Physical Models and Equilibrium Methods in Programming and Economics

Mathematics and Its Applications *(Soviet Series)*

B. S. Razumikhin
Institute for Systems Studies, Academy of Sciences, Moscow

Physical Models and Equilibrium Methods in Programming and Economics

(Revised and augmented compared to the original Russian edition and the French translation)

D. Reidel Publishing Company
A MEMBER OF THE KLUWER ACADEMIC PUBLISHERS GROUP
Dordrecht / Boston / Lancaster

Library of Congress Cataloging in Publication Data

Razumikhin, Boris Sergeevich.
Physical models and equilibrium methods in programming and economics.

(Mathematics and its applications)
Translation of: Fizicheskie modeli i metody teorii ravnovesiia v programmirovanii i ėkonomike.
"Revised and augmented English edition"–Pref., p. xi.
Includes bibliography and index.
1. Economics, Mathematical. 2. Economics–Mathematical models. 3. Equilibrium (Economics) 4. Programming (Mathematics) 5. Equilibrium. I. Title. II. Series: Mathematics and its applications (D. Reidel Publishing Company)
HB135.R3913 1984 330'.0724 83-17791
ISBN 90-277-1644-7 (cloth)

Published by D. Reidel Publishing Company
P.O. Box 17, 3300 AA Dordrecht, Holland

Sold and distributed in the U.S.A. and Canada
by Kluwer Academic Publishers
190 Old Derby Street, Hingham, MA 02043, U.S.A.

In all other countries, sold and distributed
by Kluwer Academic Publishers Group,
P.O. Box 322, 3300 AH Dordrecht, Holland

Original title: Fisičeskie modeli i metody teorii ravnovesiya
v programmirovanii i ekonomike
Translated from the Russian by M. Hazewinkel

Printed in The Netherlands

TABLE OF CONTENTS

EDITOR'S PREFACE

Approach your problems from the right end and begin with the answers. Then one day, perhaps you will find the final question.

'The Hermit Clad in Crane Feathers' in R. van Gulik's The Chinese Maze Murders.

It isn't that they can't see the solution.
It is that they can't see the problem.

G. K. Chesterton. The Scandal of Father Brown 'The Point of a Pin'.

Growing specialization and diversification have brought a host of monographs and textbooks on increasingly specialized topics. However, the "tree" of knowledge of mathematics and related field does not grow only by putting forth new branches. It also happens, quite often in fact, that branches which were thought to be completely disparate are suddenly seen to be related.

Further, the kind and level of sophistication of mathematics applied in various sciences has changed drastically in recent years: measure theory is used (non-trivially) in regional and theoretical economics; algebraic geometry interacts with physics; the Minkowsky lemma, coding theory and the structure of water meet one another in packing and covering theory; quantum fields, crystal defects and mathematical programming profit from homotopy theory; Lie algebras are relevant to filtering; and prediction and electrical engineering can use Stein spaces.

This program, Mathematics and Its Applications, is devoted to such (new) interrelations as exempla gratia:

- a central concept which plays an important role in several different mathematical and/or scientific specialized areas;
- new applications of the results and ideas from one area of scientific edeavor into another;
- influences which the results, problems and concepts of one field of enquiry have and have had on the development of another.

The Mathematics and Its Applications programme tries to make available a careful selection of books which fit the philosophy outlined above. With such books, which are stimulating rather than definitive, intriguing rather than encyclopaedic, we hope to contribute something towards better communication among the practitioners in diversified fields.

Because of the wealth of scholarly research being undertaken in the Soviet Union, Eastern Europe, and Japan, it was decided to devote special attention to work emanating from these particular regions. Thus it was decided to start three regional series under the umbrella of the main MIA programme.

The present book is concerned with the mathematics central to the decision sciences, i.e. linear and nonlinear programming, economic equilibrium and growth problems, and optimal control; from an unusual interdisciplinary point of view, however. Namely these matters are studied by means of gas- and fluid-mechanical models and this means that the principles of analytical dynamics and thermodynamics can be applied. This turns out to be quite remarkably fruitful both in terms of suggestion of algorithms (e.g. decomposition algorithms) and in terms of conceptual understanding. Thus e.g. duality in the sense of linear programming turns out to be duality between intensive and extensive variables in thermodynamics, the principle of virtual displacements of analytical mechanics is by and large the same as the Kuhn-Tucker theorem and it turns out that there is some sort of economic potential which is minimal at equilibrium. Also this setting gives a natural and suggestive interpretation to all kinds of penalty function ideas.

It seems to me that the ideas in this book have so far been exploited in only a modest way and that much more can be derived from them (notably the penalty function ones and the economic potential one).

The unreasonable effectiveness of mathematics in science ...

Eugene Wigner

Well, if you knows of a better 'ole, go to it.

Bruce Bairnsfather

What is now proved was once only imagined.

William Blake

As long as algebra and geometry proceeded along separate paths, their advance was slow and their applications limited.

But when these sciences joined company they drew from each other fresh vitality and thenceforward marched on at a rapid pace towards perfection.

Joseph Louis Lagrange

Bussum
April, 1983

Michiel Hazewinkel

PREFACE TO THE REVISED AND AUGMENTED ENGLISH EDITION

The English edition of this book is the result of substantial revisions of the Russian and French editions. The author has not only used the opportunity to remove defects from these editions, he has also introduced essential changes and additions. The most important of these revisions consists of the replacing of chapter IV (finite methods) of the Russian edition with three new chapters: "Principle of removing constraints" (Ch. IV), "The Hodograph method" (Ch. V) and "Method of displacement of elastic constraints" (Ch. VI). In chapter II there is a new section 2.7 "Models for transport problems" because the mechanical and physical models for such problems are elegant, diverse and, especially yield simple constructive devices for solving eigenvalue problems.

The book is devoted to analogies which have always played and will always play a most important role in the never ending progress of science. It seems, therefore, fitting to conclude this short preface with the words of Johannes Kepler: "And above all I value Analogies, my most faithful readers. They pertain to all secrets of nature and can be least of all neglected."

PREFACE TO THE ORIGINAL RUSSIAN EDITION

The first goal of this book is to extend the principles and methods of analytical mechanics and thermodynamics to mathematical programming and mathematical economics. To understand the results obtained, one does not need to be a sophisticated mathematician; the models from physics and mechanics which are used are simple and, in fact, are restricted to models from continuum mechanics dealing with imcompressible liquids and perfect gases. Each model represents a primal and dual problem at the same time and thus, the essential results of linear and nonlinear programming and mathematical economics acquire a physical interpretation (with the difference that in the first case one has extrinsic variables as state parameters and in the second case, intrinsic ones). The equilibrium state of the model thus defines the optimal vectors for the two associated problems and between these vectors there are simple relations resulting from the Clapeyron-Mendeleev equation. Similar results hold for models from mathematical economics which are interpreted in terms of quasi- or pseudo-static transformations of physical systems.

A substantial number of pages are devoted to numerical methods for solving mathematical programming and mathematical economics problems. These derive their mathematical description from the transition processes in the (corresponding) physical systems from an arbitrary initial state to an equilibrium state. The central idea in this part of the book is to control these processes by means of a series of temporary and time-independent constraints which permit the decomposition of the (spontaneous) time evolution towards an unknown equilibrium into several elementary processes for which a mathematical description presents no difficulties. The convergence of a corresponding algorithm then follows from the general principles of thermodynamics.

The value of ideas, concepts, and principles is measured in terms of the number of scientific disciplines in which they find useful applications. The author will have reached his goals if he succeeds in showing that the principles of analytical mechanics and thermodynamics can be effectively applied to economic research.

It is also necessary to make clear what the author understands by the words "equilibrium theory methods". The book does not pretend to expound an economic theory. Its aim is more modest: to show that certain economical-mathematical models (mostly wellknown) lead to equilibrium problems which are analogous to those which come from analytical mechanics and from thermodynamics. Equilibrium theory is a system of principles, methods, and concepts which are the basis of statics, thermodynamics, and the theory of stability.

From Archimedes and Galileo to Lyapunov and Gibbs, this theory of equilibrium has evolved from a search for conditions which equilibrium states must satisfy to the study of conditions under which a given process, that is movement, is possible. The fundamental idea is the existence of a state function the maxima and minima of which correspond to real stable equilibria or real stable movements of the system. This idea has led, in the first case, to the theory of potentials (Leibnitz, Lagrange, Helmholz, Clausius, Gibbs) and, in the second case, to the variational principle of dynamics (Maupertuis, Euler, Lagrange, Hamilton, Jacobi, Poincaré). The ideas, concepts, and methods of equilibrium theory also apply to dynamical problems which formally reduce to problems of statics. To see this, it suffices to recall the principle of d'Alembert, the variational equations of Poincaré, and the stability theory of Lyapunov.

It is possible that a number of mathematicians will find that there is too much mechanics and physics in the book. Would it not have been much simpler to present the results of this book in the language of pure mathematics while pointing out in various sections (which, for that matter, can be skipped in reading the book) the possible mechanical and thermodynamical analogues? It is even possible that, in that case, the book would have found more readers, but the author would have been less happy with his work. The time is no more that any mortal can oversee all or a great part of science. Indeed, "The consequence of the vast and rapidly growing extent of our positive knowledge was a division of labour in science right down to the minutest detail, almost reminiscent of a modern factory where one person does nothing but measure carbon filaments, while another cuts them, a third welds them in and so on. Such a division of labour certainly helps greatly to promote rapid progress in science and is indeed indispensable for it; but just as certainly it harbours great dangers. For we lose the overview of the whole, required for any mental activity aiming at discovering something essentially new or even just essentially new combinations of old ideas. In order to meet this drawback as far as possible it may be useful if from time to time a single individual who is occupied with the work of scientific detail should try to give a larger and scientifically educated public a survey of the branch of knowledge in which he is working." [1)]

This book is addressed to those who share these thoughts of that great physicist and thinker, Ludwig Boltzmann.

NOTE

1. L. Boltzmann, Populäre Schriften. Verlag Johann Ambrosius Barth, 1905, pages 198, 199. This translation is from the authorized translation: L. Boltzmann, Theoretical physics and philosophical problems, D. Reidel Publ. Co., 1974, p. 77.

INTRODUCTION

"... toutes les sciences réunies ne sont rien autre chose que l'intelligence humaine, qui reste toujours une, toujours la même, si variés que soient les sujets auxquels elle s'applique et qui n'en reçoit pas plus de changements que n'en apporte à la lumière du soleil la variété des objets qu'elle éclaire ..." Descartes

There exist a good many, sometimes excellent, books which cover the theory and methods of solving mathematical programming problems and problems from mathematical economics. This book is characterized by the fact that it contains an exposition of some of the results of these sciences as a consequence of the fundamental principles of analytical mechanics and thermodynamics. It is addressed to a wide and diverse audience, which explains the author's decision to use only a relatively narrow collection of results known to everyone interested in mechanics or physics or familiar from the standard courses taught in institutions of higher learning.

The author shares the thoughts of Lanczos on mechanics [40]:

"There is hardly any other branch of the mathematical sciences in which abstract mathematical speculation and concrete physical evidence go so beautifully together and complement each other so perfectly. It is no accident that the principles of mechanics had the greatest fascination for many of the outstanding figures of mathematics and physics. ... Analytical mechanics is much more than an efficient tool for the solution of dynamical problems that we encounter in physics and engineering." 1)

Lagrange, Laplace, Euler, and Hamilton have erected an edifice whose foundations were laid by Galileo, Leibnitz, and Newton, which is, according to Boltzmann, "a magnificent example for every physics-mathematical theory". Thus, analytical mechanics has also been a model for the creators of classical thermodynamics and one of its most important chapters: thermodynamic potential theory.

The fact that the general principles of analytical mechanics and thermodynamics play a role in all physical and chemical disciplines naturally invites us to use the same methods to study the behaviour of all material systems, with economic systems included.

The author does not pretend that this approach will lead us to a new economic theory. This book rather exemplifies an inclination which can hardly be defined better than was done by Von Neumann and Morgenstern:

> "... It is without doubt reasonable to discover what has led to progress in other sciences, and to investigate whether the application of the same principles may not lead to progress in economics also. Should the need for the application of different principles arise, it could be revealed only in the course of the actual development of economic theory. This would in itself constitute a major revolution. But since most assuredly we have not yet reached such a state - and it is by no means certain that there ever will be need for entirely different scientific principles - it would be very unwise to consider anything else than the pursuit of our problems in the manner which has resulted in the establishment of physical science." [50]. [2)]

This book describes the physical interpretations of mathematical programming problems and of certain economical models. Because the equilibrium state of the equivalent physical models defines the optimal bivector of the two (dual) associated problems, the equilibrium conditions of mechanics naturally express the fundamental duality theorem.

The models we study are subjected to rigid or elastic constraints and contain, as a moving force, an incompressible liquid or a perfect gas. One develops a general method for numerical solutions by means of the fundamental idea of controlling the transition of the physical system to an equilibrium state by the introduction of redundant and artificial constraints which are independent of time.

The reader will see physical models of various problems which one meets in linear algebra, in linear and nonlinear programming, in equilibrium theory, and in economic growth theory. He will see that maximalization and the search for equilibrium prices are, in fact, equilibrium problems for active physical systems (in the first case) or passive, i.e., isolated, systems (in the second case).

Because the search for an equilibrium of an economy with several subsystems each with their own objectives is equivalent to the problem of equilibrium for a passive physical system, the idea follows that such an economy will have a global state function which will have its maximum in the equilibrium state. In analogy to the entropy of a physical system that function does not only define the equilibrium states but also the direction in which the economy evolves spontaneously. Thus, the possibility arises in economic research of using the fundamental ideas, concepts, principles, and methods of analytical mechanics and phenomenological and statistical thermodynamics.

We would like to devote a few words to the contents of the book to help the reader decide whether he should continue to read it. Let us remark in the first place that the physical and mechanical models considered naturally lead us to a treatment of problems of mathematical programming and mathematical economics in terms of the equilibria of physical systems of which the state function is an analogue of Leibnitz's forces function or Gibbs' thermodynamic potential. As to the constraints, their analogues are the bilateral or unilateral constraints which determine the admissible parameter variations in the state of the system under consideration. In this setting, it is clear that the basic conditions for optimality and economic equilibrium must be established by means of the principles of virtual displacement, which play such a fundamental role in various sciences - from analytical mechanics, calculus of variations and thermodynamics, to the theory of relativity.

Precisely this principle, essentially, lies behind all constructions of conditions for optimality in problems of mathematical programming, to which, just as in mechanical problems there apply the words of Lagrange:

> "... and, in general, I believe it possible to predict that all the general principles which can, maybe, still be discovered in the science of equilibrium will be nothing else but the same principle of virtual velocities (displacments) viewed in different ways and only differing in the ways in which it is expressed. But

this principle is not only extremely simple in itself and very general; it also has, in addition, the unique and precious advantage that it encompasses all problems which can be posed concerning the equilibrium of solid bodies" [38].

In the literature on mathematical programming one usually finds described the method of indeterminate multipliers (Lagrange multipliers). One should not forget that Lagrange proposed this method uniquely as a rule for applying the principle of virtual displacements (velocities) to problems of mechanics for systems subject to idealized constraints.

It was also Lagrange who interpreted these multipliers, which still bear his name, in physical terms and who formulated the principle of 'removing the constraints' (_principe de la libération_) which 'has the same generality as the principle of virtual velocities' (Bertrand) [38].

Recall what has been said concerning primal and dual problems, to wit that both constitute the same equilibrium problem where only the state parameters vary independently. This situation can be considered analogous to the statics of elastic systems when the state parameters are either constraints or constituant deformations of the system; that is, extensive or intensive state variables. Such a treatment of duality not only enriches it but it also makes it more accessible. It is a curious fact that the rules which describe the physical properties of bodies subject to constraints such as Hooke's law or the law of Clapeyron-Mendeleev also connect the optimal solutions of the primal and dual problems and that the equality of the values of the 'cost' functions for the two problems is an analogue of the law of conservation of energy. [3)]

Thus the unknowns of the primal and dual problems are the parameters of the state of the system of which the equilibrium defines the optimal values of the two sets of parameters. For economic equilibrium problems this means that the prices are also parameters of the state of the economy, equally relevant as the quantitities of available goods and that a change in the external conditions will lead to a new equilibrium state in which prices and quantities of goods are connected by relations valid for all exterior conditions.

Economic systems have, as was mentioned before, a state function, analogous to entropy. The directions of growth of this function determine the possible directions of spontaneous evolution of the system and the maximum determines the equilibrium state. Consequently there arises the idea that for these economic systems there exist analogues of the thermodynamic inequalities from which there follows the principle of stability of Le Chatelier [36].

The book pays particular attention to new numerical techniques which result from the principles and methods of the general theory of equilibrium and which use the known physical properties of bodies subject to constraints which represent the optimization problems of mathematical programming, systems of equations and inequalities, and mathematical economics models.

In this connection the method of redundant constraints is central. At the basis of this method in turn lies the idea of controlling the evolution process of a physical system to an equilibrium state. This control is realized by means of a series of supplementary (redundant) constraints which are imposed and which were not present in the problem being modelled. By imposing these time-independent redundant constraints, adapted to the actual state of the system, one decomposes the big complicated process in a series of elementary transition processes between intermediate states corresponding to equilibria for the system with the additional redundant constraints. Because no work is done by the external forces when the (redundant) constraints are changed from one set to another the processes evolve spontaneously and the convergence of the 'redundant constraints algorithms' follow from the second law of thermodynamics.

We also discuss the method of a sequence of unconstrained minimizations which is also called the method of penalty functions or weight functions. When he discussed the difficulties adhering to the Rayleigh-Ritz method when applied to oscillation problems for elastic systems, Richard Courant suggested an idea for a method which may be used to overcome the difficulties which come from the presence of rigid boundaries. This idea of Courant can be formulated as follows [23]:

> "Quite generally rigid boundary conditions should be regarded as a limiting case of natural conditions in which a parameter tends to infinity. This corresponds to the physical fact that <u>rigid constraints are only an idealized limiting case of very large restoring forces</u>." 4)

This idea can be put to work in various ways. Courant

proposed to introduce in the functional to be minimized a penalty term proportional to the kinetic energy of the particles along the free boundary. 5)

Obviously the same result is obtained if one introduces forces which are proportional to the distances from the given rigid constraints (elastic constraints). Courant indicated the necessity of compromise in the practical use of penalty functions. He writes:

> "From a theoretical as well as a practical point of view it would seem worthwile to study the preferable choices of these artificial parameters". 6)

It seems thus that we owe to Courant the first statement of the method of penalty functions.

In subsequent work various minimization problems with constraints have been considered with various forms of penalty functions and under rather weak conditions convergence has been proved. For references see the bibliographies in [8] and [28]. Thus the arsenal of iterative methods has been enlarged with a procedure which associates a sequence of unconstrained minimization problems to a problem with constraints.

For some of these S.U.M.T. methods (Sequential Unconstrained Minimization Techniques), and especially for those which use a sequence of penalty functions which grow unboundedly for every finite violation of each rigid constraint, the speed of convergence diminishes rapidly with increasing size of the problem. This however does not mean that we must class penalty function methods within the class of methods which give only a first very coarse approximate solution. First of all this unfortunate property adheres only to some S.U.M.T. methods (and not the most effective ones). It is also possible to construct other methods (which reduce a constrained minimization problem to a sequence of unconstrained minimization problems 7), for instance finite ones and methods for which one can take as a penalty function one of a class of sign definite functions of violations with known properties.

There is also another much more important aspect of the problem of effective penalty function methods which goes beyond the usual mathematical framework.

In the paper cited above Courant studied methods for solving mathematical problems which have a physical origin. The mathematician is in general only marginally interested in the question of how well the mathematical model corresponds to the real physical phenomenon. If the problem is (mathematically) well posed it should be solved and all the difficulties one encounters are mathematical and are treated as mathematical problems. However one must not forget that the solution methods and also the solutions themselves are only connected to the physical phenomena via their mathematical models.

The idea of ideal constraints, absolutely rigid bodies or absolutely inextensible strings has been of considerable importance in mechanics. Indeed such constraints are expressible mathematically by equations and inequalities and the virtual work done by the corresponding reaction forces is zero. The introduction of ideal constraints as models for real constraints has been of considerable importance in the development of analytical mechanics.

It is however necessary to keep in mind that ideal constraints, absolutely rigid bodies and incompressible fluids are only models for real constraints and physical bodies. Moreover they are models which are not always useful. For instance it is not possible to explain transfer of momentum in a collision process between two bodies if they are supposed to be absolutely rigid. When considering collision problems Johann Bernouilli remarked concerning absolutely rigid bodies that they are a chimera which contradicts that general law which nature always obeys in all its manifestations, a law which one may call 'law of continuity' [22].

It would be possible to show conclusively, by means of many examples, that many initially useful concepts or models later turned out to be unsuitable and ofter their unsuitability showed itself by means of difficulties in solving the resulting mathematical problem. One should not consider the penalty parameter in the problem considered by Courant to be artificial and its choice should not be a compromise.

The method of penalty functions is much more than simply a method for the numerical solution of extremum problems subject to constraints. It is also a modelling technique; that is a method of getting away from ideal constraint hypotheses. This point of view suggests that alleviating the numerical difficulties is not the only important problem. It is even more important to choose penalty functions in such a way that they reflect sufficiently accurately the physical, economical or other properties of the real constraints in question. Thus we should turn to reality and study the real problem which gave birth to the mathematical one.

Clearly, the analogue of the method of penalty functions in nature is the property of constraints to be deformable and this is what Johann Bernouilli called 'the law of continuity'.

S.U.M.T. methods are also very effective in dealing with the variational problems of optimal control. Unfortunately the size of the book did not permit the author to delve deeply into this fascinating area. Only in the last two chapters, in connection with some dynamical problems of economics, do we consider how to solve time optimal control problems such as the problem of the minimal transition time from a given initial state to a set of given (desired) state is and the problem of a minimal time transition to a ray of maximal balanced growth.

It is not known when precisely the method of penalty functions came into being. Its originator is that person who first determined an adequate penalty for violations of rules or laws.

The famous paper of R. Courant [23] marks its mathematical beginnings. By now there is a vast collection of papers and books on the subject in which various versions of this method are developed and applied.

However, in economics penalty functions should reflect the real economic elasticities of the constraints. Here the method of penalty functions must be the mathematical apparatus for the better understanding of planning and control problems in economics, in which a central control organism and several peripheral semi-independent sub-economics work harmoniously together. Penalty function methods will certainly be very useful to determine the degree of centralization which is most effective.

NOTES

1. Loc. cit., preface, page IX.
2. J. von Neumann and O. Morgenstern, Theory of games and economic behaviour, third edition, Princetown University Press, 1953, p. 4.
3. The same result occurs in [26].
4. Loc. cit. page 8; cf. also page 13.
5. In the problem considered by Courant the penalty parameter is the density of the material at the free boundary.
6. Loc. cit. page 14.
7. See Chapter II, section 2.6; Chapter III, Section 3.5; Chapter V, VI, VII, VIII, X.

Chapter I

EQUILIBRIUM OF MECHANICAL SYSTEMS WITH LINEAR CONSTRAINTS AND LINEAR PROGRAMMING PROBLEMS

1.1. Introduction

One recognizes the analogies between problems coming from different sciences by their identical mathematical formulation. By studying these analogies one can use the achievements of one science to make progress in another. In this connection two directions can be discerned. The first is the creation of analogue and digital machines. The second is the extension of ideas, results and methods of one area of scientific inquiry to study the objects and phenomena of another. The present book has to do with the second of these directions. A minimization problem subject to constraints can obviously be treated as an equilibrium problem for a mechanical system where the role of the cost function (criterion function) is played by the potential energy and the constraints become mechanical restrictions (constraints) imposed on the system. Such a representation can be fruitfully exploited because it permits us to use the principles and methods of the equilibrium theory of mechanical systems with unilateral or bilateral constraints and the properties of bodies realizing such constraints. We shall not limit ourselves to formal analogies but will consider the concrete mechanical and physical systems which model problems of mathematical programming, problems from linear algebra and problems from mathematical economics. This will permit us to interpret the funcamental results in terms of physics and, what is more important, to consider the algorithms as mathematical descriptions of controlled transitions to equilibrium of the physical systems. This first chapter is devoted to mechanical models of systems of linear equations and linear inequalities and to models of pairs of dual linear problems. It is important to note that the primal and dual problems constitute two statements of a problem with respect to the equilibrium of the same mechanical systems, of which the constraints are expressed by the incompressibility of the fluid which fills its containers. The equilibrium conditions for such a problem express the fundamental duality theorem and the equality of the cost functions (of primal and dual problems) at equilibrium represents the law of energy conservation. The components of the optimal vectors of the primal and dual problems have a simple meaning: they define the volumes and their pressures at equilibrium. In the following chapter we shall treat the so called physical models which are obtained from the present ones by replacing the incompressible fluid by a perfect gas. These models create the possibility of using the fundamental ideas and methods of thermodynamics, not only to analyse the nature of the problems being modelled, but also to construct effective algorithms for their numerical solution.

1.2. Linear equations and inequalities

Naturally one starts to explain the method of simulation by physical models by using simple linear algebra problems, such as finding the solutions for a system of linear equations and inequalities. The models required are systems of communicating containers filled with an incompressible fluid. These models are mechanical rather than physical and useful mainly because they permit us to reduce a mathematical problem to an equilibrium problem for a mechanical system and to give an intuitive interpretation, not only of the problems themselves but also of some important theoretical results.

Consider the mechanical construct depicted in Figure 1.1. It represents a system of two communicating cylinders both of height 2ℓ and their bases have an area of a_1 and a_2 respectively. Each cylinder is divided into two parts by a piston which can move freely within the corresponding cylinder.

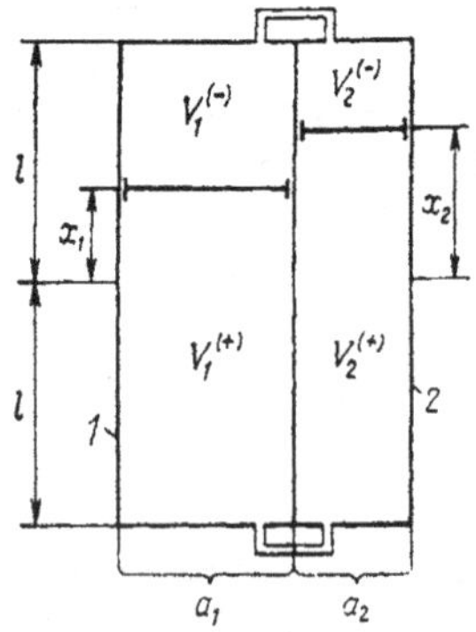

Fig. 1.1.

Let x_1 and x_2 be the coordinates defining the position of the pistons in cylinders 1 and 2. These coordinates are measured from the middle of the cylinders so that a position higher than the middle corresponds to a positive value of the corresponding coordinate. Let $V^{(+)}$ denote the sum of the volumes $V_1^{(+)}$ and $V_2^{(+)}$ beneath the two pistons and let $V^{(-)}$ denote the sum of the two volumes $V_1^{(-)}$ and $V_2^{(-)}$ above the pistons.

Then, obviously

$$V^{(+)} = V_1^{(+)} + V_2^{(+)} = a_1(\ell+x_1) + a_2(\ell+x_2)$$

$$V^{(-)} = V_1^{(-)} + V_2^{(-)} = a_1(\ell-x_1) + a_2(\ell-x_2).$$

Now let us fill the volumes $V^{(+)}$ with an amount $\bar{V}^{(+)}$ of an incompressible fluid and the volumes $V^{(-)}$ with an amount $\bar{V}^{(-)}$ of an incompressible fluid. Here $\bar{V}^{(+)}$ and $\bar{V}^{(-)}$ are chosen such that

$$\bar{V}^{(+)} + \bar{V}^{(-)} = 2\ell(a_1+a_2).$$

This condition means that $\bar{V}^{+}$ and $\bar{V}^{-}$ are chosen such that all volumes are filled completely with the incompressible fluid. In this manner we obtain a mechanical construct in which the quantities x_1 and x_2 are connected by the relation

$$V^{(+)} = \bar{V}^{(+)}$$

or

$$a_1x_1 + a_2x_2 = \bar{V}^{(+)} - \ell(a_1+a_2),$$

a relation which shows that the fluid is incompressible. We note that the construction described permits us to realize all values x_1, x_2 which satisfy the equation

$$a_1x_1 + a_2x_2 = b \tag{1.1}$$

provided $|x_1| \leqslant \ell$, $|x_2| \leqslant \ell$, if we set

$$\bar{V}^{(+)} = b + \ell(a_1+a_1), \quad \bar{V}^{(-)} = -b + \ell(a_1+a_2).$$

Thus we can consider this construct as a model for Equation (1.1).

It is necessary to note the speculative nature of the given model. We have abstracted from the real construct such properties as the inertia of solids and liquids, friction, etc., and taken into account only the main property, which is, that Equation (1.1) holds. We also consider the constraints $|x_1| \leqslant \ell$, $|x_2| \leqslant \ell$ as nonessential because the number ℓ may be as large as desired. The reader is advised to keep this in mind for all the following considerations.

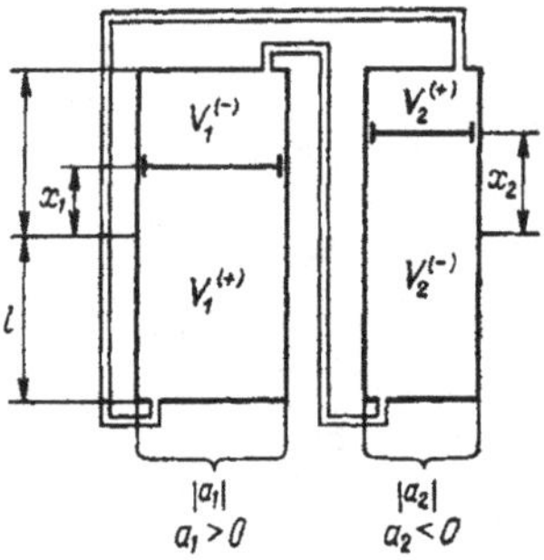

Fig. 1.2.

We now consider the problem of modelling Equation (1.1) in the case where the coefficients have different signs. Let, for example, $a_1 > 0$, $a_2 < 0$. One verifies without difficulty that a model for such an equation is represented in Figure 1.2, where the base areas of the cylinders are equal to the absolute values of the coefficients a_1 and a_2. Indeed, from Figure 1.2 we obtain

$$V^{(+)} = V_1^{(+)} + V_2^{(+)} = |a_1|(\ell+x_1) + |a_2|(\ell-x_2)$$

$$V^{(-)} = V_1^{(-)} + V_2^{(-)} = |a_1|(\ell-x_1) + |a_2|(\ell+x_2).$$

Let the quantities of incompressible liquid $\overline{V}^{(+)}$ and $\overline{V}^{(-)}$ filling the corresponding volumes $V^{(+)}$ and $V^{(-)}$ be such that

$$\overline{V}^{(+)} + \overline{V}^{(-)} = 2\ell(|a_1|+|a_2|).$$

Then, as the liquid is incompressible, the pistons will occupy the unique positions for which

$$V^{(+)} = \overline{V}^{(+)}$$

or

$$|a_1|x_1 - |a_2|x_2 = \overline{V}^+ - \ell(|a_1|+|a_2|).$$

Taking into account that $a_1 > 0$, $a_2 < 0$, we have

$$a_1x_1 + a_2x_2 = \overline{V}^+ - \ell(a_1-a_2).$$

Let us take

$$\overline{V}^+ = b + \ell(a_1-a_2)$$

$$\overline{V}^- = -b + \ell(a_1-a_2)$$

to obtain a model for Equation (1.1) with $a_1 > 0$, $a_2 < 0$. Comparing the models of Figures 1.1 and 1.2, one easily establishes the following important rule: the various volumes in the model communicate if and only if they carry the same sign (+) or (-); the sign (+) indicates a volume below the piston if the sign of the corresponding coefficient is positive, it indicates the volume above the piston in the opposite case.

In both cases the area of the base of the cylinder is equal to the absolute value of the corresponding coefficient. Let us denote with $V_1^{(+)}$, $V_1^{(-)}$ and $V_2^{(+)}$, $V_2^{(-)}$ the volumes in cylinders 1 and 2 which are marked with (+) and (-) in Figures 1.1 and 1.2. Then the rule stated above is written formally as

$$V_1^{(+)} = \begin{cases} a_1(\ell+x_1) & \text{if } a_1 \geqslant 0 \\ -a_1(\ell-x_1) & \text{if } a_1 < 0 \end{cases}$$

$$V_1^{(-)} = \begin{cases} a_1(\ell-x_1) & \text{if } a_1 \geqslant 0 \\ -a_1(\ell+x_1) & \text{if } a_1 < 0 \end{cases}$$

$$V_2^{(+)} = \begin{cases} a_2(\ell+x_2) & \text{if } a_2 \geqslant 0 \\ -a_2(\ell-x_2) & \text{if } a_2 < 0 \end{cases}$$

$$V_2^{(-)} = \begin{cases} a_2(\ell-x_2) & \text{if } a_2 \geqslant 0 \\ -a_2(\ell+x_2) & \text{if } a_2 < 0. \end{cases}$$

It is not difficult to imagine the physical model for a linear equation with many unknowns (Figure 1.3).

$$\sum_{i=1}^{n} a_i x_i = b. \tag{1.2}$$

In this case the rule for establishing the signs of volumes $V_i^{(+)}$ and $V_i^{(-)}$ is given exactly as in the case of the models of Figures 1.1 and 1.2. That is

$$V_i^{(+)} = \begin{cases} a_i(\ell+x_i), & a_i \geqslant 0 \\ -a_i(\ell-x_i), & a_i < 0 \end{cases}$$

$$V_i^{(-)} = \begin{cases} a_i(\ell-x_i), & a_i \geqslant 0 \\ -a_i(\ell+x_i), & a_i < 0. \end{cases}$$

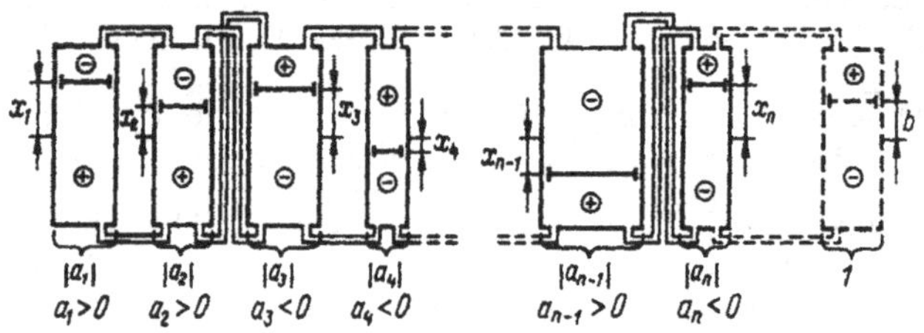

Fig. 1.3.

It follows that

$$V^{(+)} = \sum_{i=1}^{n} V_i^{(+)} = \ell\Big(\sum_{a_i>0} a_i\Big) - \ell\Big(\sum_{a_i<0} a_i\Big) + \sum a_i x_i =$$

$$= \ell \sum_{i=1}^{n} |a_i| + \sum_{i=1}^{n} a_i x_i$$

$$V^{(-)} = \sum_{i=1}^{n} V_i^{(-)} = \ell\Big(\sum_{a_i>0} a_i\Big) - \ell\Big(\sum_{a_i<0} a_i\Big) - \sum a_i x_i =$$

$$= \ell\Big(\sum_{i=1}^{n} |a_i|\Big) - \sum_{i=1}^{n} a_i x_i .$$

Filling the communicating volumes $V_1^{(+)}$, $V_2^{(+)}$, ..., $V_n^{(+)}$ with an incompressible liquid in the quantity

$$\bar{V}^{(+)} = \ell \sum_{i=1}^{n} |a_i| + b$$

and volumes $V_1^{(-)}$, $V_2^{(-)}$, ..., $V_n^{(-)}$ with an incompressible liquid to the amount of

$$\bar{V}^{(-)} = \ell \sum_{i=1}^{n} |a_i| - b$$

we obtain a model for equation (1.2) such that from the incompressibility conditions $V^{(+)} = \bar{V}^{(+)}$ and $V^{(-)} = \bar{V}^{(-)}$ there results the validity of (1.2) for all possible states of the model.

It is useful for the following to indicate a modification of the model of Figure 1.3 which corresponds to the inhomogeneous equation

$$\sum_{i=1}^{n-1} a_i x_i = b.$$

This equation can be written in the form

$$\sum_{i=1}^{n} a_i x_i = 0 \qquad (1.3)$$

where

$$a_n x_n = -b.$$

The model of Equation (1.3) is distinguished from the model of Equation (1.2) by the fact that in this case the piston of the n-th cylinder is fixed and either the n-th coordinate $x_n = b$ and the base area is 1, or $x_n = 1$ and the base area is $|b|$.

Obviously, if b is considered as a variable quantity, then the model transforms itself into a construction which gives the values of the linear form

$$\sum_{i=1}^{n-1} a_i x_i.$$

It is now clear that the modelling of a linear inequality of the form

$$\sum_{i=1}^{n} a_i x_i \leqslant b \qquad (1.4)$$

also presents no difficulties. Indeed, introducing an additional nonnegative variable x_{n+1}, inequality (1.4) reduces to an equality

$$\sum_{i=1}^{n+1} a_i x_i = b$$

where

$$a_{n+1} = 1, \; x_{n+1} \geqslant 0.$$

The constraint $x_{n+1} \geqslant 0$ (nonnegativeness) can be modelled by a stop in the middle of the cylinder limiting the possible positions of the corresponding piston.

1.3. Systems of linear equations and inequalities

It is now really very simple to construct a model for a system of linear equations and inequalities. Obviously a model of a system of m linear equations in n unknowns is a system of m models for each of the m equations connected in such a way that the movable pistons which indicate the values of one and the same unknown are in the same position. It is natural to realize such a connection by attaching these pistons to a rigid rod.

It is clear now that the model for a system of linear equations like

$$\sum_{i=1}^{n} a_{si} x_i = b_s, \quad s = 1,2,\dots,m \tag{1.5}$$

is as pictures by Figure 1.4 and that it is a union of the m models of the m equations making up the system.
The element of the given model which corresponds to the i-th column and s-th row is a cylinder with a base area equal to the absolute value of the coefficient a_{si}. The total volume of this cylinder, equal to $2\ell|a_{si}|$, is divided by the movable piston into two parts $V_{si}^{(+)}$ and $V_{si}^{(-)}$. Figure 1.5 depicts these parts in the cases $a_{si} > 0$ and $a_{si} < 0$. From the rule assigning the signs (+) and (-) there follow the formulas

$$V_{si}^{(+)} = \begin{cases} a_{si}(\ell + x_i) & \text{for } a_{si} \geqslant 0 \\ -a_{si}(\ell - x_i) & \text{for } a_{si} < 0 \end{cases}$$

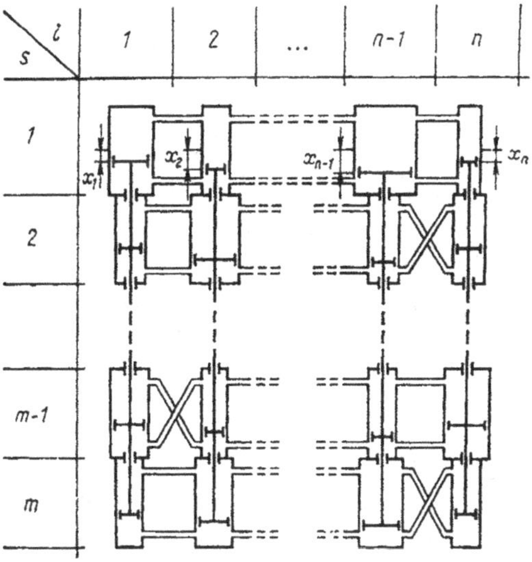
i
s
1
2
...
n-1
n
1
2
m-1
m
x_1
x_2
x_{n-1}
x_n

Fig. 1.4.

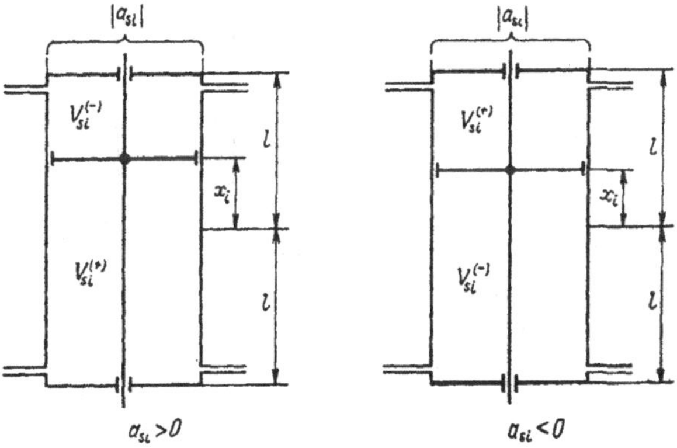
$|a_{si}|$
$V_{si}^{(-)}$
$V_{si}^{(+)}$
x_i
l
l
$a_{si}>0$
$|a_{si}|$
$V_{si}^{(+)}$
$V_{si}^{(-)}$
x_i
l
l
$a_{si}<0$

Fig. 1.5.

$$V_{si}^{(-)} = \begin{cases} a_{si}(\ell - x_i) & \text{for } a_{si} \geqslant 0 \\ -a_{si}(\ell + x_i) & \text{for } a_{si} < 0 \end{cases}$$

$$\left.\begin{aligned} V_s^{(+)} &= \sum_{i=1}^{n} V_{si}^{(+)} = \ell \sum_{i=1}^{n} |a_{si}| + \sum_{i=1}^{n} a_{si}x_i \\ V_s^{(-)} &= \sum_{i=1}^{n} V_{si}^{(-)} = \ell \sum_{i=1}^{n} |a_{si}| - \sum_{i=1}^{n} a_{si}x_i . \end{aligned}\right\} \quad (1.6)$$

Therefore, the admissible states of the models are solutions of the system 1.5 if

$$V_s^{(+)} = \bar{V}_s^{(+)}, \quad V_s^{(-)} = \bar{V}_s^{(-)}, \quad s = 1,2,\ldots,m$$

where, clearly

$$\begin{aligned} \bar{V}_s^{(+)} &= \ell \sum_{i=1}^{n} |a_{si}| + b_s \\ \bar{V}_s^{(-)} &= \ell \sum_{i=1}^{n} |a_{si}| - b_s \end{aligned} \qquad s = 1,2,\ldots,m. \qquad (1.7)$$

In vector-matrix form the system of Equation (1.5) has the form

$$Ax = b$$

where

$$A = \begin{pmatrix} a_{11} & a_{12} & \cdots & a_{1n} \\ a_{21} & a_{22} & \cdots & a_{2n} \\ \cdots & \cdots & \cdots & \cdots \\ a_{m1} & a_{m2} & \cdots & a_{mn} \end{pmatrix}, \quad x = \begin{pmatrix} x_1 \\ x_2 \\ \vdots \\ x_m \end{pmatrix}, \quad b = \begin{pmatrix} b_1 \\ b_2 \\ \vdots \\ b_m \end{pmatrix}.$$

Let us denote with $a_s = (a_{s1}, a_{s2}, \ldots, a_{sn})$ the s-th rowvector, and introducing as the norm $\|a\|$ of vector a the sum of the absolute values of its components we get

$$\|a_s\| = \sum_{i=1}^{n} |a_{si}| .$$

With these notations formulas (1.6) and (1.7) take the form

$$\begin{cases} V_s^{(+)} = \ell\|a_s\| + (a_s, x) \\ V_s^{(-)} = \ell\|a_s\| - (a_s, x) \end{cases} \qquad s = 1,\ldots,m \qquad (1.8)$$

$$\left\{\begin{array}{l} \overline{V}_s^{(+)} = \ell \| a_s \| + b_s \\ \overline{V}_s^{(-)} = \ell \| a_s \| - b_s \end{array}\right. \qquad s = 1,\ldots,m. \tag{1.9}$$

The case of systems consisting of m_1 inequalities and $m-m_1$ equalities of the form

$$\sum_{i=1}^{n} a_{si}x_i \leqslant b_s, \quad s = 1,2,\ldots,m_1$$

$$\sum_{i=1}^{n} a_{si}x_i = b_s, \quad s = m_1+1,\ldots,m$$

reduces to the case already considered by means of introducing additional unknowns $x_{n+1},\ldots,x_{n+m_1}$, which are constrained to be nonnegative.

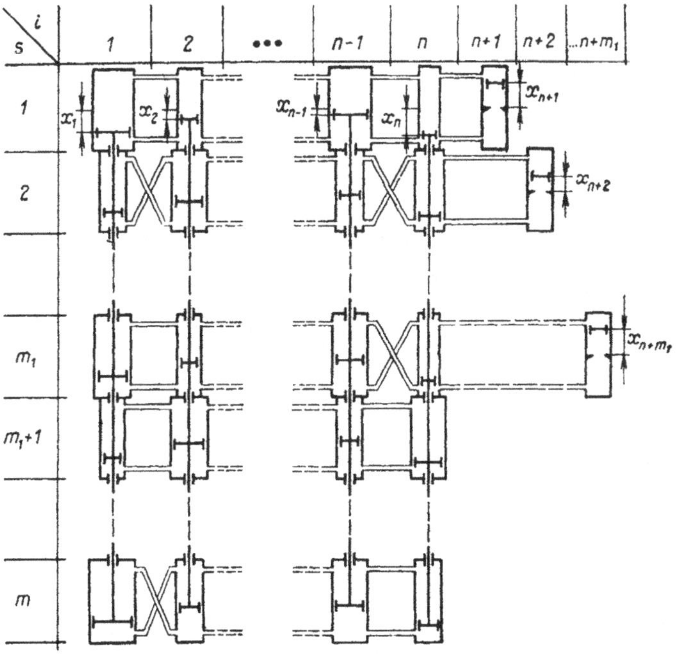

Fig. 1.6.

$$\sum_{i=1}^{n} a_{si}x_i + x_{n+s} = b_s, \quad s = 1,\dots,m_1$$
$$\sum_{i=1}^{n} a_{si}x_i = b_s, \quad s = m_1+1,\dots,n \tag{1.10}$$
$$x_{n+s} \geqslant 0, \quad s = 1,\dots,m_1$$

Conditions $x_{n+s} \geqslant 0$ are guaranteed to be satisfied by means of stops in the middle of the corresponding cylinder. The model of system (1.10) is depicted in Figure 1.6.

In the following it will be more convenient to start with a different form of system of linear equations and inequalities (1.10), that is,

$$\sum_{i=1}^{n} a_{si}x_i - \xi_s = 0, \quad s = 1,\dots,m \tag{1.11}$$

$$\xi_s = \begin{cases} \leqslant b_s & \text{for } s = 1,\dots,m_1 \\ = b_s & \text{for } s = m_1+1,\dots,m. \end{cases} \tag{1.12}$$

Setting

$$A_1 = \begin{pmatrix} a_{11} & a_{12} & \dots & a_{1n} & -1 & 0 & \dots & 0 \\ a_{21} & a_{22} & \dots & a_{2n} & 0 & -1 & \dots & 0 \\ \cdot & \cdot & \cdot & \cdot & \cdot & \cdot & \cdot & \cdot \\ a_{m1} & a_{m2} & \dots & a_{mn} & 0 & 0 & \dots & -1 \end{pmatrix}$$

$$u = \begin{pmatrix} u_1 \\ u_2 \\ \vdots \\ u_{n+m} \end{pmatrix}, \quad u_i = \begin{cases} x_i & \text{for } i = 1,\dots,n \\ \xi_s & \text{for } i = n+s,\ s = 1,\dots,m \end{cases} \tag{1.13}$$

the system (1.11), (1.12) can be rewritten in the form

$$A_1u = 0. \tag{1.14}$$

The norm of rowvector a_s of matrix A_1 is equal to

$$\|a_s\| = 1 + \sum_{i=1}^{n} |a_{si}|. \tag{1.15}$$

By analogy with (1.8) and (1.9), the quantities $V_s^{(+)}$, $V_s^{(-)}$, $\bar{V}_s^{(+)}$, $\bar{V}_s^{(-)}$ take the form

$$V_s^{(+)} = \ell\|a_s\| + (a_s,u) = \ell\|a_s\| + \sum_{i=1}^{n} a_{si}x_i - \xi_s$$
$$V_s^{(-)} = \ell\|a_s\| - (a_s,u) = \ell\|a_s\| - \sum_{i=1}^{n} a_{si}x_i + \xi_s \tag{1.16}$$

$$\bar{V}_s^{(+)} = \bar{V}_s^{(-)} = \ell\|a_s\|, \quad s = 1,\dots,m. \tag{1.17}$$

Clearly the permissible states for the model of Figure 1.7

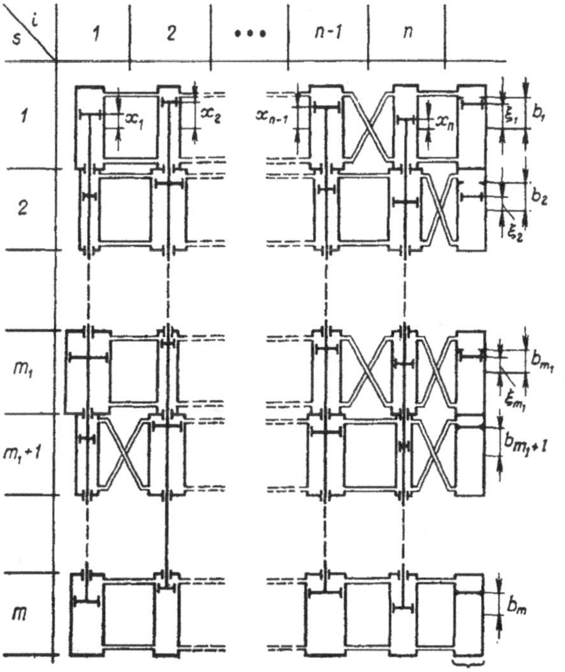

Fig. 1.7.

satisfy system (1.11) - (1.12) if the following conditions hold

$$V_s^{(+)} = \bar{V}_s^{(+)}, \ V_s^{(-)} = \bar{V}_s^{(-)}, \quad s = 1,\ldots,m.$$

Thus the models we have considered above are mechanical systems consisting of solid bodies which admit translation movements. The displacements of these bodies are subject to linear constraints which are bilateral when they correspond to equations and unilateral when their mathematical expression is an inequality. The incompressible fluid which fills the various volumes realizes these constraints which are imposed on the system. In section 1.4 we shall study the general problems of linear programming for which the models just constructed define the set of admissible points.

Remark. As is well known, it is possible that the system of linear equations and linear inequalities (1.11)-(1.12) has no solutions. In the case of the model described above, it is then precisely the incompressiblity of the fluid which prevents the system from assuming a position of rigid bars (connecting the pistons) for which the conditions

$$\bar{V}_s^{(+)} = V_s^{(+)}, \ \bar{V}_s^{(-)} = V_s^{(-)}, \quad s = 1,2,\ldots,m$$

are fulfilled. Thus, every such model of a linear system of equations and inequalities necessarily presupposes the existence of a solution of these linear equations and inequalities.

In the sequel we shall see that by replacing the incompressible fluid by a perfect gas one succeeds in modelling inconsistent systems of linear equations and inequalities, and that the equilibrium state of such a model of volumes filled with a perfect gas corresponds to a solution which minimizes an error norm.

1.4. Linear programming problems. Duality theorems

Many problems in the theory of decision and control reduce to a problem of finding the maximum or minimum of a linear function on a set which is defined by a system of linear equations and inequalities. The problems thus arrived at are called linear programming problems, and they are the mathematical counterpart of a great number of important problems in which one looks for a most efficient allocation of resources in economics, industry and administrative affairs.

The theory of linear programming, of which the basic fundamentals can be found in the works of L. V. Kantorovic [5, 6, 33], has a vast bibliography containing a large collection of treatises, monographs and articles. This permits us to assume that the reader is acquainted with the elements of this theory and enables us to restrict ourselves to the basic facts and to pass directly to the study of the physical modelling method and its applications.

The problems of mathematical programming in general, and linear programming in particular, cannot be separated from their numerical aspects. It is therefore important that the method of physical modelling has in fact given rise to a great number of algorithms for the numerical solution of mathematical programming and economics problems. The algorithms described below are derived naturally from the physical properties of the corresponding models and the convergence of the resulting calculation procedures is physically obvious.

The most important property of the models for the optimum problems under discission lies in the fact that they are models for a corresponding dual pair of problems and that, in fact, the equilibrium state defines the optimal vector for both the primal and dual problems.

We shall proceed on the basis of the following formulation of a primal and dual linear programming model. Let there be given two sets of indices

$$M = \{1,2,\dots,m\}, \quad N = \{1,2,\dots,n\}$$

and let each of them be divided into two disjoint subsets

$$M_1 = \{1,2,\dots,m_1\}, \quad M_2 = \{m_1+1,\dots,m\}$$

$$N_1 = \{1,2,\dots,n_1\}, \quad N_2 = \{n_1+1,\dots,n\}.$$

Given this, the general linear programming problem can be described as follows. Find an n-dimensional vector $x = (x_1,\dots,x_n)$ which satisfies the conditions

$$\sum_{i=1}^{n} p_i x_i \rightarrow \max$$

$$\sum_{i=1}^{n} a_{si} x_i \leqslant b_s \quad \text{for} \quad s \in M_1$$

$$\sum_{i=1}^{n} a_{si}x_i = b_s \quad \text{for} \quad s \in M_2 \tag{1.18}$$

$$x_i \geqslant 0 \quad \text{for} \quad i \in N_2.$$

The quantities x_i for $i \in N_1$ may be both positive or negative (or zero).

The following problem is called the dual to the problem just formulated. Find an m-dimensional vector $w = (w_1,\dots,w_m)$ which satisfies the conditions

$$\sum_{s=1}^{m} b_s w_s \to \text{minimum}$$

$$\sum_{s=1}^{m} a_{si}w_s = p_i \quad \text{for} \quad i \in N_1$$

$$\sum_{s=1}^{m} a_{si}w_s \geqslant p_i \quad \text{for} \quad i \in N_2 \tag{1.19}$$

$$w_s \geqslant 0 \quad \text{for} \quad s \in M_1.$$

The quantities w_s for $s \in M_2$ can be both positive or negative (or zero).

Introducing quantities $\xi_1,\dots,\xi_m$ for the primal (direct) problem and $\eta_1,\dots,\eta_n$ for the dual one, we can rewrite these problems in the following form. The primal problem:

$$\sum_{i=1}^{n} p_i x_i \to \text{maximum} \tag{1.20}$$

$$\sum_{i=1}^{n} a_{si}x_i - \xi_s = 0 \quad \text{for} \quad s \in M$$

$$x_i \geqslant 0 \qquad \text{for} \quad i \in N_2 \tag{1.21}$$

$$\xi_s = \begin{cases} \leqslant b_s & \text{for} \quad s \in M_1 \\ = b_s & \text{for} \quad s \in M_2 \end{cases}$$

and the dual problem

$$\sum_{i=1}^{m} b_s w_s \to \text{minimum} \tag{1.22}$$

$$\sum_{s=1}^{m} a_{si} w_s - \eta_i = 0 \quad \text{for} \quad i \in N$$

$$w_s \geqslant 0 \qquad \text{for} \quad s \in M_1 \tag{1.23}$$

$$\eta_i = \begin{cases} = p_i & \text{for} \quad i \in N_1 \\ \geqslant p_i & \text{for} \quad i \in N_2. \end{cases}$$

We shall call the linear forms in (1.20) and (1.22) the objective functions for the primal and dual problems.

Recall that the n-vector x and the m-vector w are called feasible (admissible) if vector x satisfies conditions (1.21) and vector w satisfies conditions (1.23).

The vectors x and w, which satisfy, respectively, conditions (1.20)-(1.21) and (1.22)-(1.23), are called the optimal vectors for the primal and dual problems.

Fundamental for the theory of linear programming are the duality theorems which tie together the primal and dual problems. We shall now state the fundamental duality theorems but not discuss their proofs which the reader can find in books by Dantzig [24], Gale [29], Karlin [34], and other treatises on linear programming.

THEOREM 1.1. <u>If $(x_1,\ldots,x_n)$ is a feasible point for the primal problem and $(w_1,\ldots,w_m)$ is a feasible point for the dual problem and we have</u>

$$\sum_{i=1}^{m} a_{si} w_s - p_i = 0 \qquad \text{for} \quad i \in N_1 \tag{1.24}$$

$$\sum_{s=1}^{m} a_{si} w_s - p_i = R_i \geqslant 0 \quad \text{for} \quad i \in N_2 \tag{1.25}$$

<u>then the condition</u>

$$R_i = 0 \quad \text{if} \quad x_i > 0 \tag{1.26}$$

<u>is a necessary and sufficient condition for the optimality of both vectors.</u>

THEOREM 1.2. <u>If there exists a feasible point for the primal problem, and if the objective function is bounded on the set</u>

of feasible points for the primal problem, then there exists an optimal solution.

THEOREM 1.3. *If there exists an optimal solution for the primal problem then there also exists an optimal solution for the dual problem. Moreover, the value of the objective function of the primal problem at an arbitrary feasible point is no larger than the value of the objective function of the dual problem at any of its feasible points and, in addition, for optimal solutions the objective functions of both problems have equal value, i.e.*

$$\max \sum_{i=1}^{n} p_i x_i = \min \sum_{s=1}^{m} b_s w_s . \qquad (1.27)$$

THEOREM 1.4. *Optimal solutions* x^* *and* w^* *of the primal and dual problems have the following properties*

1) *If in formula* (1.21) $\xi_s^* < b_s$ *for some* $s \in M_1$ *then* $w_s = 0$
2) *If in formula* (1.23) $\eta_i^* > p_i$ *for some* $i \in N_2$ *then* $x_i = 0$
3) *If* $x_i^* > 0$ *for some* $i \in N_2$ *then* $\eta_i^* = p_i$
4) *If* $w_s^* > 0$ *for some* $s \in M_1$ *then* $\xi_s^* = b_s$.

Theorem 1.4. means that if in one of the pair of problems an optimal solution satisfies a constraint with a strict inequality, then the component of the solution of the dual problem corresponding to this constraint is equal to zero and, inversely, if some component which must satisfy a nonnegative constraint of an optimal vector for one of the two dual problems is strictly positive, then the constraint corresponding to this component is satisfied with equality.

We now proceed to consider a mechanical model for a linear programming problem.

If in the models for systems of linear equations and inequalities depicted by figures 1.6 or 1.7 we insert stops restricting the possible positions of the pistons in the columns $n_1+1,\ldots,n$ of coaxial cylinders to positions for which $x_i \geqslant 0$ $(i \in N_2)$, then we have a mechanical system for which the possible states define precisely the feasible points of a corresponding linear programming problem. Thus, one obtains two models for the constraints (1.18) or (1.21), respectively, for general linear programming problems. Both these models are passive mechanical systems in the sense that they are free of external forces acting on them.

Now suppose that the bars (connecting the pistons) of the model of the system of constraints (1.21) are subject to forces $p_1, p_2, \ldots, p_n$. Here force p_i is directed in the same direction as x_i if $p_i > 0$ and in the opposite direction if $p_i < 0$.

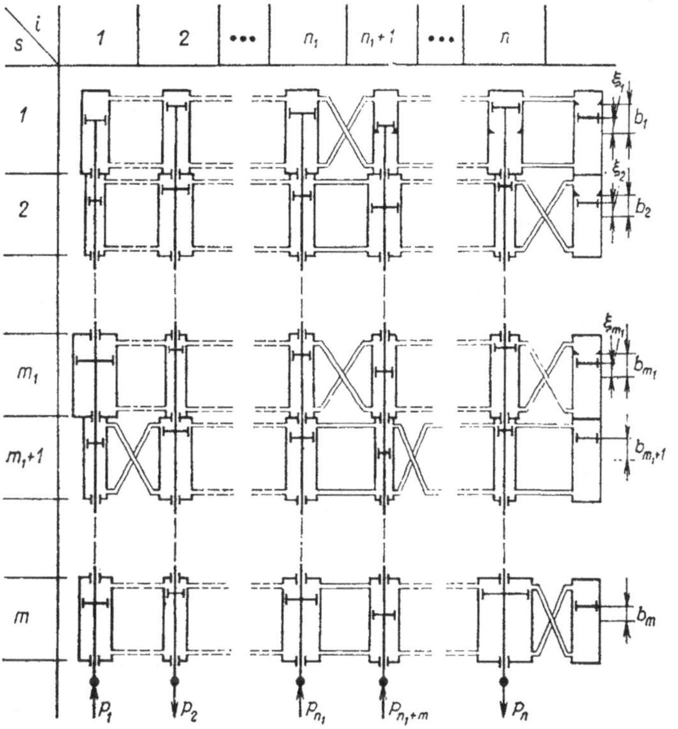

Fig. 1.8.

The potential energy of this mechanical system, depicted in Figure 1.8, is equal to $-\sum_{i=1}^{n} p_i x_i$ and it assumes its minimal value at the equilibrium point x^* which then clearly maximizes the objective function $\sum_{i=1}^{n} p_i x_i$ [1)]. This concludes the construction of a mechanical model for the linear programming problem (1.20)-(1.21).

It remains to show that the equilibrium states of the model constructed, as depicted in Figure 1.8, are indeed optimal solutions for the general linear programming problem. To this end, let us first of all obtain the conditions for equilibrium of the model in question, i.e. the mechanical system subject to unilateral and bilateral constraints (1.21). Recall that ideal constraints are said to be bilateral if their analytical expression is an equality and unilaterial if that expression is an inequality. We note that by enlarging the dimension of the system, that is by introducing the free coordinates $x_{n+1},\ldots,x_{n+k}$, it is possible to rewrite the unilaterial constraints

$$g_s(x_1,\ldots,x_n) \geqslant 0, \quad s = 1,\ldots,k$$

in the following form

$$g_s(x_1,\ldots,x_n) - x_{n+s} = 0, \quad x_{n+s} \geqslant 0, \quad s = 1,2,\ldots,k$$

In this last formulation, it is useful that the unilateral constraints have a particularly simple form:

$$x_{n+s} \geqslant 0, \quad s = 1,\ldots,k.$$

Let us return to the illustration of the model of linear programming problem (Figure 1.8). In the equilibrium state the exterior forces $p_1,\ldots,p_n$, acting on the bars of the model, must be balanced by the internal pressures in volumes $V_s^{(+)}$ and $V_s^{(-)}$.

Let us denote with $q_s^{(+)}$ and $q_s^{(-)}$ the respective pressures in volumes $V_s^{(+)}$ and $V_s^{(-)}$. Obviously, the model will be in equilibrium if and only iff each of its various movable solid parts is in equilibrium, that is, if and only if this is the case for each of the n bars on which the forces $p_1,\ldots,p_n$ act and also the case for the m_1 free pistons whose positions define the quantities $\xi_1,\ldots,\xi_{m1}$.

From Figure 1.9, which depicts the i-th system of coaxial cylinders of the model, it is clear, that in the case $i \in N_1$, that is, in the case where no unilateral constraint $x_i \geqslant 0$ applies, the condition for equilibrium has the form

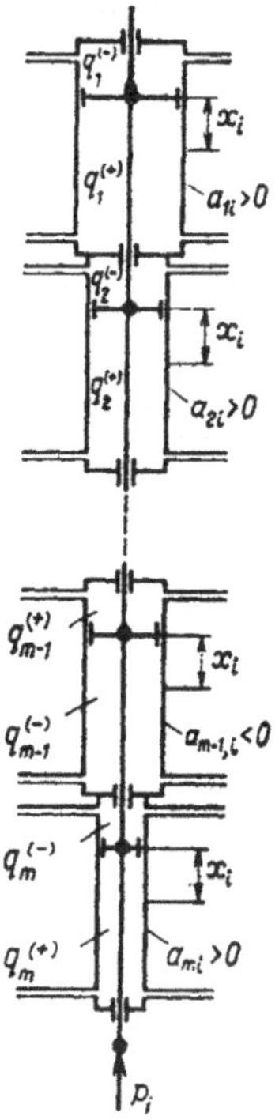

Fig. 1.9.

$$\sum_{s=1}^{m} a_{si}(q_s^{(-)}-q_s^{(+)}) - p_i = 0, \quad i \in N_1. \tag{1.28}$$

Here $a_{si}(q_s^{(-)}-q_s^{(+)})$ is the force acting on the s-th piston fixed to the i-th bar. In the case $i \in N_2$, that is in the case where the movements of the i-th bar are restricted by unilateral constraint $x_i \geqslant 0$, the condition for equilibrium has the following form

$$\sum_{s=1}^{m} a_{si}(q_s^{(-)}-q_s^{(+)}) - p_i = \begin{cases} =0 \text{ if } x_i^* > 0 \\ \geqslant 0 \text{ if } x_i^* = 0 \end{cases}, \quad i \in N_2 \tag{1.29}$$

where x_i is the position of the i-th bar at equilibrium.

The inequality in the case $x_i^* = 0$ means that the possibly nonnegative value of the force on the left-hand side of (1.29) is compensated by a reaction force coming from the stop (the unilateral constraint) which forces x_i to be $\geqslant 0$ (cf. conditions (1.24) and (1.25) of Theorem 1.1). The conditions for equilibrium for the pistons, which indicate the values of the quantities ξ_s for $s \in M_1$, are clearly

$$q_s^{(-)} - q_s^{(+)} = \begin{cases} = 0 & \text{for } \xi_s < b_s \\ \geqslant 0 & \text{for } \xi_s = b_s \end{cases} \tag{1.30}$$

Conditions (1.28), (1.29), and (1.30) are a set of equilibrium conditions for this model of a general linear programming problem. In an equilibrium state let us write

$$w_s^* = q_s^{(-)} - q_s^{(+)}, \quad s = 1,2,\dots,m$$

then the equilibrium conditions can be written as follows:

$$\sum_{s=1}^{m} a_{si} w_s^* - p_i = 0 \quad \text{for} \quad i \in N_1 \tag{1.31}$$

$$\sum_{s=1}^{m} a_{si} w_s^* - p_i = \begin{cases} = 0 & \text{if } x_i^* > 0 \\ > 0 & \text{if } x_i^* = 0 \end{cases} \quad \text{for } i \in N_2 \tag{1.32}$$

$$w_s^* = \begin{cases} = 0 & \text{if } \xi_s < b_s \\ \geqslant 0 & \text{if } \xi_s = b_s \end{cases} \quad \text{for } s \in M_1 \tag{1.33}$$

Comparing (1.31) and (1.32) with conditions (1.24)-(1.26) of Theorem 1.1 and (1.33) with conditions 1) and 4) of Theorem 1.4, we obtain the following important result.

<u>The pressure differences</u> $q_s^{(-)} - q_s^{(+)}$ <u>of the pressures in</u> $V_s^{(-)}$ <u>and</u> $V_s^{(+)}$ <u>of the model of the primal problem in an equilibrium state are the compoments of an optimal vector for the dual problem and the equilibrium conditions of the model express the fundamental duality theorem.</u>

And thus we have established that the model we have constructed represents a dual pair of linear programming problems and that an equilibrium state defines a pair of optimal vectors for the primal and dual problems.

From equilibrium conditions (1.31)-(1.33) one also

easily obtains the results of Theorem 1.3. Multiply the respective equations and inequalities (1.31) and (1.32) with x_i^* and sum the results. This obviously gives the equality

$$\sum_{i=1}^{n} \sum_{s=1}^{m} a_{si} w_s^* x_i^* - \sum_{i=1}^{n} p_i x_i^* = 0.$$

Further, using constraints (1.21), let us write this equality in the form

$$\sum_{s=1}^{m} w_s^* b_s - \sum_{s=1}^{m_1} w_s^*(b_s - \xi_s) = \sum_{i=1}^{n} p_i x_i^*. \tag{1.34}$$

Now use conditions (1.33) which imply

$$w_s^*(b_s - \xi_s) = 0, \quad s \in M_1 \tag{1.35}$$

From (1.34) and (1.35) we obtain the well-known result

$$\sum_{s=1}^{m} w_s^* b_s = \sum_{i=1}^{n} p_i x_i^*$$

which expresses the equality of the optimal values of the objective functions of the primal and dual problems.

Thus, the first duality theorem is equivalent to the statement that vector (x^*, w^*) of an equilibrium state of our model of a linear programming problem is an optimal vector of the dual pair of problems (1.20)-(1.21) and (1.22)-(1.23) and, consequently, the equilibrium conditions (1.31)-(1.33) are necessary and sufficient conditions for the optimality of this vector. Because the linear programming problem is equivalent to an equilibrium problem of a mechanical system, it follows that the necessary and sufficient conditions for optimality derive from the principles of analytical mechanics. The proof of this statement is based on the fundamental theorem of mechanics which expresses the principle of virtual displacements (Lagrange [38]; Appell [19]).

THEOREM 1.5. *A necessary and sufficient condition that the system be in equilibrium in the position* $x_1, \ldots, x_n$ *is that for all possible displacements compatible with the constraints in a neighbourhood of state* x, *the sum of the amount of work performed by the given forces be zero or negative; it must* be zero in the case of displacements of inessential constraints only and zero or negative in the case of displacements of essential constraints.

Some extra explanations may be useful. Let us consider a mechanical system whose movements are subject to m_1 unilateral constraints of the form

$$g_s(x_1,\ldots,x_n) \leqslant 0, \quad s = 1,\ldots,m_1$$

and m_2 bilateral constraints

$$g_s(x_1,\ldots,x_n) = 0, \quad s = 1,\ldots,m_2.$$

The unilaterial constraint

$$g_s(x_1,\ldots,x_n) \leqslant 0$$

is called active at the state x if there is equality

$$g_s(x_1,\ldots,x_n) = 0$$

and passive if

$$g_s(x_1^*,\ldots,x_n^*) < 0.$$

In Theorem 1.5, one considers only those unilateral constraints which are active at x*, because it is clear that the passive unilateral constraints do not constitute a restriction for small possible diaplacements. One must also remember that the phrase 'the given forces' refers to those forces applied to the system and which are not reaction forces deriving from the constraints.

Thus let $x_1^*,\ldots,x_n^*$, $\xi_1^*,\ldots,\xi_{m_1}^*$ be an equilibrium state of the model satisfying equilibrium conditions (1.31)-(1.33) and let the following constraints be active

$$x_i \geqslant 0 \quad \text{for} \quad i \in N_2^{(1)} \subset N_2$$

$$\xi_s \leqslant b_s \quad \text{for} \quad s \in M_1^{(1)} \subset M_1.$$

This means that

$$\begin{aligned} x_i &= 0 \quad \text{for} \quad i \in N_2^{(1)} \\ \xi_s &= b_s \quad \text{for} \quad s \in M_1^{(1)}. \end{aligned} \tag{1.36}$$

From the constraints (1.21), viewed as an analytical expression of the unilateral and bilateral ideal constraints, and conditions (1.36) it follows that the possible displacements δx and $\delta\xi$ are subject to

$$\sum_{i=1}^{n} a_{si}\delta x_i - \delta\xi_s = 0 \qquad s \in M \tag{1.37}$$

$$\delta\xi_s = 0 \qquad s \in M_2 \tag{1.38}$$

$$\delta x_i \geqslant 0 \qquad i \in N_2^{(1)} \subset N_2 \tag{1.39}$$

$$\delta\xi_s \leqslant 0 \qquad s \in M_1^{(1)} \subset M_1. \tag{1.40}$$

Multiplying equilibrium conditions (1.31)-(1.32) with the respective x_i, summing over i and taking (1.39) into account we obtain

$$\sum_{s=1}^{m} w_s^* \sum_{i=1}^{n} a_{si}\delta x_i - \sum_{i=1}^{n} p_i\delta x_i \geqslant 0$$

or, according to (1.37) and (1.38),

$$\sum_{s\in M_1} w_s^*\delta\xi_s - \sum_{i=1}^{n} p_i\delta x_i \geqslant 0. \tag{1.41}$$

Multiplying condition (1.33) with the respective $\delta\xi_s$ for $s \in M_1$, and taking (1.40) into account, we obtain

$$\sum_{s\in M_1^{(1)}} w_s^*\delta\xi_s \leqslant 0, \qquad \sum_{s\in M_1\setminus M_1^{(1)}} w_s^*\delta\xi_s = 0 \tag{1.42}$$

It follows from (1.41) and (1.42) that

$$\delta\left(\sum_{i=1}^{n} p_i x_i\right) = \sum_{i=1}^{n} p_i\delta x_i \leqslant \sum_{s\in M_1} w_s^*\delta\xi_s \leqslant 0$$

Thus, Theorem 1.1 follows from the equilibrium conditions

of the model and the condition

$$\delta\left(\sum_{i=1}^{n} p_i x_i\right) \leqslant 0,$$

which expresses that the objective function reaches its maximum in the equilibrium state. Theorem 1.4 now clearly follows from equilibrium conditions (1.31)-(1.33).

Let us now elucidate the mechanical meaning of the optimal value equality of the objective functions of a primal and dual pair of programming problems. Let x^* be an optimal vector of the linear programming problem (1.18). One easily verifies that the vector λx^*, where $\lambda \geqslant 0$ is a scalar, is an optimal vector for the problem

$$\sum_{i=1}^{n} p_i x_i \rightarrow \text{maximum}$$

$$\begin{aligned} &\sum_{i=1}^{n} a_{si} x_i \leqslant b_s \quad && s \in M_2 \\ &\sum_{i=1}^{n} a_{si} x_i = b_s \quad && s \in M_2 \\ &x_i \geqslant 0 \quad && s \in N_2 \end{aligned} \qquad (1.43)$$

for every $\lambda \geqslant 0$, and the optimal vector w^* of the problem dual to (1.18) is independent of λ. That this assertion is correct follows immediately from equilibrium conditions (1.31)-(1.33) [2] which remain valid after substituting λx^* for x^*.

The independence of vector w^* with respect to parameter λ also follows from the form of problem (1.19) in which the substitution of λb_s for b_s only results in a multiplication of the objective function with a positive factor.

Consider now the model of Figure 1.10 for the problem (1.43), in which parameter λ is a variable quantity and $-b_s$ is the (n+1)st column vector of the matrix of coefficients. In the model depicted by Figure 1.10, the (n+1)st column is a system of coaxial cylinders whose base areas are equal to $|b_1|,\ldots,|b_m|$. The last column, the (n+2)nd, contains m_1 cylinders whose base areas are equal to one.

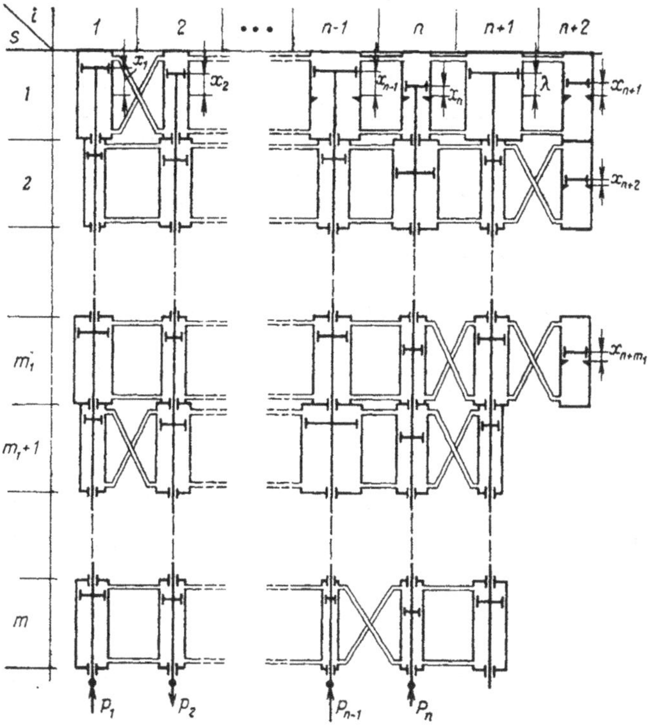

Fig. 1.10

As in the case of Figure 1.6, the positions of the pistons in these cylinders indicate the values of the free variables

$$x_{n+s} = - \sum_{i=1}^{n} a_{si}x_i + b_s, \quad s \in M_1.$$

Obviously if we fix the (n+1)st bar in the position $\lambda = 1$ we obtain the model for the linear programming problem (1.18). Suppose that the model is in an equilibrium state x^* for $\lambda = 1$.

Parameter λ determines the external conditions and if this parameter varies the system must either perform work on the exterior bodies or the exterior bodies must supply work to the system. Correspondingly, the potential energy of the physical model must either increase or decrease. Let us now implement the transition process of the model from an equilibrium state for $\lambda = 1$ to an equilibrium state for $\lambda = 0$, by varying the magnitude of parameter λ sufficiently slowly so that every intermediate state can be assumed to be an equilibrium state for the corresponding value of parameter λ. Such processes are called quasi static and they play a fundamental role in thermodynamics. Below (in Chapter II) we shall explain the basic concepts of thermodynamics, but, meanwhile let us just remark that in mechanics a process is called quasi static if the total energy of the sytem at every moment in time is only a function of the coordinates with respect to a fixed set of axes defining the position of the bodies making up the system.

Because vector w^* is independent of parameter λ, the work received by the system (model) during a quasi static transition from state x^* to the origin is equal to

$$\sum_{s=1}^{m} b_s w_s^*.$$

The potential energy, meanwhile, varies from $-\sum_{i=1}^{n} p_i x_i^*$ to zero. Thus, the equality

$$\sum_{i=1}^{n} p_i x_i^* = \sum_{s=1}^{m} b_s w_s^*$$

expresses that well-known theorem of analytical mechanics [3] which Maurice Levy formulates in the following manner

THEOREM 1.6. The potential energy is the maximum amount of effective work which can be obtained using only the interior forces of the system and without using the velocities acquired by the points of the system.

What has been said above shows the deep links which exist between the basics of linear programming and the fundamental principles of the analytical mechanics of systems with bilateral and unilateral constraints. It seems likely that establishing these links together with their applications to the theory of decision will bring extra depth to accounts of the fundamentals of linear programming.

Remark. If one considers problem (1.22)-(1.23) as original then, of course, problem (1.20)-(1.21) is dual. Then in an equilibrium state of the model of (1.22)-(1.23) the positions of the pistons define the components of the optimal vector w^* of problem (1.22)-(1.23) and the pressure differences in the respective volumes $V_i^{(+)}$ and $V_i^{(-)}$ define the components of the optimal vector for problem (1.20)-(1.21). The equilibrium conditions in this case are similarly obtained and they take the form

$$\sum_{i=1}^{n} a_{si}x_i^* - b_s = 0 \quad s \in M_2 \tag{1.44}$$

$$\sum_{i=1}^{n} a_{si}x_i^* - b_s = \begin{cases} = 0 & \text{if } w_s^* > 0 \\ \leqslant 0 & \text{if } w_s^* = 0 \end{cases}, \quad s \in M_1$$

$$x_i^* = \begin{cases} = 0 & \text{if } \eta_i > p_i \\ \geqslant 0 & \text{if } \eta_i = p_i \end{cases}, \; i \in N_2. \tag{1.45}$$

It is now important to note that conditions (1.31)-(1.33) and conditions (1.44)-(1.46) are equivalent, as both are equilibrium conditions for mechanical systems both of which constitute an exact model for one and the same pair of dual linear programs. Thus, conditions (1.31)-(1.33) and (1.44)-(1.46) are two analytical expressions of sufficient and necessary conditions for optimality of a feasible bivector 4) (x^*,w^*) for a dual pair of linear programming problems.

NOTES

1. The potential energy is normalized by setting it equal to zero for the state of system $x_1 = \ldots = x_n = 0$ (even when that state is not feasible).
2. The optimality of $x = 0$ for $\lambda = 0$ follows from Theorem 1.2.
3. The statement of the theorem below expresses the principle of conservation of energy. It should be remarked that the same interpretation of the fundamental duality relation has been found by Dennis [26] in his studies of the analogy between problems of mathematical programming and the theory of electrical networks.
4. A bivector is a pair of vectors of a different nature, defining the state of a system. An example of a bivector in mechanics is the pair (R,M) where R is the resultant force vector of all the forces acting on the system and M is the resultant moment of these forces with respect to some centre.

Chapter II

EQUILIBRIUM OF PHYSICAL SYSTEMS AND LINEAR PROGRAMMING PROBLEMS

2.1. Introduction

In Chapter I we saw that the general linear programming problem is equivalent to an equilibrium problem for certain mechanical systems. We also saw that the mechanical model was a model for both the primal and the dual linear programming problems and that the equilibrium state of the model determined the optimal vector for both these problems. In this way the fundamental results of the theory found an intuitive mechanical interpretation and in fact turned out to be consequences of the basic principles of mechanics. In this chapter we proceed with the consideration of models in which the active substance subjected to the various constraints is not an incompressible fluid but an ideal gas. Below we shall see that switching to such models opens up great possibilities for the construction of algorithms for solving linear and nonlinear programming problems.

Let us return once more to the problem of finding the solutions of an algebraic equation with constant coefficients.

$$\sum_{i=1}^{n} a_i x_i = b. \tag{2.1}$$

A model of this problem is pictured in Figure 1.3. In Chapter I we convinced ourselves that this model admits all states which satisfy Equation (2.1) if the volumes $V^{(+)}$ and $V^{(-)}$ are filled with an incompressible fluid in the amount of $\overline{V}^{(+)}$ and $\overline{V}^{(-)}$ respectively, cf. Section 1.2. Now we shall consider what happens if the model is equipped with a thermostat at temperature T and we fill the volumes $V^{(+)}$ and $V^{(-)}$ with an ideal gas in such quantities that for all states $(x_1,\dots,x_n)$ satisfying (2.1) the pressures in the two volumes are equal and equal to an arbitrary pregiven positive value q_0.

Let us denote with $\mu^{(+)}$ and $\mu^{(-)}$ the number of mols [1)] of the ideal gas we are looking for in the respective volumes $V^{(+)}$ and $V^{(-)}$ and recalling that the pressure, volume, absolute temperature and number of mols are connected by means of the Clapeyron-Mendeleev equation

$$q_0\bar{V}^{(+)} = \mu^{(+)}RT, \quad q_0\bar{V}^{(-)} = \mu^{(-)}RT$$

(where R is the universal gas constant) one finds from this that

$$\mu^{(+)} = \frac{q_0\bar{V}^{(+)}}{RT}, \quad \mu^{(-)} = \frac{q_0\bar{V}^{(-)}}{RT}.$$

In the following we shall limit ourselves to the consideration of isothermal processes and we shall assume that for every state transition in the model the temperature of the gas remains constant. In the model with containers filled with an incompressible liquid the only possible states were those for which the Equation (2.1) was fulfilled, precisely because the liquid was incompressible. But now all states are possible but Equation (2.1) will be satisfied only in an equilibrium state for which the conditions

$$V^+ = \bar{V}^{(+)} \quad \text{or} \quad V^- = \bar{V}^{(-)}$$

hold. It follows in fact from the Clapeyron-Mendeleev equation that it is only under these conditions that the pressures below and above each piston are equal and all pistons are in a state of equilibrium.

An analogous statement also holds in the case of a system of linear equations or inequalities of which a model is depected by Figure 1.7. If in the volumes $V_s^{(+)}$ and $V_s^{(-)}$, $s = 1,\ldots,m$ (see formula (1.16)) of this model one replaces the incompressible fluid by an ideal gas in the amount of mols given by the equations

$$\mu_s^{(+)} = \frac{q_0 V_s^{(+)}}{RT}, \quad \mu_s^{(-)} = \frac{q_0 V_s^{(-)}}{RT}, \tag{2.2}$$

where $\bar{V}_s^{(+)}$ and $\bar{V}_s^{(-)}$ are given by formula (1.17), then one finds a model of the system (1.11) such that the set of equilibrium states of the model coincides with the set of solutions of (1.11)-(1.12). However every state $x_1^{(0)},\ldots,x_n^{(0)}, \xi_1^{(0)},\ldots,\xi_{m_1}^{(0)}$ is still physically possible.

Proceeding similarly with respect to a linear programming model, i.e. by replacing the incompressible fluid in the volumes $V^{(+)}$ and $V^{(-)}$ with an ideal gas in the amounts given by (2.2) we obtain a physical model such that the equilibrium states are determined by the constraints (1.21) and the magnitudes of the active forces $p_1,\ldots,p_n$ and the physical properties of the gas constrained by the restrictions to which the system is subjected. Precisely the physical properties of the active medium will play an important role in the following in the formulation of numerical algorithms for the solution of linear programming problems and problems from mathematical economics.

Does an equilibrium state of such a model indeed define an optimal solution of a mathematical programming problem? How does the state function look which attains its maximum for equilibrium states? These are the questions which we shall try to elucidate in the following. For now we can only say that in an equilibrium state some thermodynamical potential attains an extremum. Therefore we should leave behind for the moment the subject of physical models for extremum problems and it seems necessary to recall some facts from physics and in particular from that part of physics which is called thermodynamics.

2.2. Some concepts from thermodynamics

In the following we shall deal with physical models of important mathematical programming problems and we shall see that these problems are equivalent to certain problems on equilibrium states of the corresponding physical systems. It is therefore necessary to remind the reader of some concepts from physics. We restrict ourselves to a highly condensed basic outline of equilibrium theory which is the basic core of thermodynamics.

The concepts of system and state. A system is some isolated collection of bodies which may be in mutual interaction and which may also be in interaction with bodies not belonging to the isolated collection. The interactions may take various forms: for example the form of transport of motion, energy or mass, or it may take the form of chemical reactions.

Thermodynamics considers systems of bodies of sufficient size so that aggregated or averaged quantitites can be considered to characterize the properties of the systems.

Examples of such quantities are volume, pressure, density, concentration, temperature, generalized coordinates, velocities, inertial masses entering the systems etcetera. The indicated quantities and indicators characterize the system under consideration and have meaning only in so far as they can be measured directly or indirectly; they are called the state parameters of the system.

Equilibrium states and state equations. A state of a physical system is called an equilibrium state if all the state parameters of the system remain constant for arbitrarily long times as long as the exterior conditions do not change. Equilibrium is an important concept in thermodynamics. We have already remarked that one or another state parameter has meaning only if some direct or indirect method is indicated for measuring it. Measurements take place by means of contact between the measuring device and the bodies of the systems. Therefore the measuring device just indicates the value of a parameter of its own state, and consequently the possibility of measuring requires the following equilibrium postulate to hold: If the body and the device which is in contact with the body are isolated from external influences, then an equilibrium establishes itself between them and this equilibrium is maintained as long as necessary and moreover this equilibrium does not depend on the initial state of the device. Thus thermodynamics studies essentially only processes which are sufficiently slow, that is processes whose velocity is small compared to the average speed with which a local equilibrium establishes itself.

In general a physical system admits a whole family of equilibrium states. And it appears that for any of the possible equilibrium states the state parameters are not independent but are connected by certain relations which are usually determined experimentally. Such a relation, when expressed in analytical form, is called a state equation. Examples of such state equations are: the Clapeyron-Mendeleev equation for an ideal gas, the Van der Waals equation for a real gas, and Hooke's law for elastic bodies.

The characteristic variables of a physical system can be divided into two classes depending on whether they depend on the dimensions and mass of the system or not: intensive and extensive variables. If a system which is in equilibrium is divided into parts by means of impermeable walls then each part will remain in equilibrium. An equilibrium state of a homogeneous system is therefore an intrinsic property and it determines parameter values which are independent of the dimensions of the system. Such parameters are called intensive. Examples of intensive parameters are temperature, pressure, concentration. On the other hand there are parameters whose values are proportial to the size or mass of the system when it is divided into parts without disturbing the equilibrium. These are called extensive variables, and examples are volume, mass, energy and entropy.

The reason we pay attention to this division of the parameters into two groups is that in the physical models of extremum problems this corresponds to the division of quantities into primal and dual ones. It is thus that the physical model of the primal problem also becomes a model of the dual problem and that the optimal vectors of the two problems are connected by the Clapeyron-Mendeleev state equation.

Conditions for equilibrium. The theory of equilibrium rests on the concept of (thermodynamic) potential introduced by Gibbs. In thermodynamics as in mechanics, a potential is a state function such that changes in its value reflect the amount of work done by the system and such that the maximum or minimum values define the equilibrium states. Below we shall see that the conditions for equilibrium of a given state of a physical system stated by means of minimum conditions on some state function follow from the basic principles of thermodynamics.

The first principle of thermodynamics is the law of the conservation of energy and it can be stated in the form

$$dU + \delta A = \delta Q, \tag{2.3}$$

where U is the internal energy, A the amount of work and Q the amount of heat. Let us consider a system which evolves from a state defined by the vector X to the state $X + \Delta X$. Equation (2.3) signifies that for each transition of the system from state X to state $X + \Delta X$ the sum of the change in internal energy and the amount of work done by the system in this transition is equal to the amount of heat received by the system from external sources. The internal energy turns out to be a function of the state and its changes do not depend on the precise path followed from X to $X + \Delta X$. Therefore the change in internal energy is a total differential which is denoted dU. The quantities A and Q in general do depend on the precise process (path) involved and their changes are therefore denoted δA and δQ. Let Y be a vector of parameters describing the exterior bodies which are in interaction with the system and let the components of the vector Y be such that the state X is an equilibrium state in the sense of the definition given above. Let us suppose that the vector $Y + \Delta Y$ defines exterior conditions for which $X + \Delta X$ is an equilibrium state. If the exterior conditions change continuously and infinitely slowly from the state Y to the state $Y + \Delta Y$ then the system under consideration likewise changes continuously and infinitely slowly from state X to state $X + \Delta X$, passing through a continuous sequence of intermediate equilibrium states. We thus obtain idealized processes which play an important role in thermodynamics; they are called quasistatic processes.

Consider two processes under which the system evolves from X to $X + \Delta X$. Let one of them be quasistatic and the other a real process, that is a process corresponding to a change of exterior conditions in finite time (Figure 2.1).

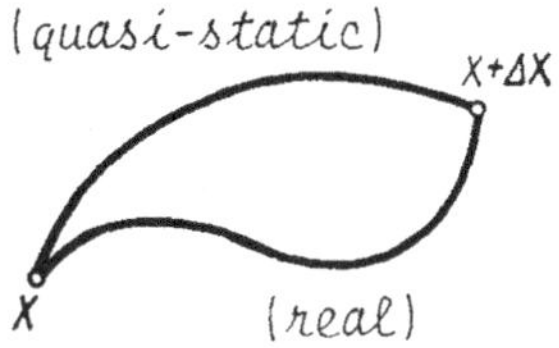

Fig. 2.1.

Denote with δA_{qs} and δQ_{qs} the corresponding amounts of work and heat in the case of the quasistatic transition and with δA_{re} and δQ_{re} the values of these quantities in case of a real transition from the state X to the state $X + \Delta X$. For the quasistatic process the law of the conservation of energy can be written in the form

$$dU + \delta A_{qs} = \delta Q_{qs}. \tag{2.4}$$

Using (2.4) let us write down the negative difference of the work done by the system in the transition from X to $X + \Delta X$ in the two processes.

$$-(\delta A_{qs} - \delta A_{re}) = dU - \delta Q_{qs} + \delta A_{re}. \tag{2.5}$$

The amount of work δA_{re} done by the system consists of the work done by the forces of pressure under the changes of volumes of the various parts of the system and the work done by generalized forces (acting on the system). Thus,

$$\delta A_{re} = \sum_{s=1}^{m} q_s \delta V_s + \sum_{i=1}^{n} p_i \delta x_i \tag{2.6}$$

where $V_1,\ldots,V_m$ are the volumes occurring in the system, $q_1,\ldots,q_m$ the pressures in these volumes; $x_1,\ldots,x_n$ are the generalized coordinates defining the positions of the solid parts making up the system and $p_1,\ldots,p_n$ are the generalized forces corresponding to the generalized coordinates $x_1,\ldots,x_n$.

We now use the second principle of thermodynamics which in the formulation of Clausius says the following: a process in which there is no other change than an exchange of heat from a warmer body to a colder is irreversible; or in other words, heat cannot spontaneously go from a colder body to a warmer one without some other change in the system.

By means of the concept of entropy, introduced by Clausius, this second principle can be given a mathematical form. Suppose that the system goes through a cyclic process such that it comes in thermic contact with a sequence of heat sources (baths) at temperatures $T_1,T_2,\ldots,T_N$. Let $Q_1,Q_2,\ldots,Q_N$ be the amounts of heat exchanged with the sources. Let us say that $Q_i > 0$ if Q_i is an amount of heat obtained from the source with temperature T_i and $Q_i < 0$ if Q_i is an amount of heat supplied by the system. From the second principle of thermodynamics there follows the inequality

$$\sum_{i=1}^{n} \frac{Q_i}{T_i} \leqslant 0 \tag{2.7}$$

or more precisely

$$\sum_{i=1}^{N} \frac{Q_i}{T_i} = \begin{cases} = 0 & \text{if the cycle is reversible} \\ < 0 & \text{if the cycle is irreversible} \end{cases}$$

In the case where the sources are distributed continuously obviously

$$\oint \frac{dQ}{T} = \begin{cases} = 0 & \text{for a reversible cycle} \\ < 0 & \text{for an irreversible cycle.} \end{cases} \tag{2.8}$$

The inequality (2.7) gave Clausius the idea of constructing

a new state function which he called entropy.

Indeed, let X_0 be an arbitrary state which will be taken as a starting point, and let X_1 be some other state. The entropy is then equal to the integral

$$S = \int_{X_0}^{X_1} \frac{dQ_{rev}}{T}$$

calculated along the trajectory of a reversible process going from X_0 to X_1. By (2.8) this integral is independent of the particular reversible process chosen.

Let σ be some process under which the system goes from the state X_1 to the state $X_1 + \Delta X$. The obvious additivity of entropy and the inequality (2.8) then give

$$S(X_1+\Delta X) - S(X_1) \geqslant \int_{X_1}^{X_1+\Delta X} \frac{dQ}{T} \qquad (2.9)$$

or more precisely

$$S(X_1+\Delta X) - S(X_1) = \begin{cases} = \int_{X_1}^{X_1+\Delta X} \frac{dQ}{T} & \text{if } \sigma \text{ is reversible} \\ > \int_{X_1}^{X_1+\Delta X} \frac{dQ}{T} & \text{if } \sigma \text{ is irreversible.} \end{cases}$$

In the case of an isolated system (that is if during the process σ $dQ = 0$) we have from (2.9) that

$$S(X_1+ X) \geqslant S(X_1). \qquad (2.10)$$

This last inequality says that for any process which takes place in an isolated system the entropy of the final state cannot be smaller than the entropy of the initial state.

From the inequality (2.10) the following important result follows:

Within the set of states of an isolated system which are possible under the given constraints the stable equilibrium states are those for which the entropy of the system is maximal.

Now let us turn to the equality (2.5). From the definition of entropy we have

$$\delta Q_{qs} = TdS \qquad (2.11)$$

and from (2.5), (2.6), and (2.11) it follows that

$$-(\delta A_{qs}-\delta A_{re}) = dU - TdS + \sum_{s=1}^{m} q_s\delta V_s + \sum_{i=1}^{n} p_i\delta x_i. \quad (2.12)$$

Our aim is the construction of a general thermodynamic potential. This goal will have been reached if we can describe the conditions for which the left hand side of equality (2.12) is a complete differential of some state function ϕ, that is conditions under which equality (2.12) is equivalent to the following

$$d\phi = -(\delta A_{qs}-\delta A_{re}) = dU - TdS + \sum_{s=1}^{m} q_s dV_s + \sum_{i=1}^{n} p_i dx_i. \quad (2.13)$$

After integration along the trajectory of a transition of the the system from state X_0 to state X this takes the form

$$\phi - \phi_0 = U - U_0 - \int_{X_0}^{X} TdS + \sum_{s=1}^{m} \int_{X_0}^{X} q_s dV_s + \sum_{i=1}^{n} \int_{X_0}^{X} p_i dx_i. \quad (2.14)$$

These transformations from (2.12) to (2.13) and (2.14) are certainly permissble if the physical conditions are such that the integrals which enter in the right hand side of equality (2.14) can be calculated (unambiguously). Then the function ϕ is a state function and is called a general thermodynamic potential. For example the integrals in the right hand side of (2.14) can be defined if the physical conditions guarantee for each process that the following functional relations hold

$$T = T(S) \quad (2.15)$$

$$q_s = q_s(V_s), \quad s = 1,2,\ldots,m \quad (2.16)$$

$$p_i = \frac{\partial\pi}{\partial x_i}, \quad i = 1,2,\ldots,n \quad (2.17)$$

where the function $\pi(x_1,\ldots,x_n)$, if it exists, is called the force function of the generalized forces.

From (2.14) under the conditions (2.15)-(2.17) one easily obtains all well known thermodynamic potentials. For instance in the case of an isothermic system (T = constant), setting

$$\phi_0 = U_0 = S_0 = 0$$

we obtain

$$\phi = U - TS + \sum_{i=1}^{n} \int_{X_0}^{X} q_s(V_s)\, dV_s + (\pi - \pi_0).$$

As is well known the function

$$F = U - TS \tag{2.18}$$

is called the Helmholtz free energy and in the case of a passive isochoric (= constant volume) process ($dV_s = 0$, $s = 1,\ldots,m$; $P_i = 0$, $i = 1,\ldots,n$) we find

$$\phi = F$$

It now remains to show that the states in which the potential ϕ assumes its minima are equilibrium states. In fact from the first and second principles of thermodynamics, that is from the conditions

$$\delta A_{qs} = \delta Q_{qs} - dU, \quad \delta Q_{re} = \delta Q_{re} - dU$$

$$\delta Q_{qs} = TdS, \qquad \delta Q_{re} \leqslant TdS$$

it follows that

$$-\delta\phi = \delta A_{qs} - \delta A_{re} \geqslant 0 \tag{2.19}$$

or $\delta\phi \leqslant 0$.

Let X^* be a state in which the potential ϕ is minimal, that is $\phi(X^*) = \min \phi(X)$. This means that for every variation in the parameters compatible with the constraints in a neighbourhood of X^* we have the inequality

$$\delta\phi = \phi(X^* + \delta X) - \phi(X^*) \geqslant 0.$$

But according to (2.19) no such process compatible with the constraints can occur spontaneously. This means that X^* is an equilibrium state. And thus the equilibrium states of a physical system are defined by the condition

$$U - U_0 - \int_{X_0}^{X} T_d S + \sum_{s=1}^{m} \int_{X_0}^{X} q_s dV_s + \sum_{i=1}^{n} \int_{X_0}^{X} P_i dx_i \to \min.$$

To conclude we mention the particular case of an active system for which the physical conditions are such that only isochoric and isothermal processes are possible.

$$T = \text{const.}, \quad dV_s = 0, \quad s = 1,2,\dots,m.$$

Under these conditions for equilibrium take the form

$$F + \sum_{i=1}^{n} \int_{X_0}^{X} p_i dx_i \rightarrow \min \qquad (2.20)$$

where F, given by formula (2.18), is the Helmholtz free energy. Precisely such processes will be considered as physical models of extremum problems.

Because the volumes of our models will be filled with an ideal gas it is useful to quote the expressions for the free energy and the entropy for one mol of an ideal gas in a volume V (cf. for eample [36]):

$$F = c_V T + w - T(c_V \text{In}T + R\ln V + a)$$
$$S = c_V \text{In}T + R\ln V + a_1 \qquad (2.21)$$

where c_V is the heat capacity at constant volume, w, a, and a_1 are integration constants, T is the absolute temperature and R is the universal gas constant. For isothermic processes clearly

$$F = c_1 - RT\ln V, \quad S = c_2 + R\ln V \qquad (2.22)$$

where c_1 and c_2 are new arbitrary constants. From (2.22) it follows that in the case of an isothermic system the states X for which the free energy is minimal and the states with maximal entropy are the same.

2.3. Physical models of dual pairs of systems of linear equations and inequalities. Alternative theorems

Nonhomogeneous equations and inequalities. The study of the equilibrium problem of the physical model of a system of linear equations and inequalities of general type

$$\sum_{i=1}^{n} a_{si}x_i \leqslant b_s, \quad s \in M_1$$
$$\sum_{i=1}^{n} a_{si}x_i = b_2, \quad s \in M_2 \qquad (2.23)$$
$$x_i \geqslant 0, \quad i \in N_2$$

with, as previously, $M_1 = \{1,2,\ldots,m_1\}$, $M_2 = \{m_1+1,\ldots,m\}$, $N_1 = \{1,\ldots,n_1\}$, $N_2 = \{n_1+1,\ldots,n\}$ enables one to obtain a whole series of important results of the theory of linear equations and inequalities.

In this particular section we shall be concerned with the question of the existence of solutions of such systems and to that end we shall consider physical models of such systems of linear equations and inequalities consisting of containers filled with an ideal gas. Indeed to elucidate questions concerning the existence of solutions the models of containers filled with an incompressible liquid do not suffice because these models exist only when there are solutions; compare the remark in Section 1.3. The physical models with containers filled with an ideal gas are very useful for the study of problems of existence of solutions as in the equilibrium state of such a model a certain positive function of the errors assumes its minimum. This function has the physical meaning of free energy. Thus a solution of the system (2.23) exists if in an equilibrium state the free energy of its physical model is equal to zero, and such a solution does not exist if in the equilibrium state the free energy is different from zero and positive. In both cases equilibrium states exist.

The physical model of problem (2.23) is constructed as was described in Section 1.2 and is depicted by Figure 1.7 [2)]. This model is a passive physical system and its thermodynamic potential is the Helmholtz free energy such as is obtained from (2.20) for $p_i = 0$, $i = 1,\ldots,n$. Let us calculate the free energy assuming for simplicity that the state transition processes in the system take place at constant temperature. Obviously the amount of free energy is equal to the sum of the quantities of free energy of the ideal gases in the 2m systems of containers $V_s^{(+)}$ and $V_s^{(-)}$, $s = 1,\ldots,m$. Consequently according to (2.22) we shall have

$$F = C - RT \sum_{s=1} (\bar{V}_s^{(+)} \ln V_s^{(+)} + \bar{V}_s^{(-)} \ln V_s^{(-)}).$$

Using formula (2.2) one obtains

$$F = C - q_0 \sum_{s=1}^{m} (\bar{V}_s^{(+)} \ln V_s^{(+)} + \bar{V}_s^{(-)} \ln V_s^{(-)}).$$

The arbitrary constant C can always be chosen in such a way that the following condition holds

$$F = 0 \text{ if } V_s^{(+)} = \bar{V}_s^{(+)},\ V_s^{(-)} = \bar{V}_s^{(-)}, \quad s = 1,2,\ldots,m.$$

To see this it suffices to set

$$C = q_0 \sum_{s=1}^{m} (\bar{V}_s^{(+)} \ln \bar{V}_s^{(+)} + \bar{V}_s^{(-)} \ln \bar{V}_s^{(-)}).$$

Then clearly

$$F = q_0 \sum_{s=1}^{m} (\bar{V}_s^{(+)} \ln \frac{\bar{V}_s^{(+)}}{V_s^{(+)}} + \bar{V}_s^{(-)} \ln \frac{\bar{V}_s^{(-)}}{V_s^{(-)}}).$$

Let us denote with $y_1,\ldots,y_m$, the errors in system (2.23), that is the quantities

$$y_s = \sum_{i=1} a_{si} x_i - \xi_s$$

where

$$\xi_s = \begin{cases} \leqslant b_s & \text{for } s \in M_1 \\ = b_s & \text{for } s \in M_2. \end{cases}$$

Further from formulas (1.16)-(1.17) it follows that

$$V_s^{(+)} = \bar{V}_s^{(+)} + y_s, \quad V_s^{(-)} = \bar{V}_s^{(-)} - y_s, \quad s = 1,\ldots,m \tag{2.25}$$

and the expression for the free energy takes the form

$$F(y_1,\ldots,y_m) = q_0 \sum_{i=1}^{m} (\bar{V}_s^{(+)} \ln \frac{\bar{V}_s^{(+)}}{\bar{V}_s^{(+)}+y_s} + \bar{V}_s^{(-)} \ln \frac{\bar{V}_s^{(-)}}{\bar{V}_s^{(-)}-y_s}). \tag{2.26}$$

The function $F(y_1,\ldots,y_m)$ is a positive definite function of the quantities $y_1,\ldots,y_m$ and strongly convex as a consequence of the inequality

$$\frac{\partial^2 F}{\partial y_s^2} = \frac{q_0 \bar{V}_s^{(+)}}{(\bar{V}_s^{(+)}+y_s)^2} + \frac{q_0 \bar{V}_s^{(-)}}{(\bar{V}_s^{(-)}-y_s)^2} > 0, \quad s = 1,\dots,m.$$

The conditions for equilibrium are clearly the particular case of the conditions (1.31)-(1.33) for which $p_i = 0$, $i = 1,\dots,n$ and take the form

$$\sum_{s=1}^{m} a_{si}\bar{w}_s = 0, \quad i \in N_1 \tag{2.27}$$

$$\sum_{s=1} a_{si}\bar{w}_s = \begin{cases} = 0 & \text{for } \bar{x}_i > 0 \\ \geqslant 0 & \text{for } \bar{x}_i = 0 \end{cases}, \quad i \in N_2 \tag{2.28}$$

$$\bar{w}_s = \begin{cases} = 0 & \text{for } \xi_s < b_s \\ \geqslant 0 & \text{for } \xi_s = b_s \end{cases}, \quad s \in M_1 \tag{2.29}$$

where

$$\bar{w}_s = q_s^{(-)} - q_s^{(+)}, \quad s = 1,2,\dots,m.$$

Thus in an equilibrium state $\bar{x}_1,\dots,\bar{x}_n$, $\bar{\xi}_1,\dots,\bar{\xi}_{m_1}$ of the physical model of system (2.23) the pressure differences $q_s^{(-)} - q_s^{(+)}$ satisfy conditions (2.27)-(2.29) independently of whether the system (2.23) admits solutions or not.

We shall now show that the conditions for the existence of a solution of the system under consideration are the conditions

$$\bar{w}_s = 0, \quad s = 1,\dots,m.$$

Let $V_s^{(+)}$, $V_s^{(-)}$, $\bar{V}_s^{(+)}$, $\bar{V}_s^{(-)}$ be defined by formulas (1.16)-(1.17). Then from the relations

$$q_s^{(+)}V_s^{(+)} = q_0\bar{V}_s^{(+)}, \quad q_s^{(-)}V_s^{(-)} = q_0\bar{V}_s^{(-)}, \quad s = 1,2,\dots,m$$

which express the law of Boyle-Mariotte we find

$$\bar{w}_s = q_s^{(-)} - q_s^{(+)} = q_0 \frac{\bar{V}_s^{(-)}V_s^{(+)} - V_s^{(-)}\bar{V}_s^{(+)}}{V_s^{(+)}V_s^{(-)}}, \quad s = 1,2,\dots,m$$

and, using formula (2.25) and (1.17) we obtain

$$\bar{w}_s = 2q_0 \; \ell\|a_s\| \frac{y_s}{\ell^2\|a_s\|^2 - y_s^2}, \quad s = 1,2,\ldots,m. \tag{2.30}$$

From (2.30) it follows that $y_s = 0$, $s = 1,\ldots,m$, if and only if $\bar{w}_s = 0$, $s = 1,\ldots,m$. In this way we have shown that the conditions $\bar{w}_s = 0$, $s = 1,\ldots,m$ are necessary and sufficient for the existence of a solution of system (2.23).

Let us now pay some attention to the simplification of relation (2.30). First of all the parameter ℓ must be sufficiently large such that the solutions which interest us of the system (2.23) or the vectors $\bar{x}$ which interest us which minimize the norms of the errors, satisfy the conditions

$$|\bar{x}_i| < \ell, \quad i = 1,\ldots,n.$$

Moreover the equilibrium state $\bar{x}$ does not depend on the magnitude of the parameter $q_0 > 0$. These reflections can be used to simplify formula (2.30) significantly. Let us introduce a new parameter of the model $\tilde{q}_0$ which is connected to q_0 and ℓ by the equality

$$\tilde{q}_0 = \frac{2q_0}{\ell}.$$

Substituting this in formula (2.30) gives the result

$$\bar{w}_s = \tilde{q}_0 \frac{\ell^2\|a_s\| y_s}{\ell^2\|a_s\|^2 - y_s^2}, \quad s = 1,2,\ldots,m. \tag{2.31}$$

Now it is clearly possible to go to the limit as $\ell \to \infty$. And then the formula connecting $\bar{w}_s$ and y_s takes the form

$$\bar{w}_s = \tilde{q}_0 \frac{y_s}{\|a_s\|}, \quad s = 1,\ldots,m. \tag{2.32}$$

One easily verifies that as $\ell \to \infty$ the free energy F becomes a well defined positive definite form in the quantities $y_1,\ldots,y_m$. The quantities $y_1,\ldots,y_m$ are defined by the formulas (2.24), which contain the free variables $\xi_1,\ldots,\xi_m$ subject to the conditions

$$\xi_s \begin{cases} \leqslant b_s & \text{for} \quad s \in M_1 \\ = b_s & \text{for} \quad s \in M_2. \end{cases}$$

Let $\bar{\xi}_1,\ldots,\bar{\xi}_m$ be the values of the quantities $\xi_1,\ldots,\xi_m$ in an equilibrium state of the model. These quantities can of course not be considered independent because their values in an equilibrium state $\bar{x}_1,\ldots,\bar{x}_n$ obviously satisfy the formulas

$$\bar{\xi}_s = \begin{cases} \sum_{i=1}^{n} a_{si}\bar{x}_i & \text{for} \quad \sum_{i=1}^{n} a_{si}\bar{x}_i < b_s, \; s \in M_1 \\ b_s & \text{for} \quad \sum_{i=1}^{n} a_{si}x_i \geqslant b_s, \; s \in M_1 \\ b_s & \text{for} \quad s \in M_2. \end{cases} \tag{2.33}$$

Denoting with

$$z_s = \sum_{i=1}^{n} a_{si}x_i - b_s, \quad s = 1,\ldots,m$$

we obtain from (2.33)

$$y_s = \begin{cases} 0 & \text{for} \quad z_s < 0, \quad s \in M_1 \\ z_s & \text{for} \quad z_s \geqslant 0, \quad s \in M_1 \\ z_s & \text{for} \quad s \in M_2 \end{cases}$$

or

$$y_s = \begin{cases} z_s \,\underline{1}\,[z_s], & s \in M_1 \\ z_s, & s \in M_2 \end{cases} \tag{2.34}$$

where

$$\underline{1}\,[z] = \begin{cases} 0 & \text{for} \quad z < 0 \\ z & \text{for} \quad z \geqslant 0. \end{cases}$$

In this way formula (2.32) takes its final form

$$\bar{w}_s = \begin{cases} q_0\|a_s\|^{-1} \, z_s \,\underline{1}\,[z_s], & s \in M_1 \\ q_0\|a_s\|^{-1} \, z_s, & s \in M_2 \end{cases} \tag{2.35}$$

Thus the physical system depicted by Figure 1.7 is a model for the system of linear equations and inequalities (2.23) in the following sense: if the system is solvable then the equilibrium states $\bar{x}$ are solutions of the system (2.23), and if the system is not solvable then the vector $\bar{x}$ (of an equilibrium state) minimizes a positive definite function of the errors $y_1,\ldots,y_m$ of the system. In particular an equilibrium state defines a bivector $(\bar{x},\bar{w})$ of extensive $(\bar{x})$ and intensive $(\bar{w})$ variables of the physical model (compare Section 2.2), connected by Equations (2.35), which follow from the state equation of Clapeyron-Mendeleev. Therefore the following assertion holds:

THEOREM 2.1. *There exists a bivector $(\bar{x},\bar{w})$ which satisfies the conditions (2.27)-(2.29) and (2.35) such that moreover if the system (2.23) is solvable then $\bar{x}$ is a solution of it and $\bar{w}$ is the zero vector, and if (2.23) is not solvable then $\bar{x}$ minimizes the positive definite function (2.26) of the errors in the system and $\bar{w}$ is a nonzero vector.*

Proof. The existence of the vector $\bar{x}$ follows from the existence of the minimum of the (positive) definite function of errors in (2.23). And the existence of the vector $\bar{w}$ then follows from the Clapeyron-Mendeleev state equation. It is clear that the minimal free energy is equal to zero for every solution of the system if such a solution exists. In that case the assertion $\bar{w} = 0$ follows from formula (2.35).

In all cases the vector of intensive variables $\bar{w}$ in the physical model of the system (2.23) is a solution of the following system of equations and inequalities which derives from (2.27)-(2.29)

$$\sum_{s=1}^{m} a_{si} w_s = 0, \quad i \in N_1$$

$$\sum_{s=1}^{m} a_{si} w_s \geq 0, \quad i \in N_2 \tag{2.36}$$

$$w_s \geq 0, \quad s \in M_1.$$

Using the above we can now easily establish the fundamental theorem of systems of equations and inequalities in the following most general form.

THEOREM 2.2. One and one only of the following alternatives holds: either the system (2.23)

$$\sum_{i=1}^{n} a_{si}x_i \leqslant b_s, \quad s \in M_1$$

$$\sum_{i=1}^{n} a_{si}x_i = b_s, \quad s \in M_2$$

$$x_i \geqslant 0, \quad i \in N_2$$

has a solution, or there is a solution of the system

$$\sum_{s=1}^{m} a_{si}w_s = 0, \quad i \in N_1$$

$$\sum_{s=1}^{m} a_{si}w_s \geqslant 0, \quad i \in N_2 \qquad (2.37)$$

$$\sum_{s=1}^{m} b_s w_s < 0, \quad w_s \geqslant 0 \quad \text{for} \quad i \in M_1.$$

Proof. Let us start with proving that there do not exist two vectors $x^0 = (x_1^0,\dots,x_n^0)$ and $w^0 = (w_1^0,\dots,w_m^0)$ such that the first vector satisfies the conditions (2.23) and the second one the conditions (2.37). Suppose on the contrary that such vectors did exist. Multiply the first n conditions of (2.37) with $x_1^0,\dots,x_n^0$ respectively and sum the results. One obtains

$$\sum_{i\in N_1} x_i^0 \sum_{s=1}^{m} a_{si}w_s + \sum_{i\in N_2} x_i^0 \sum_{s=1}^{m} a_{si}w_s \geqslant 0 \qquad (2.38)$$

because the first sum is zero by (2.37) and the second non-negative by (2.37) and the conditions $x_1^0 \geqslant 0$ for $i \in N_2$. Now multiply the first m conditions of (2.23) with $w_1^0,\dots,w_m^0$ respectively and sum. This gives

$$\sum_{s\in M_1} w_s^0 \sum_{i=1}^{n} a_{si} x_i^0 + \sum_{s\in M_2} w_s^0 \sum_{i=1}^{n} a_{si} x_i^0 \leqslant$$
$$\leqslant \sum_{s\in M_1} b_s w_s^0 + \sum_{s\in M_2} b_s w_s^0 = \sum_{s=1}^{m} b_s w_s^0. \qquad (2.39)$$

Indeed, this inequality holds because of conditions (2.23) and the condition $w_s^0 \geqslant 0$ for $s \in M_1$. The left hand sides of the equalities (2.38) and (2.39) are equal as can be seen by interchanging the summation order. Therefore (2.38) and (2.39) would mean that

$$\sum_{s=1}^{m} b_s w_s^0 \geqslant 0$$

which contradicts the (n+1)st condition of (2.37). Consequently the systems (2.23) and (2.37) cannot be solved simultaneously.

It remains to show that one of the systems (2.23) and (2.37) does have a solution. To that end we consider an equilibrium state of the physical model of system (2.23); existence of such an equilibrium state is independent of whether (2.23) is solvable or not.

The vector $\bar{x}$ satisfies the equilibrium conditions (2.27)-(2.29). There are two possibilities. 1) In an equilibrium state the free energy is equal to zero. This means that a solution of (2.27)-(2.29) is necessarily the zero vector and consequently that the system (2.23) has a solution. 2) In an equilibrium state $\bar{x}$ the free energy is strictly positive and that means that the system (2.23) is not solvable. In the first case it was shown that the system (2.37) was not solvable. To conclude the proof of the theorem it remains to see that the system (2.37) is solvable in the second case. In fact in this case, because of Theorem 2.1 there exists a non trivial solution $\bar{w}$ of system (2.36) together with the equilibrium conditions (2.27)-(2.29). Multiply the n conditions of (2.27)-(2.28) with $\bar{x}_1,\ldots,\bar{x}_n$ respectively and sum. This yields

$$\sum_{s=1}^{m} \bar{w}_s \sum_{i=1}^{n} a_{si} \bar{x}_i = 0.$$

This equality is obviously equivalent to the following

$$\sum_{s=1}^{m} \bar{w}_s \left(\sum_{i=1}^{n} a_{si} \bar{x}_i - b_s \right) + \sum_{s=1}^{m} \bar{w}_s b_s = 0$$

or, according to the definition of the quantities z_s, we have

$$\sum_{s=1}^{m} \bar{w}_s b_s + \sum_{s=1}^{m} \bar{w}_s \bar{z}_s = 0.$$

In this second case under consideration at least one of the error quantities $y_1, \ldots, y_m$ is different from zero and consequently because of (2.35)

$$\sum_{s=1}^{m} \bar{w}_s \bar{z}_s = \tilde{q}_0 \Big(\sum_{s \in M_1} \|a_s\|^{-1} \bar{z}_s^2 \underset{=}{1} [z_s] + \sum_{s \in M_2} \|a_s\|^{-1} \bar{z}_s^2 \Big) > 0.$$

And with this last inequality it follows from (2.40) that

$$\sum_{s=1}^{m} b_s \bar{w}_s < 0.$$

So that $\bar{w}_1, \ldots, \bar{w}_m$ is indeed a solution of (2.37). This proves the theorem.

It is interesting to interpret Theorem (2.2) in physical terms. If there exists a solution x^* of a system (2.23) then, clearly, the vector λx^*, where $\lambda \geqslant 0$ is a solution of the system

$$\begin{aligned} &\sum_{i=1}^{n} a_{si} x_i \leqslant \lambda b_s, \quad s \in M_1 \\ &\sum_{i=1}^{n} a_{si} x_i = \lambda b_s, \quad s \in M_2. \end{aligned} \tag{2.41}$$

In this way as it was done in Section 1.4 of Chapter I we can let the model evolve according to a quasistatic process by varying the magnitude on the parameter λ infinitely slowly from zero to one. Under this procedure the physical model passes from the equilibrium state $x = 0$ to the equilibrium state x^* through all the intermediate equilibrium states λx^*. The work done under these changes in the exterior conditions is given by the linear form $(b, w(\lambda))$ as in Section 1.4. This work, clearly, is equal to zero because $w(\lambda) = 0$ for all $\lambda \geqslant 0$.

Let us now consider the case where the system (2.23) does not have a solution. In this case the minimum of the free energy of the physical model attained in an equilibrium state $\bar{x}$ is different from zero and positive and the vector $\bar{w}$ defined by the formulas (2.35) is different from zero. Let us consider again a quasi static process taking place in the model (see Figure 1.10 for $p_1 = p_2 = \ldots = p_n = 0$) of infinitely slow changes in the parameter λ from zero to one. Clearly for $\lambda = 0$ the system (2.41) has the solution $\bar{x}(0) = 0$ and because this solution is at the same time an equilibrium state we have $\bar{w}(0) = 0$. Now under the quasi static process just described the vector of extensive variables changes according to the linear rule $\bar{x}(\lambda) = \lambda\bar{x}$. And the vector of intensive variables $\bar{w}(\lambda) = \lambda\bar{w}$ changes similarly, and the free energy of the physical model grows from zero to some strictly positive magnitude. By virtue of the law of the conservation of energy the change in free energy is equal to the work done by the exterior forces P_0 acting on the bar (compare Figure 1.10) of which the position defines the magnitude of the parameter λ, during the movements of this bar from the position $\lambda = 0$ to the position $\lambda = 1$. As is known the transition forces of the model from the state $x = 0$ to the state $\bar{x}$ will be quasi static if for each λ the exterior force P_0 is balanced by the internal forces of the system. This means that the following equality holds

$$P_0(\lambda) + \sum_{s=1}^{m} b_s \bar{w}_s(\lambda) = 0$$

or

$$P_0(\lambda) = -\lambda \sum_{s=1}^{m} b_s \bar{w}_s .$$

By virtue of the law of the conservation of energy the strictly positive amount $F(\bar{x})$ of free energy of the physical model in state $\bar{x}$ is equal to the work done by the force $P_0(\lambda)$ under the changes of λ from zero to one. One therefere has the equation

$$\int_0^1 P_0(\lambda)d\lambda = -\tfrac{1}{2} \sum_{s=1} b_s \bar{w}_s = F(\bar{x}) > 0$$

from which it follows that a vector $\bar{w}$ which satisfies the conditions (2.36) also satisfies the inequality

$$\sum_{s=1}^{m} b_s \bar{w}_s < 0$$

which means that it is a solution of the system (2.37). And thus the fact that the basic Theorem 2.2 holds, follows from the equilibrium conditions of the physical model of (2.23) and the law of the conservation of energy.

Remark. Theorem 2.2 remains true if in the system (2.37) the condition $\sum_{s=1}^{m} b_s \bar{w}_s < 0$ is replaced by the condition

$$\sum_{s=1}^{m} b_s \bar{w}_s = -c \tag{2.42}$$

where c is an arbitrary given strictly positive number. Indeed from (2.40) and (2.35) it follows that

$$\sum_{s=1}^{m} b_s w_s = - \sum_{s=1}^{m} \bar{w}_s \bar{z}_s =$$

$$= -q_0 [\sum_{s \in M_1} \|a_s\|^{-1} \bar{z}_s^{-2} \underset{=}{1} [\bar{z}_s] + \sum_{s \in M_2} \|a_s\|^{-1} \bar{z}_s^{-2}]$$

where

$$\bar{z}_s = \sum_{i=1}^{n} a_{si} \bar{x}_i - b_s, \quad s = 1,\dots,m.$$

The quantities $\bar{z}_1,\dots,\bar{z}_m$, just as the equantities $\bar{x}_1,\dots,\bar{x}_n$ do not depend on the magnitude of the parameter q_0 which can be chosen to be equal to

$$\tilde{q}_0 = c(\sum_{s \in M_1} \|a_s\|^{-1} \bar{z}_s^{-2} \underset{=}{1} [\bar{z}_s] + \sum_{s \in M_2} \|a_s\|^{-1} \bar{z}_s^{-2})^{-1}.$$

Then if the system (2.23) has no solutions the quantities $\bar{w}_1,\dots,\bar{w}_m$ will be the components of a vector $\bar{w}$ for which both conditions (2.37) and (2.42) hold. The theorem we have proced generalizes several important theorems of the theory of linear equations and inequalities (compare [5]).

<u>COROLLARY</u> 2.2.1. (solvability of linear equations). <u>One and one only of the following alternatives holds: either the system</u>

$$\sum_{i=1}^{n} a_{si} x_i = b_s, \; s = 1,\dots,m$$

<u>has a solution or there is a solution of the system</u>

$$\sum_{s=1}^{m} a_{si} w_s = 0, \quad i = 1,\dots,n$$

$$\sum_{s=1}^{m} b_s w_s = -c,$$

<u>where c is a strictly positive number</u>.

This result is the particular case of Theorem 2.2 for which $M_1 = \emptyset$, $M_2 = \emptyset$.

COROLLARY 2.2.2. (Nonnegative solutions of systems of linear equations). *One and one only of the following alternatives holds: either the system*

$$\sum_{i=1}^{n} a_{si}x_i = b_s, \quad s = 1,\ldots,m$$

has a nonnegative solution or there exists a solution of the system of inequalities

$$\sum_{s=1}^{m} a_{si}w_s \geqslant 0, \; i = 1,\ldots,n$$

$$\sum_{s=1}^{m} b_s w_s < 0.$$

This assertion which is called a separating hyperplane theorem, coincides with the assertion of Theorem 2.2 in the case where $M_1 = \emptyset$ and $N_1 = \emptyset$.

COROLLARY 2.2.3. (Solutions of linear inequalities). *One and one only of the following alternatives holds: either the system of inequalities*

$$\sum_{i=1}^{n} a_{si}x_i \leqslant b_s, \quad s = 1,\ldots,m$$

has a solution or there exists a nonnegative solution of the system of equations

$$\sum_{s=1}^{m} a_{si}w_s = 0, \quad i = 1,\ldots,n$$

$$\sum_{s=1}^{m} b_s w_s = -1.$$

This assertion coincides with that of Theorem 2.2 in the case where $M_2 = \emptyset$ and $N_2 = \emptyset$ (compare also the remark below Theorem 2.2).

COROLLARY 2.2.4. (Nonnegative solutions of linear inequalities). *One and one only of the following alternatives holds: the system of inequalities*

$$\sum_{i=1}^{n} a_{si}x_i \leqslant b_s, \quad s = 1,\ldots,n$$

$$x_i \geqslant 0, \quad i = 1,\ldots,m$$

has a solution, or there is a solution for the system

$$\sum_{s=1}^{m} a_{si}w_s \geqslant 0, \quad i = 1,\ldots,n$$

$$\sum_{s=1}^{m} b_s w_s < 0, \quad w_s \geqslant 0, \quad s = 1,\ldots,m.$$

In the case $M_2 = \emptyset$, $N_1 = \emptyset$ this Corollary 2.2.4 is precisely the assertion of Theorem 2.2.

Homogeneous equations and inequalities. To conclude we consider some theorems on homogeneous systems of equations and inequalities. We shall start with a theorem on semi positive solutions of homogeneous equations. Recall some necessary definitions. (1) The vector $x = (x_1,\ldots,x_n)$ is called nonnegative if $x_i \geqslant 0$, $i = 1,\ldots,n$. The symbol $\geqq 0$ denotes the property of being nonnegative. (2) The vector x is called semi positive if $x_i \geqslant 0$, $i = 1,\ldots,n$ and if there exists at least one value of the index i such that $x_i > 0$. The property of being semi positive is denoted by means of the symbol $\geqslant 0$. (3) The vector x is called positive if $x_i > 0$, $i = 1,\ldots,n$ and this property is denoted by means of the symbol $\gg 0$.

THEOREM 2.3. One and only one of the following alternatives holds: either the system

$$\sum_{i=1}^{n} a_{si}x_i = 0, \quad s = 1,\ldots,m \tag{2.34}$$

has a semi positive solution or there exists a solution of the system of inequalities

$$\sum_{s=1}^{m} a_{si}w_s > 0, \quad i = 1,\ldots,n. \tag{2.44}$$

Proof. Assuming the opposite it is simple to prove the incompatibility of the two alternatives. Let both systems be solvable and let $x^0 \geqslant 0$ be a solution of (2.43) and $w^{(0)}$ a solution of system (2.44). Now multiply the equations (2.43) with respectively $w_1^{(0)},\dots,w_m^{(0)}$ and the relations (2.44) with respectively $x_1^{(0)},\dots,x_n^{(0)}$. Sum the results to obtain, using (2.43)

$$\sum_{i=1}^{m} w_s^{(0)} \sum_{i=1}^{n} a_{si}x_i^{(0)} = 0 = \sum_{i=1}^{n} x_i^{(0)} \sum_{s=1}^{m} a_{si}w_s^{(0)}.$$

On the other hand because x^0 is semi positive we have from (2.44) that

$$\sum_{i=1}^{n} x_i^{(0)} \sum_{s=1}^{m} a_{si}w_s^{(0)} > 0$$

and the contradiction thus obtained proves that the two alternatives are not compatible. To conclude the proof it remains to show that if one of the systems (2.43) or (2.44) is not solvable, then the other is solvable. Conditions (2.44) mean that there is a positive number $\varepsilon > 0$ sufficiently small such that the following inequalities hold

$$\sum_{s=1}^{m} a_{si}w_s \geqslant \varepsilon, \quad i = 1,\dots,n. \tag{2.45}$$

Setting $w_s = -v_s$ we rewrite (2.45) in the more convenient form

$$\sum_{s=1}^{m} a_{si}v_s \leqslant -\varepsilon, \quad i = 1,\dots,n. \tag{2.46}$$

Now we can use Corollary 2.2.3. of Theorem 2.2 which says that the alternative belonging to (2.46) is the system

$$\begin{gathered}\sum_{i=1}^{n} a_{si}\xi_i = 0 \quad s = 1,\dots,m \\ \xi_i \geqslant 0 \quad i = 1,\dots,n, \quad \varepsilon \sum_{i=1}^{n} \xi_i = 1.\end{gathered} \tag{2.47}$$

By virtue of Corollary 2.2.3 there exists a solution of (2.47) if there does not exist a solution of (2.46) or the equivalent system (2.44). Thus if there does not exist a solution to (2.44) then there exists a vector ξ satisfying the conditions of (2.47). Because of the conditions

$$\xi_i \geqslant 0, \quad i = 1,\dots,n, \quad \sum_{i=1}^{n} \xi_i = \varepsilon^{-1} > 0$$

the vector ξ is semi positive and clearly $x = \xi \geqslant 0$ is a solution of system 2.43. This proves the theorem.

THEOREM 2.4. One and one only of the following alternatives holds: either the system of inequalities

$$\sum_{i=1}^{n} a_{si} x_i \leqslant 0, \quad s = 1,\dots,m \tag{2.48}$$

has a semi positive solution or there is a nonnegative solution of the system of inequalities

$$\sum_{s=1}^{m} a_{si} w_s > 0, \quad i = 1,\dots,n. \tag{2.49}$$

Proof. Again the incompatibility of the two alternatives is proved by assuming the opposite. Suppose that $x^0 \geqslant 0$ and $w^0 \geqq 0$ are solutions of respectively system (2.48) and (2.49). Then from (2.48) it follows that

$$\sum_{i=1}^{m} w_s^{(0)} \sum_{i=1}^{n} a_{si} x_i^{(0)} \leqslant 0$$

and on the other hand from (2.49) one obtains

$$\sum_{i=1}^{n} x_i^{(0)} \sum_{s=1}^{m} a_{si} w_s^{(0)} = \sum_{s=1}^{m} w_s^{(0)} \sum_{i=1}^{n} a_{si} x_i^{(0)} > 0$$

The two inequalities thus obtained are contradictory and establish the incompatibility of the two alternatives. It reamins to show that a semi positive vector $x \geqslant 0$ exists which satisfies the conditions (2.48) if the system (2.49) has no nonnegative solutions. As in the proof of Theorem 2.3 we write (2.49) in the form

$$\sum_{s=1}^{m} -a_{si} w_s \leqslant -\varepsilon, \quad i = 1,\dots,n \tag{2.50}$$

where ε is a sufficiently small positive number.

Now use Corollary 2.2.4. of Theorem 2.2. The alternative system belonging to (2.50) obviously is the system of inequalities

$$\sum_{i=1}^{n} a_{si} \xi_i \leqslant 0, \quad s = 1,\dots,m \tag{2.51}$$

$$\varepsilon \sum_{i=1}^{n} \xi_i > 0, \quad \xi_i \geqslant 0, \quad i = 1,\dots,n.$$

Consequently if the system (2.49) has no nonnegative solution then there exists a semi positive vector $x = \xi \geqslant 0$ which satisfies the conditions (2.48). And in fact the system (2.51) coincides with (2.48) and the condition (2.52) guarantees the semi positiveness of the vector ξ. This proves the theorem.

The proofs of these theorems on alternatives for homogeneous systems of equations and inequalities is based on a reduction of such systems to inhomogeneous ones. This permits the use of the fundamental Theorem 2.2 for their proofs and on the other hand this reduction also permits to give Theorems 2.3 and 2.4 a similar physical interpretation.

2.4. A physical model for linear programming problems. Equilibrium conditions

Consider the physical model for the linear programming problem (1.20)-(1.21)

$$\sum_{i=1}^{n} p_i x_i \to \max$$

$$\sum_{i=1}^{n} a_{si} x_i - \xi_s = 0, \quad s \in M = \{1,\dots,m\}$$

$$x_i \geqslant 0, \quad i \in N_2 = \{n_1+1,\dots,n\} \tag{2.53}$$

$$\xi_s = \begin{cases} \leqslant b_s, & s \in M_1 = \{1,2,\dots,m_1\} \\ = 0, & s \in M_2 = \{m_1+1,\dots,m\}. \end{cases}$$

The general description of this model was given in Section 1.4. In an equilibrium state of this model (see Figure 1.8) the thermodynamic potential is minimized. This thermodynamic potential is the sum of the free energy of the ideal gases in the volumes $V_s^{(+)}$ and $V_s^{(-)}$ and the work done by the system on exterior bodies (formula 2.20) in Section 2.2).

For the physical systems under consideration this work done on exterior bodies is equal to minus the amount of work done by the exterior forces $p_1,\dots,p_n$ acting on the bars connecting the pistons, in deplacing these bars.
Thus the second term in the expression (2.30) for the thermodynamic potential has the form

$$\sum_{i=1}^{n} \int_{x^{(0)}}^{x} P_i dx_i = -\sum_{i=1}^{n} p_i (x_i - x_i^{(0)}). \tag{2.54}$$

And an expression for the free energy of the physical model

was obtained above in Section 2.3 (see 2.26)). Setting $x^{(0)} = 0$ in (2.54) we obtain according to (2.20) an expression for the thermodynamical potential of the model

$$\phi = -\sum_{i=1}^{n} p_i x_i + q_0 \sum_{s=1}^{m} [\bar{V}_s^{(+)} \ln \frac{\bar{V}_s^{(+)}}{\bar{V}_s^{(+)} + y_s} + \bar{V}_s^{(-)} \ln \frac{\bar{V}_s^{(-)}}{\bar{V}_s^{(-)} - y_s}], \tag{2.55}$$

where

$$y_s = \sum_{i=1}^{n} a_{si} x_i - \xi_s, \quad s = 1,\dots,m. \tag{2.56}$$

And thus an equilibrium state of the model turns out to be a solution of the following problem:

$$-\sum_{i=1}^{n} p_i x_i + q_0 \sum_{x=1}^{m} [\bar{V}_s^{(+)} \ln \frac{\bar{V}_s^{(+)}}{\bar{V}_s^{(+)} + y_s} + \bar{V}_s^{(-)} \ln \frac{\bar{V}_s^{(-)}}{\bar{V}_s^{(-)} - y_s}] \to \min \tag{2.57}$$

under the constraints $x_i \geqslant 0$, $i \in N_2$.

We are now faced with the problem of explaining the connection between the general linear programming problem and the equilibrium problem described above (see Figure 1.8). The solutions of these problems are clearly different given the fact that the physics of an ideal gas and an incompressible liquid are different. On the other hand the solutions of the equilibrium problem of the physical model will be arbitrarily near solutions of the linear programming model for sufficiently large values of the parameter q_0 or $\tilde{q}_0$. This happens because in the region of very high pressures the properties of an ideal gas resemble those of an incompressible liquid.

It is perfectly clear that the equilibrium conditions of the model of Figure 1.8 remain as in the case of the model with volumes filled with an incompressible liquid. In the case of the physical model the quantities $q_s^{(+)}$ and $q_s^{(-)}$ occurring in the equilibrium conditions (1.31)-(1.33) are the pressures of the ideal gases in the volumes $V_s^{(+)}$ and $V_s^{(-)}$, and the quantities $w_1,\dots,w_m$ represent the differences $q_s^{(-)} - q_s^{(+)}$, $s = 1,\dots,m_n$ of these pressures. The only difference will be in the consequences and the solutions of these equilibrium conditions because, as follows from (2.32) the value of the quantity w_s will be different from zero if and only if

$$V_s^{(+)} \neq \bar{V}_s^{(+)} \quad \text{and} \quad V_s^{(-)} \neq \bar{V}_s^{(-)}$$

or in other words when $y_s \neq 0$; that is when the corresponding constraint of the linear programming problem (2.53) is not fulfilled.

Let $\bar{x}_1,\ldots,\bar{x}_n$, $\bar{\xi}_1,\ldots,\bar{\xi}_m$ be the coordinates of an equilibrium state for the physical model of problem (2.53). The quantities $\bar{\xi}_1,\ldots,\bar{\xi}_m$ as was shown in Section 2.8 cannot be considered independent and are defined by the formulas (2.33). The equilibrium state $\bar{x}$ satisfies the following conditions obtained in Section 1.4:

$$\sum_{i=1}^{m} a_{si}\bar{w}_s - p_i = 0, \; i \in N_1 \tag{2.58}$$

$$\sum_{i=1}^{m} a_{si}\bar{w}_s - p \begin{cases} = 0 & \text{for } \bar{x}_i > 0 \\ \geqslant 0 & \text{for } \bar{x}_i = 0 \end{cases}, \; i \in N_2 \tag{2.59}$$

$$w_s \begin{cases} = 0 & \text{for } \bar{\xi}_s < b_s \\ \geqslant 0 & \text{for } \bar{\xi}_s = b_s \end{cases}, \; s \in M_1. \tag{2.60}$$

In Section 1.4 of Chapter I we obtained from these equilibrium conditions the basic results of duality theory. Let us see what results can be obtained by analogous considerations in the case of models with volumes containing ideal gases. Nultiplying the equations and inequalities (2.58) and (2.59) with $\bar{x}_1,\ldots,\bar{x}_n$ respectively and summing the results one obtains

$$\sum_{s=1}^{m} \bar{w}_s \sum_{i=1}^{n} a_{si}\bar{x}_i = \sum_{i=1}^{n} p_i\bar{x}_i$$

or, according to (2.56)

$$\sum_{s=1}^{m} \bar{w}_s(\bar{\xi}_s+\bar{y}_s) = \sum_{i=1}^{n} p_i\bar{x}_i$$

where

$$\bar{y}_s = \sum_{i=1}^{n} a_{si}\bar{x}_i - \bar{\xi}_s, \quad s = 1,\ldots,m.$$

From the last group of equilibrium conditions (2.60) it follows that

$$\sum_{s=1}^{m} \bar{w}_s \bar{\xi}_s = \sum_{s=1}^{m} \bar{w}_s b_s$$

and then the equality (2.61) takes the form

$$\sum_{s=1}^{m} \bar{w}_s b_s + \sum_{s=1}^{m} \bar{w}_s \bar{y}_s = \sum_{i=1}^{n} p_i \bar{x}_i . \qquad (2.62)$$

We note that the quantities $\xi_1, \ldots, \xi_{m_1}$ do not uccur in relation (2.62). Using formulas (2.34)' and (2.35) the equality (2.62) can be rewritten in the form

$$\sum_{s=1}^{m} \bar{w}_s b_s + \sum_{s=1}^{m} \bar{w}_s \bar{z}_s = \sum_{s=1}^{n} p_i \bar{x}_i \qquad (2.63)$$

where according to (2.35)

$$\bar{w}_s \begin{cases} = 0 \text{ for } \bar{z}_s < 0 \\ \geqslant 0 \text{ for } \bar{z}_s \geqslant 0 \end{cases}, \quad s \in M_1$$

$$\bar{z}_s = \sum_{i=1}^{n} a_{si} \bar{x}_i - b_s, \quad s = 1, \ldots, m \qquad (2.64)$$

$$\sum_{s=1}^{m} \bar{w}_s \bar{z}_s = q_0 \{ \sum_{s=1}^{m_1}{}' \frac{\bar{z}_s^2 \, 1[\bar{z}_s]}{\|a_s\|} + \sum_{s=m_1+1}^{m} \frac{\bar{z}_s^2}{\|a_s\|} \} > 0.$$

Equation (2.63) is an analogue of a well known result from duality theory. The transition from a model whose volumes are filled with an incompressible liquid to a model with volumes containing an ideal gas, is equivalent to the replacement of rigid constraints with elastic ones. And in this way the free energy of the gas is a measure of elasticity for the constraints or a penalty function which measures to what extent these constraints are violated. We shall explain the economic meaning of Equation (2.63) in the case of the well known planning problem.

Suppose that some firm produces n kinds of goods $A_1, \ldots, A_n$ by means of m kinds of recources $B_1, \ldots, B_m$ which are available in the respective quantities $b_1, \ldots, b_m$. Let there be given prices $p_1, \ldots, p_n$ of the n goods and a production matrix (a_{si}) of dimensions $m \times n$. The element a_{si} of the production matrix (a_{si}) represents the amount of resource B_s which is necessary to produce one unit of product A_i. We shall now use the word 'plan' to denote a nonnegative n-dimensional vector $x = (x_1,\dots,x_n)$ whose components represent the amounts of products manufactured during a given period of time.

The production of x_i units of product A_i requires the investment of respectively $a_{1i}x_i, a_{2i}x_i,\dots,a_{mi}x_i$ units of resources $B_1,\dots,B_m$. The quantity

$$\sum_{i=1}^{n} p_i x_i$$

is the value of the assortment of goods $(x_1,\dots,x_n)$. An optimal plan is a vector $x^* = (x_1^*,\dots,x_n^*)$ which satisfies the conditions

$$\sum_{i=1}^{n} p_i x_i \to \max \tag{2.65}$$

$$\begin{aligned} \sum_{i=1}^{n} a_{si}x_i &\leqslant b_s, \quad s = 1,\dots,m \\ x_i &\geqslant 0, \quad i = 1,\dots,n. \end{aligned} \tag{2.66}$$

The problem described by (2.65)-(2.66) is a linear programming problem and the equilibrium conditions of its model are obtained from (2.56)-(2.60) by taking $M_1 = \{1,\dots,m\}$, $M_2 = \emptyset$, $N_1 = \emptyset$, $N_2 = \{1,\dots,n\}$. They take the form

$$\sum_{s=1}^{m} a_{si}w_s - p_i \begin{cases} = 0 & \text{for } x_i > 0 \\ \geqslant 0 & \text{for } x_i = 0 \end{cases}, \quad i = 1,\dots,n \tag{2.67}$$

$$w_s \begin{cases} = 0 & \text{for } \xi_s < b_s \\ \geqslant 0 & \text{for } \xi_s = b_s \end{cases}, \quad s = 1,\dots,m.$$

If the volumes in the model of problem (2.65)-(2.66) are filled with an incompressible liquid then conditions (2.67) and (2.68) are fulfilled by an optimal plan x^* and an optimal vector of dual prices of resources w^*, and one then has in virtue of the duality theorem

$$\sum_{s=1}^{m} w_s b_s = \sum_{i=1}^{n} p_i x_i^*$$

that is the maximal value is equal to the cost of the resources used. If the volumes $V_s^{(+)}$ and $V_s^{(-)}$ of our model contain ideal gases in the quantities defined by formula (2.2), then an equilibrium state $\bar{x}$, $\bar{w}$ of that physical model, which also satisfies the conditions (2.67), (2.68), represents an approximately optimal plan for the problem (2.65)-(2.66) an an approximately optimal solution of the dual problem of (2.65)-(2.66). One then has the inequalities

$$\sum_{i=1}^{n} p_i \bar{x}_i > \sum_{i=1}^{n} p_i x_i^*, \quad \sum_{s=1}^{m} b_s \bar{w}_s \geqslant \sum_{s=1}^{m} b_s w_s^*. \tag{2.69}$$

The inequality (2.69) holds because the vector $\bar{x}$ is not a feasible solution of the primal problem but the vector $\bar{w}$ is a feasible vector for the dual problem in virtue of the equilibrium conditions. The first inequality is obtained from the equilibrium conditions (2.67), (2.68) by setting $\bar{x} = x^* + \Delta x$ and by replacing $\bar{w}_s$, $s = 1,\dots,m$ by linear forms in the quantities $\Delta x_1,\dots, \Delta x_n$ according to (2.35). Multiplying the equilibrium conditions with $\Delta x_1,\dots,\Delta x_n$ respectively and summing the results one easily shows that the scalar product $(p,\Delta x)$ is equal to a positive definite quadratic form and that the matrix of this form is the Gramm matrix consisting of the column vectors of the matrix of constraints. We leave it to the reader as an exercise to carry out the necessary formal verifications and content ourselves with a different, less formal but more simple demonstration. Suppose that the assertion $(p,\bar{x}) > (p,x^*)$ is false and that instead $(p,\bar{x}) \leqslant (p,x^*)$. From the last inequality it follows that in the half space $(p,x) \geqslant (p,\bar{x})$ there exists a nonempty set of feasible vectors (for instance the vector x^* belongs to this half space) and for each of these feasible vectors x the free energy is zero. Also $(p,x) \geqslant (p,\bar{x})$. Consequently in $\bar{x}$ the thermodynamic potential (2.55) of the physical model is not minimal. That means that $\bar{x}$ is not an equilibrium state. The contradiction thus obtained shows that the inequality $(p,\bar{x}) > (p,x^*)$ holds.

Let us return to problem (2.65)-(2.66) with elastic constraints, that is violations of its constraints are permitted. We remark that according to (2.64) the quantities $\bar{z}_s \underline{\underline{1}} [\bar{z}_s]$ represent the excess quantities of resources needed to fulfil plan $\bar{x}$ and that the quantities $\bar{w}_s\bar{z}_s$ are the costs of these amounts of resources. The equality (2.63), a consequence of the equilibrium conditions, retains also in the case of elastic constraints, the same economic meaning of an equality between the value $(p,\bar{x})$ and the cost of the resources actually used. One easily verifies that the dual prices $\bar{w}_1,\ldots,\bar{w}_m$ are the components of the optimal solution of the problem dual to the following linear programming problem

$$\sum_{i=1}^{n} p_i x_i \rightarrow \max$$

$$\sum_{i=1}^{n} a_{si} x_i \leqslant b_s + \bar{z}_s \underset{=}{1} [\bar{z}_s], \quad s = 1,\ldots,m$$

$$x_i \geqslant 0 \ , \quad i = 1,\ldots,n.$$

Let us finally consider a particular case which has a very rich economic interpretation. In the pages that follow its importance will become clear. One searches the equilibrium state of a physical model obtained by the conditions (2.53) for $b_s = 0$, $s = 1,\ldots,m$. The fundamental equation (2.63) then looks like

$$\sum_{s=1}^{m} \bar{w}_s \bar{z}_s = \sum_{i=1}^{n} p_i x_i \qquad (2.63')$$

and $\bar{z}_1,\ldots,\bar{z}_m$ are the amounts of resources actually used: $\bar{z}_s = \sum_{i=1}^{n} a_{si}\bar{x}_i$, $s = 1,\ldots,m$. Equation (2.63') is an analogue of (1.27), the fundamtal duality theorem if $\bar{x}$ is interpreted as an optimal program under the assumption that the economic system can buy the resources necessary at prices $\bar{w}_1,\ldots,\bar{w}_m$.

In the particular case under consideration the physical equilibrium problem represents an economic situation where the economic system has no supply of the resources and buys them from example by means of the money obtained by the sale of its products at the given prices $p_1,\ldots,p_n$. In this case the equality (2.63') expresses the following well known economic result: revenue = cost of resources used. One notes that the planning problem is not represented approximately but is represented by an exact model which permits a new approach to planning and centralized rational allocation of resources. In this model the amounts of resources bought and the amounts of goods produced are determined by the prices of the latter and also by the penalty function (the Helmholtz) free energy of the model). The choice of the latter is the prerogative of the centre of decision making. There is more. One perceives the necessity of generalizing the physical model by replacing the scalar parameter q_0 by an m-vector $q = (q^{(1)},\dots,q^{(m)})$ of parameters and then the function to be minimized takes the form

$$-\sum_{i=1}^{n} p_i x_i + \sum_{s=1}^{m} q^{(s)}[\bar{V}_s^{(+)} \ln \frac{\bar{V}_s^{(+)}}{\bar{V}_s^{(+)}+y_s} + \bar{V}_s^{(-)} \ln \frac{\bar{V}_s^{(-)}}{\bar{V}_s^{(-)}-y_s}] .$$

The control of the allocation of resources is effected not by means of the price mechanism but by means of the choice of the parameters $q^{(1)},\dots,q^{(m)}$. Each term of the second sum defines the quantity y_s of a given resource which the economic system under consideration desires to buy and the amount of money it is prepared to pay, and the sum itself represents evidently the maximal cost of the resources necessary to the system.

One verifies without trouble that for the right choice of values for the parameters $q^{(1)},\dots,q^{(m)}$ for the model of problem (2.53) for $b_s = 0$, $s = 1,\dots,m$, the equilibrium state $\bar{x}(q^{(1)},\dots,q^{(m)})$ is an optimal vector for this problem with given positive $b_1,\dots,b_m \neq 0$; that is $\bar{x}(q^{(1)},\dots,q^{(m)}) = x^*$. Indeed let $q_0^{(1)},\dots,q_0^{(m)}$ be initial positive values of the parameters and let $\bar{x}^{(0)}$ be the equilibrium vector of the model for these initial values. The following 'better' values can then be calculated by the simple formulas

$$q_1^{(s)} = \frac{q_0^{(s)}}{b_s} \bar{z}_s^{(0)}, \quad s = 1,\dots,m$$

and by

$$q_{\alpha+1}^{(s)} = \frac{q_\alpha^{(s)}}{b_s} \bar{z}_s^{(\alpha)}, \quad s = 1,\dots,m$$

where

$$\bar{z}_\alpha^{(s)} = \sum_{i=1}^{n} a_{si}\bar{x}_i^{(\alpha)}$$

if $\bar{x}^{(\alpha)}$ is the equilibrium vector of the model with parameters $q_\alpha^{(1)},\dots,q_\alpha^{(m)}$. In Chapter V we shall discuss an algorithm for finding the equilibrium of a model with given parameters. Note that the conditions $b_s > 0$, $s = 1,\dots,m$ are definitely necessary (for the procedure to make sense). The reader is advised to show the convergence to x^* of the sequence $\bar{x}^{(0)}, \bar{x}^{(1)}, \dots$.

It is important that this approach generalizes to the case of centralized allocation of resources in an economy with several agents each with its own objective.

Let there be an economy composed of k linear units and a decision centre which has resources available in the amounts $b_1,\dots,b_m$. One generalizes without trouble the formulas listed above to

$$q_{\alpha+1}^{(s)} = q_\alpha^{(s)} \frac{\sum_{\nu=1}^{k} \|a_s^{(\nu)}\|}{k\cdot b_s} \sum_{\nu=1}^{k} \|a_s^{(\nu)}\|^{-1} z_s^{(\alpha)\nu}, \quad s = 1,\dots,m, \quad \alpha = 0,1,\dots .$$

where ν labels the noncentral units. It is important to note that in this treatment of allocation of resources it is not necessary that the central organism knows the objective of each economic unit.

2.5. Penalty methods

By using an ideal gas as an active medium in the volume of our models we have replaced linear programming problems with problems of minimization of thermodynamic potentials of active physical systems. In such problems the Helmholtz free energy of the ideal gas is a positive definite function of the violations of the constraints and consequently can be considered as a penalty function of the errors with respect to those constraints. Thus we are naturally led to an important method for maximalization problems with constraints, a class of methods which are known as penalty methods. [3, 6, 22, 24, 25, 34, 37, 42].

A sketch of the essential idea behind penalty methods (methods of successive unconstrained maximalizations) was expounded by the author of this book in the introduction.

In this section we shall first review that variant of the method which is usually called the method of penalty functions and then proceed to a consideration of the basic contents of this section.

Let us consider the following problem

$$f(x_1,\dots,x_n) \to \max \tag{2.70}$$

in a domain $Q \subset E^n$, defined by the conditions

$$\begin{aligned} g_s(x_1,\dots,x_n) &\leqslant 0, \quad s = 1,\dots,m_1 \\ g_s(x_1,\dots,x_n) &= 0, \quad s = m_1+1,\dots,m. \end{aligned} \tag{2.71}$$

The system of restrictions (2.71) can in many ways be described by a single constraint

$$g(x_1,\dots,x_n) \leqslant 0.$$

To obtain this it suffices, for example, to take for $g(x_1,\dots,x_n)$ the function

$$g(x_1,\dots,x_n) = \max\{g_1,\dots,g_{m_1}, g^2_{m_1+1},\dots,g^2_m\}.$$

Another example of such a function is a logarithmic penalty function of the form

$$g(x_1,\dots,x_n) = \sum_{s=1}^{m} \left(A_s \ln \frac{A_s}{A_s+y_s} + B_s \ln \frac{B_s}{B_s-y_s}\right) \tag{2.72}$$

where

$$y_s = \begin{cases} g_s(x_1,\dots,x_n) \underline{\underline{1}} [g_s(x_1,\dots,x_n)], & s = 1,\dots,m_1 \\ g_s(x_1,\dots,x_n), & s = m_1+1,\dots,m \end{cases}$$

and $A_1,\dots,A_m$, $B_1,\dots,B_m$ are sufficiently large positive constants. It is easy to verify that the function (2.72) is well defined (exists) in the region

$$-A_s < y_s < B_s$$

that it is differentiable if the $g_1,\dots,g_m$ are differentiable, that it is identically zero in the domain Q (as a function of $x_1,\dots,x_n$) and that it is strictly positive outside that domain.

We shall call a function $\phi(g(x)) = \psi(x)$ a penalty function if it is convex in E^n and such that

$$\phi(t) \begin{cases} = 0 & \text{for } t \leqslant 0 \\ > 0 & \text{for } t > 0 \end{cases}, \tag{2.73}$$

$$\lim_{t\to\infty} \phi(t) = +\infty. \tag{2.74}$$

We shall also consider sequences of penalty functions $\phi_1(x), \psi_2(x), \ldots$ satisfying the following property

$$\lim_{k\to\infty} \psi_k(x) = \lim_{k\to\infty} \phi_k(g(x)) = +\infty \tag{2.75}$$

for every x in the domain $\{x | g(x) > 0\}$.
If $\psi(x)$ is a penalty function then an example of such a sequence is $K_0\psi(x)$, $K_1\psi(x)$, ... where K_0, K_1 is a monotonic unbounded growing sequence of positive real numbers. The penalty function method consists of the reduction of problem (2.70)-(2.71) to the solution up to accuracy ε_k with $\lim_{k\to\infty} \varepsilon_k = 0$ of a sequence of maximization problems without constraints

$$f(x) - \psi_k(x) \to \max, \quad k = 0,1,2,\ldots .$$

That is it consists of the determination of a sequence of vectors $x^{(0)}, x^{(1)}, \ldots$ such that

$$f(x^{(k)}) - \psi_k(x^{(k)}) \geqslant d_k + \varepsilon_k$$

$$d_k = \sup_{x\in E_k} [f(x) - \psi_k(x)] \tag{2.76}$$

$$\lim_{k\to\infty} \varepsilon_k = 0.$$

The following theorem is proved in [22]. It establishes conditions under which the limit of such a sequence of vectors is a solution of problem (2.70)-(2.71).

THEOREM 2.5. *Let* $f(x)$ *and* $g(x)$ *be continuous, let the set* Q *be nonempty,* $d_k \leqslant k < +\infty$, $\psi_k(x) \geqslant 0$ *for all* x *and* k, $\lim_{k\to\infty} \psi_k(x) = 0$ *on a set* R *which is everywhere dense in* Q, *and*

$\lim_{k\to\infty} \psi_k(t) = +\infty$ *for* $t > 0$. *Then a sequence* $x^{(0)}, x^{(1)}, \ldots$ *satisfying the conditions* (2.76) *is such that* $f(x^{(k)}) \geq f^* = \sup_{x\in Q} f(x)$, $\lim_{k\to\infty} g(x^{(k)}) \leq 0$. *If the restriction* $g(x) \leq 0$ *is well posed* [3] *and* $f(x)$ *satisfies a Lipshitz condition in some open neighbourhood of* Q *then* $\lim_{k\to\infty} f(x^{(k)}) = f^*$ *and* $\lim_{k\to\infty} \rho(x^{(k)}), Q) = 0$.

It is possible to generalize the penalty function method, see for example [24, 37], by considering sequences of convex functions $\psi_k(x) = \phi_k(g(x))$ satisfying the conditions

$$\lim_{k\to\infty} \phi_k(t) = \begin{cases} +\infty & \text{if } t > 0 \\ C > 0 & \text{if } t = 0 \\ 0 & \text{if } t < 0 \end{cases} \tag{2.77}$$

$$\phi_k(t) > 0 \quad \text{for} \quad t > 0. \tag{2.78}$$

An example of such functions are the $\phi_k(g(x)) = Ce^{\lambda_k g(x)}$, where $\lambda_k > 0$ and $\lim_{k\to\infty} \lambda_k = \infty$. Using such functions one obtains a maximizing sequence $x^{(0)}, x^{(1)}, \ldots$, such that each element is a feasible vector. Such functions are usually called violation functions.

There are cases where it is expedient to conserve part of the constraints. In such cases penalty function methods permit us to reduce the problem to a series of maximalization problems in the presence of only that part of the constraints which does not affect the solution. A generalization of Theorem 2.6 to cover this case is not difficult to establish. Examples of restrictions which can be retained using penalty function methods are positivity constraints on some of the variables.

Suppose that the penalty function $\phi_s(g_s)$ corresponding to one of the constraints (2.71) contains a parameter. That is suppose that it is a function of the parameter λ_s and of $g_s(x)$. Let for every $\lambda_s > 0$

$$\phi_s(\lambda_s, g_s(x)) = \begin{cases} = 0 & \text{for } \begin{cases} g_s(x) \leqslant 0, & s \in M_1 \\ g_s(x) = 0, & s \in M_2 \end{cases} \\ > 0 & \text{for } \begin{cases} g_s(x) > 0, & s \in M_1 \\ g_s(x) \neq 0, & s \in M_2 \end{cases} \end{cases}$$

$$\lim_{t \to \infty} \phi_s(\lambda_s, t) = \infty, \quad s \in M$$

$$\lim_{\lambda_s \to \infty} \phi_s(\lambda_s, g_s(x)) = \infty, \quad \text{for} \quad x \in Q, \; s \in M.$$

Further let $\bar{x}(\lambda_1^{(0)}, \ldots, \lambda_m^{(0)})$ be a solution of the equation

$$\text{grad } (f(x) - \sum_{s=1}^{m} \phi_s(\lambda_s^{(0)}, g_s(x))) = 0 \tag{2.79}$$

for some fixed values $\lambda_s^{(0)} > 0$, $s \in M$. Let us consider some arbitrary continuous vector function $\lambda(\tau) = (\lambda_1(\tau), \ldots, \lambda_m(\tau))$ to the space E^m which satisfies the following conditions

$$\lambda_s(0) = \lambda_s^{(0)} > 0$$

$$\lambda_s(\tau) > 0 \quad \text{for} \quad \tau \geqslant 0 \tag{2.80}$$

$$\lim_{\tau \to \infty} \lambda_s(\tau) = \infty, \quad s \in M.$$

Let $\bar{x}(\lambda(\tau))$ be a vector function satisfying equation (2.79) for all $\tau \geqslant 0$. It follows from theorem 2.6 that

$$\lim_{\tau \to \infty} \bar{x}(\lambda(\tau)) = x^*$$

where x is an optimal vector for the problem (2.70)-(2.71). Thus the hodograph of the vector function $\bar{x}(\lambda(\tau))$ is a curve in E^n joining the point $\bar{x}(\lambda^{(0)})$ to the optimal point x^*. Obviously different vector functions $\lambda(\tau)$ will give rise to different trajectories joining the point $\bar{x}(\lambda^{(0)})$ to the point x^*. See Figure 2.2.

Fig. 2.2.

Let $\bar{x}(\lambda(\tau))$ be the trajectory resulting from a vector function $\lambda(\tau)$, satisfying conditions (2.80). The function

$$f(x) - \sum_{s=1}^{m} \phi_s(\lambda_s(\tau), g_s(x))$$

can be considered as the potential of some physical system S, of which the states are defined by the vector x. The parameter τ for such a system determines the exterior conditions. A vector $\bar{x}(\lambda(\tau))$, satisfying equation (2.79) for a fixed value of τ is an equilibrium state of the physical system S under these constant exterior conditions. If the exterior conditions change infinitesimally slowly represented by infinitely slow changes in the parameter τ from the value τ_0 to $\tau_1 > \tau_0$, then the system S goes through a quasi static transition from the state $\bar{x}(\tau_0)$ to the state $\bar{x}(\tau_1)$ passing through a sequence of intermediate equilibrium states $\bar{x}(\tau)$. Thus the trajectory $\bar{x}(\lambda(\tau))$ for $\tau_0 < \tau < +\infty$ represents itself a known quasi static (reversible) transition process of the system X from the state $\bar{x}(\lambda(\tau_0)$ to a state $x^* = \bar{x}(\lambda(\infty))$, which enjoys the extremal properties of reversible thermodynamic processes. However precisely these systems acting on an exterior medium which attain their maximum by means of reversible processes do not satisfy the criterion of minimal time by means of which numerical experts judge such or such an algorithm. This leads to the necessity of constructing processes which are quite different from quasi static ones. An example of such a process is a descent process in which there is given a sequence of values of the parameter τ and one uses some kind of gradient method to go from the state $\bar{x}(\tau_\alpha)$ to the state $\bar{x}(\tau_{\alpha+1})$. A picture of the structure of such a process is Figure 2.3.

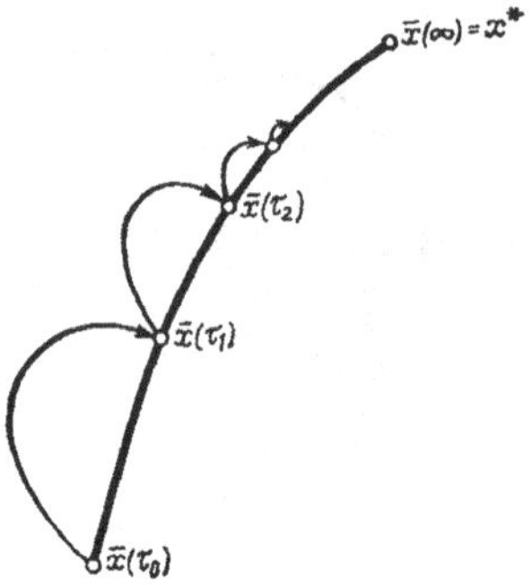

Fig. 2.3.

A physical interpretation of the structure of various descent processes, using ideas from the penalty function method, suggests various ways to solve the problem of speeding up the procedure. The fact that different sequences of penalty functions give rise to different gradient processes means that the penalty functions can be and should be considered as functions which control the calculation process and their choice is restricted solely by conditions (2.73)-(2.75). Therefore it is natural to consider the calculation processes by means of penalty functions as control processes for which it makes sense to pose the time optimal control problem. Further, once a sequence of penalty functions satisfying condition (2.75) is given, additional improvements in calculation speed may be obtained by using known ways of predicting gradient processes [28]. The numerical aspects of mathematical programming problems will be discussed in the chapter below. So in this section we content ourselves with a few general remarks.

In the literature on mathematical programming one finds now optimistic statements and then pessimistic ones on the question of the efficacy of penalty function methods. While recognizing the universality and simplicity of the method some authors are still led to the conclusion that penalty function methods can be only successfully used as a kind of initial stage for other methods. This opinion seems to be based on the fact that the speed of convergence of gradient methods decreases as the number of the sequence of problems to be solved increases because of property (2.75) of a sequence of penalty functions. It suffices to penetrate a bit more deeply into the theory of duality to see that these authors are wrong.

The essence of the method of penalty functions consists in replacing rigid constraints by elastic ones which generate a field of elastic reaction forces. The error in the optimal vector for the primal problem is then defined by the deformations in the constraints at equilibrium and the error in the optimal vector of the dual problem is measured by the difference of the gradients of the objective function in the optimal point x^* and the equilibrium point $\bar{x}$. That entails a result of considerable importance. If the components of the n-vector $(\bar{x}-x^*)$ are first order small then the components of the m-vector $(\bar{w}-w^*)$ are higher order small. Recall that there are simple relations between $\bar{x}$ and $\bar{w}$ which result from the laws which determine the elastic forces resulting from elastic deformations (Hooke's law and the Clapeyron-Mendeleev equation for example). In Section 2.6 and in Chapters IV, V, VI, VIII these questions will be treated in more detail and also from different points of view.

Let us finally mention once more, perhaps the most important property of the method of penalty functions. This is that it is not only a method of solving extremum problems with constraints but it is also a method which mathematically models real objects of study. Indeed the equality and inequality constraints are the analytical counterpart of ideal two sided or one sided restrictions (connections) and these connecting restrictions are deformable and, as in nature, the resulting reaction forces are a function of the magnitude of the deformation. Thus the equality and inequality constraints describe only the geometrical properties of the inactive (non constrained) restrictions and the penalty function can be interpreted as the deformation energy of the constraints which are deformed. For example the exact solution of an equilibrium problem for a system of elastic bodies is considered as an 'approximate' solution of an equilibrium problem of a system of perfectly rigid bodies, obtained by penalty function methods.

Another example is constituted by the considerations in this chapter concerning equilibria of physical models of containers filled with an ideal gas. The solution to these problems can also be viewed as approximate solutions obtained by penalty function methods of the analogous equilibrium problem of the same model of containers filled with an incompressible liquid. As we saw, in this last case the penalty function represents the Helmholtz free energy of the ideal gas, that is it characterizes the physical properties of the substance used to realize the constraints.

Thus the basic problem of the method is not the artificial problem of numerical efficacy but the problem of choice of a finite penalty function which describes best the properties of the real restraints following from physical, social or economic circumstances.

To conclude let us devote a few words to problems of economic planning and control. Here penalty function methods constitute the mathematical way of dealing with a centralized economy with central control and partically independent economic subsystems. In this setting the central control of the economy furnishes production plans and decides the allocation of recources and the penalty function penalizes violations of the plans and the constraints in resources. The choice of the penalty function, that is the degree of elasticity in the plans and resource restrictions clearly reflects the degree of central control of the economy. A high level of central control of the economy is reasonable if and only if there is sufficiently complete, reliable and continuous information on the state of the economy and in the circumstance that this information can be digested. There are also methods to construct a system of unconstrained maximalizations which are different from those described in the first part of this section. Below we discuss the ideas behind those methods.

The method of reducing to a sequence of problems by minimizing the norm of the errors for an incompatible system of equations and inequalities.

Let $x^* = (x_1^*,\dots,x_n^*)$ be an optimal vector for the problem (2.70)-(2.71) and let $\lambda^{(0)}$ be a number satisfying

$$f(x^*) < \lambda^{(0)}.$$

Then there exists obviously no solution to the system of equations and inequalities

$$\begin{aligned} &g_s(x) < 0, \quad s = 1,\dots,m_1 \\ &g_s(x) = 0, \quad s = m_1+1,\dots,m \qquad\qquad (2.81) \\ &f(x)-\lambda^{(0)} = 0. \end{aligned}$$

Let $F(y_1,\ldots,y_{m+1})$ be a positive function strictly convex in the $y_1,\ldots,y_{m+1}$, which are defined by

$$y_s = \begin{cases} g_s(x) \;\underline{\underline{1}}\; [g_s(x)], & s = 1,\ldots,m_1 \\ g_s(x), & s = m_1+1,\ldots,m \\ f(x) - \lambda, & s = m+1 \end{cases}$$

and suppose that $F(y_1,\ldots,y_{m+1})$ has continuous derivatives. Thus

$$F(y_1,\ldots,y_m,y_{m+1}) = \Phi(x_1,\ldots,x_n,\lambda).$$

The incompatibility of (2.81) for $\lambda = \lambda^0$ means that the minimum of $\Phi(x;\lambda^{(0)})$ is strictly positive and is assumed in a point $x^{(0)}$ which necessarily does not lie in Q and also does not satisfy $f(x) - \lambda^{(0)} = 0$. It is easy to show also that

$$f(x^*) < f(x^{(0)}) < \lambda^{(0)}. \tag{2.82}$$

Now let $x^{(1)}$ be a point in which the function $\Phi(x,\lambda^{(1)})$ assumes its minimum, where $\lambda^{(1)} = f(x^{(0)})$. By virtue of the inequality (2.82) clearly $x^{(1)} \notin Q$ and the point $x^{(1)}$ does not lie in the surface $f(x) - \lambda^{(1)} = 0$ and by analogy with (2.82) one has the inequality

$$f(x^*) < f(x^{(1)}) < \lambda^{(1)} = f(x^{(0)}) < \lambda^{(0)}.$$

How to continue is clear and we thus obtain a sequence of unconstrained minimization problems

$$\Phi(x,\lambda^{(\alpha)}) \to \min, \quad \alpha = 0,1,2,\ldots$$

where $\lambda^{(\alpha)} = f(x^{(\alpha-1)})$ and $x^{(\alpha-1)}$ is the optimal vector of the problem $\Phi(x,\lambda^{(\alpha-1)}) \to \min$.

The sequence of vectors $x^{(0)},x^{(1)},\ldots$ satisfies the conditions

$$\lambda^{(0)} > f(x^{(0)}) > f(x^{(1)}) > \dots$$

$$f(x^{(\alpha)}) > f(x^*) \qquad (2.83)$$

$$\lim_{\alpha\to\infty} x^{(\alpha)} = x^*.$$

The properties (2.83) hold under the same conditions on the functions $g(x)$, $g_1(x),\dots,g_m(x)$ which were required in Theorem 2.5. In Section 3.5 of Chapter III we shall apply this method to linear programming problems. The physical meaning of the method is that one essentially uses a method of reducing an equilibrium problem for an active physical system to a sequence of equilibrium problems for isolated (passive) systems. To that end we extend the physical model by adding to it a physical model for the equation $f(x) - \lambda = 0$, where λ is the (n+1)st component of the state. The algorithm then presents itself as a two step cyclic process of subsequent maximalizations of the entropy $S(x_1,\dots,x_n,\lambda)$ where in the first stage the maximum is found of $S(x_1,\dots,x_n,\lambda)$ for the parameters $x_1,\dots,x_n$ for a fixed value of λ, and in the second step the value of λ is found which maximizes the entropy for fixed values of the state parameters $x_1,\dots,x_n$. The sequence $S(x^{(\alpha)},\lambda^{(\alpha)})$, $\alpha = 1,2,\dots$ is clearly monotonically increasing and

$$\lim_{\alpha\to\infty} S(x^{(\alpha)},\lambda^{(\alpha)}) = S(x^*,f(x^*)) = 0.$$

For the description of another method see Section 3.5. Chapter X is essentially devoted to a development of this method and an extension to problems of optimal control.

<u>The reduction method by parametrization of the constraints of a maximalization problem with constraints</u>

Let $x^{(0)}$ be a point where the function

$$f(x) - \sum_{s=1}^{m} \psi_s(x)$$

assumes its maximum. Here

$$\psi_s(x) = \begin{cases} \phi_s(g_s(x))\ \underline{1}\ [g_s(x)], & = 1,\dots,m_1 \\ \phi_s(g_s(x)), & s = m_1+1,\dots,m \end{cases}$$

and $\phi_1(t),\ldots,\phi_m(t)$ are penalty functions which satisfy the conditions (2.73), (2.74) and which are such that the maximum of function (2.84) exists. The vector $x^{(0)}$ is an approximate solution of problem (2.70)-(2.71). On the other hand that same vector $x^{(0)}$ is an exact solution of the problem

$$f(x) \to \max$$

under the conditions

$$g_s(x) \leq g_s(x^{(0)}) \ \underline{1}\ [g_s(x^{(0)})], \qquad s = 1,\ldots,m_1$$

$$g_s(x) = g_s(x^{(0)}), \qquad s = m_1+1,\ldots,m.$$

This means that the approximate solution of problem (2.70)-(2.71) obtained by changing that problem to the unconstrained maximalization problem

$$f(x) - \sum_{s=1}^{m} \psi_s(x) \to \max \tag{2.85}$$

is an exact solution of some other maximalization problem with precisely the same objective function but on a domain defined by a different selection of functions from one-parameter families of functions $G_s(x;c_s) = g_s(x) - c_s$, $s = 1,\ldots,m$. The inverse assertion is also correct and this underlies what is proposed in the following. Given a choice of penalty functions satisfying condtions (2.73)-(2.74) then there exists an m-dimensional vector $c^* = (c_1^*,\ldots,c_m^*)$ such that the optimal vector of problem (2.85) with

$$\phi_s(x) = \begin{cases} \phi_s(G_s(x,c_s^*)) \overset{1}{=} [G_s(x,c_s^*)] , & s = 1,\ldots,m_1 \\ \phi_s(G_s(x,c_s^*)), & s = m_1+1,\ldots,m \end{cases}$$

is also an optimal vector for problem (2.70)-(2.71).

There are several ways to determine the vector c^*. One of them involves solving the following sequence of unconstrained maximalization problems.

$$f(x) - \sum_{s=1}^{m} \phi_s(y_s^{(\alpha)}) \to \max$$

$$y_s^{(\alpha)} = \begin{cases} G_s(x, c_s^{(\alpha)}) \underline{1} [G_s(x, c_s^{(\alpha)})], & s = 1, \ldots, m_1 \\ G_s(x, c_s^{(\alpha)}), & s = m_1+1, \ldots, m \end{cases}$$

$$c_s^{(\alpha)} = G_s(x^{(\alpha)}), \quad \alpha = 0, 1, 2, \ldots$$

$$c_s^{(0)} = 0, \qquad s = 1, \ldots, m$$

where $x_s^{(\alpha)}$ is an optimal vector of the problem

$$f(x) = \sum_{s=1}^{m} \phi_s(y_s^{(\alpha-1)}) \rightarrow \max.$$

Corresponding to the sequence of problems (2.86) there is a sequence of optimal vectors $x^{(0)}, x^{(1)}, \ldots$ and a sequence of vectors $c^{(0)}, c^{(1)}, \ldots$ which satisfy the conditions

$$\lim_{\alpha\to\infty} c^{(\alpha)} = c \quad \text{and} \quad \lim_{\alpha\to\infty} x^{(\alpha)} = x^*.$$

Let us remark that for a linear programming problem the sequence (2.86) is finite and that the number of problems in this sequence does not exceed the number of constraints.

Underlying the two methods set out above for solving extremal problems with constraints is the idea of penalty functions. But is important to note that for these methods it is not necessary to use a sequence of penalty functions which satisfies condition (2.75).

The idea for a final method which we shall see again in Chapter V for linear programming problems can be arrived at by reflecting on certain remarks of Bertrand concerning the solution of Lagrange's problem on equilibria of bodies connected by non extensible strings [38]: "One understands in fact that once an equilibrium has established itself and the string has assumed a certain invariant length, it is not important anymore whether the length of the string is required to be constant or not" 4).

Boltzmann is absolutely right when he remarks that all ideas are prepared, somewhat anticipated, vaguely outlined by someone else before someone finally arrives who puts in all together into a coherent whole.

The problems of finding an equilibrium for a system subject to deformable exterior constraints can be reduced to an equilibrium problem of the same system subject to nondeformable constraints if one displaces the exterior bodies precisely as far as these exterior bodies constituting the constraints are deformed.

There is also another possibility of constructing problems on equilibria of systems subject to elastic constraints equivalent to equilibrium problems of the same system subject to indeformable constraints. This consists in giving finite fixed deplacements of the exterior bodies and subsequently choosing such values for the parameters characterizing the degree of elasticity of the bodies constituting the constraints that the deformations of these bodies at equilibrium are equal to the given finite values. This simple and purely mechanical idea turns out to be very fruitful in the theory of extremal problems at it leads to a new method of obtaining a sequence of unconstrained problems which is free from the disadvantages of the methods which are based on a sequence of penalty functions satisfying conditions (2.73)-(2.75). This method will be describing in Chapters VI and VIII for linear and nonlinear programming problems.

Thus conditions (2.73)-(2.75) only define one class of functions by means of which one can construct sequences which satisfy (2.75). The remarks outlined above attest that condition (2.73) is not necessary and that it is possible to enlarge considerably the class of functions which can be used in methods of sequences of unconstrained minimizations.

Consider for example the problem of minimizing the convex function $f(x_1,\ldots,x_n)$ on a convex set

$$g_s(x_1,\ldots,x_n) \leqslant 0, \quad s = 1,2,\ldots,m$$

which is assumed to contain an interior point; that is it is assumed that the Slater condition holds. We remark that large classes of problems which do not satisfy this condition can be reduced to problems of this type. As an example of these we point to linear programming problems whose constraints contain linear equations of the type

$$\sum_{i=1}^{n} a_{si}x_i = b_s .$$

To reduce such a problem to a problem with only inequality constraints it suffices to find an equilibrium state $\bar{x}(\tilde{q}_0)$ of its physical model for some arbitrary fixed positive value of the parameter $\tilde{q}_0$. In the state $\bar{x}(\tilde{q}_0)$ the error variables for the equality constraints are different from zero and the condition $(a_s, x) = b_s$ can be replaced by $(a_s, x) \leqslant b_s$ if $(a_s, \bar{x}(q_0)) - b_s > 0$ or the condition $-(a_s, x) \leqslant -b_s$ if $(a_s, \bar{x}(\tilde{q}_0)) - b_s < 0$. This is still true for nonlinear programming problems with the exception of a very restricted class of problems. Thus we can assume that our problem is reduced to one in a form where the set of feasible points contains an interior point. For such problems one can take as penalty functions, functions $\phi(g_s)$ which satisfy the conditions

$$\phi(g_s) \begin{cases} > 0 & \text{for } g_s > 0 \\ = 0 & \text{for } g_s = 0 \\ < 0 & \text{for } g_s < 0 \end{cases}$$

and use for the construction of a sequence of unconstrained minimization problems of such functions $\phi_1(g_s)$, $\phi_2(g_s), \ldots$ satisfying the conditions

$$\lim_{k\to\infty} \phi_k(g_s) = \begin{cases} +\infty & \text{for } g_s > 0 \\ 0 & \text{for } g_s \leqslant 0. \end{cases}$$

Obviously sequences which satisfy conditions (2.73)-(2.74) belong to the class of sequences satisfying this condition. However this generalization is not purely formal becase among such penalty functions there exist choices of functions bounded in finite neighbourhood of the optimal solution such that the optimal solution of the minimization problem is the exact unconstrained minimum of the function

$$L(x_1, \ldots, x_n) = f(x_1, \ldots, x_n) + \sum_{s=1}^{m} \phi_s(g_s(x_1, \ldots, x_n)).$$

Obviously the condition which must be satisfied by the indicated optimal choice of the penalty functions is that the gradient of the function $L(x_1^*, \ldots, x_n^*)$ be zero in the optimal solution $(x_1, \ldots, x_n)$ of the constrained minimization problem. Some methods of constructing optimal families of penalty functions will be described in Chapters V and VIII.

2.6. Some properties of approximate solutions of dual problems of linear programming problems

In the following chapter we shall study numerical methods for solving linear programming problems and for that we shall make essential use of the properties of the physical models for those problems. Therefore we shall assume in the present section that the equilibrium state vector $\bar{x}(\tilde{q}_0)$ of the physical model with parameter $\tilde{q}_0$ of the linear problem can be calculated as precisely as wanted. Here we shall consider a number of general results which will play an important role in the discussion of numerical aspects of mathematical programming. First of all let us turn to formula (2.35) which was obtained by a limiting process from formula (2.31) which in turn followed from the state equation for an ideal gas. These two formulas clearly connect approximately the optimal vectors $\bar{x}(q_0)$ and $\bar{w}(q_0)$ of the primal and dual linear programming problem.

In this section we consider two-sided estimates for approximate solutions to dual pairs of linear programming problems in the form (1.20)-(1.21) and (1.22)-(1.23) under the condition that <u>all</u> constraints to which the variable quantities $x_1,\ldots,x_n$ are subject are elastic. This means that in conditions (1.21) $N_2 = \emptyset$ and in conditions (1.23) $M_1 = \emptyset$, i.e. the 'non-negativity constraints' are included among the 'inequality constraints'.

Such a model is depicted in Figure 1.8, and, obviously, this restriction does not diminish the generality of the problems under consideration.

Thus, once one has obtained an approximate solution $\bar{x}(\tilde{q}_0)$ of the primal problem in the guise of an equilibrium state of its physical model with parameter $\tilde{q}_0$, then an approximate solution of the dual problem can be calculated by a formula which follows from the law of Boyle-Mariott

$$\bar{w}_s(\tilde{q}_0) = \begin{cases} \|a_s\|^{-1}\, \tilde{q}_0 \bar{z}_s\, \underline{\underline{1}}\,[\bar{z}_s], & s \in M_1 \\ \|a_s\|^{-1}\, \tilde{q}_0 \bar{z}_s, & s \in M_2 \end{cases} \tag{2.87}$$

where according to (2.64) and (1.15)

$$\bar{z}_s = \sum_{i=1}^{n} a_{si}\bar{x}_i(\tilde{q}_0) - b_s, \quad s = 1,\ldots,m$$

$$\|a_s\| = 1 + \sum_{i=1}^{n} |a_{si}|, \qquad s = 1,\ldots,m.$$

Here it is important to pay attention to the fact that the vector $\bar{x}(\tilde{q}_0)$ is not a feasible vector for the linear programming problem and that moreover the inequality (2.69) is valid

$$\sum_{i=1}^{n} p_i \bar{x}_i(\tilde{q}_0) > \sum_{i=1}^{n} p_i x_i^*$$

where x^* is an optimal vector. On the other hand the vector $\bar{w}(\tilde{q}_0)$ calculated by means of formula (2.87) is a feasible vector of the dual problem by virtue of the equilibrium conditions (2.58) (2 60) and consequently the following condition is fulfilled

$$\sum_{s=1}^{m} b_s \bar{w}_s(\tilde{q}_0) \geqslant \sum_{s=1}^{m} b_s w_s^*$$

where w^* is an optimal vector of the dual problem.

Consider now the physical model with parameter $\tilde{q}_0$ of the dual problem (1.22)-(1.23). The equilibrium conditions for this model (1.33)-(1.46) obviously remain the same independently of whether its containers are filled with an incompressible liquid or a perfect gas. In Section 1.4 it was shown that if the containers in the models of the primal and dual problems contain an incompressible liquid, then these models are equivalent in the sense that the extensive and intensive values in both cases define an optimal bivector of the dual pair of the linear programming problems. In the case that the containers of these two models are filled with an ideal gas we have no ground for the analogous assertion and below we shall convince ourselves that the bivector $(\bar{x},\bar{w})$ of an equilibrium state of the physical model of the primal problem with parameter value $\tilde{q}_0$ is necessarily different from the bivector $(\bar{\bar{x}},\bar{\bar{w}})$ of an equilibrium state of the physical model of the dual problem. In fact, repeating the arguments which led us to the formula (2.35) we easily obtain an analogous formula linking the components of the vector $\bar{\bar{x}}$ to the components of the vector $\bar{\bar{w}}$:

$$\bar{\bar{x}}(\tilde{q}_0) = \begin{cases} \|a_i\|^{-1} \, \tilde{q}_0 \bar{\bar{\xi}}_i, & i \in N_1 \\ \|a_i\|^{-1} \, q_0 \bar{\bar{\xi}}_i \doteq [\bar{\bar{\xi}}_i], & i \in N_2 \end{cases} \tag{2.88}$$

where

$$\bar{\bar{\xi}}_i = p_i - \sum_{s=1}^{m} a_{si}\bar{\bar{w}}_s(\tilde{q}_0)$$

$$\|a_i\| = 1 + \sum_{s=1}^{m} |a_{si}|.$$

This time it is the vector $\bar{\bar{w}}(q_0)$ which is not a feasible vector for problem (1.22)-(1.23) and the vector $\bar{\bar{x}}(q_0)$ is a feasible vector for the primal problem (1.20)-(1.21) in virtue of the equilibrium conditions (1.44)-(1.46). Therefore the following inequalities hold [5]

$$\sum_{s=1}^{m} b_s\bar{\bar{w}}_s(\tilde{q}_0) < \sum_{s=1}^{m} b_s w_s \tag{2.89}$$

$$\sum_{i=1}^{n} p_i\bar{\bar{x}}_i(\tilde{q}_0) < \sum_{i=1}^{n} p_i x_i^*. \tag{2.90}$$

Moreover from equation (2.63) there follows the inequality

$$\sum_{s=1}^{m} b_s\bar{w}_s < \sum_{i=1}^{n} p_i\bar{x}_i \tag{2.91}$$

because according to (2.87)

$$\sum_{s=1}^{m} \bar{z}_s\bar{w}_s > 0.$$

Further, from the equilibrium conditions of the physical model of the dual problem (1.22)-(1.23) one easily obtains an analogue of the duality theorem:

$$\sum_{i=1}^{n} p_i\bar{\bar{x}}_i - \sum_{i=1}^{n} \bar{\bar{\xi}}_i\bar{\bar{x}}_i = \sum_{s=1}^{m} b_s\bar{\bar{w}}_s. \tag{2.92}$$

From (2.88) it follows that

$$\sum_{i=1}^{n} \bar{\bar{\xi}}_i\bar{\bar{x}}_i > 0.$$

Therefore from the inequality (2.92) we have

$$\sum_{i=1}^{n} p_i\bar{\bar{x}}_i > \sum_{s=1}^{m} b_s\bar{\bar{w}}_s . \tag{2.93}$$

The results thus obtained show that the following assertion is true.

THEOREM 2.6. <u>The equilibrium states of the physical models of the primal and dual linear programming problem for each value of the parameter $\tilde{q}_0$ define respectively a pair of vectors $\bar{x}(\tilde{q}_0) \in E^n$, $\bar{w}(\tilde{q}_0) \in E^m$ and a pair of vectors $\bar{\bar{x}}(\tilde{q}_0) \in E^n$, $\bar{\bar{w}}(\tilde{q}_0) \in E^m$ which satisfy the conditions</u>

1. $$\sum_{i=1}^{n} p_i \bar{x}_i > \sum_{s=1}^{m} b_s \bar{w}_s \geqslant \sum_{i=1}^{n} p_i x_i^* = \sum_{s=1}^{m} b_s w_s^* \geqslant \sum_{i=1}^{n} p_i \bar{\bar{x}}_i > \sum_{s=1}^{m} b_s \bar{\bar{w}}_s \qquad (2.94)$$

2. <u>The vectors $\bar{x}(q_0)$ and $\bar{\bar{w}}(q_0)$ are not feasible and the vectors $\bar{\bar{x}}(q_0)$ and $\bar{w}(q_0)$ are feasible vectors for respectively the primal and the dual linear programming problem</u>

3. $$\lim_{\tilde{q}_0 \to \infty} \bar{x}(q_0) = \lim_{\tilde{q}_0 \to \infty} \bar{\bar{x}}(q_0) = x^*$$
$$\lim_{\tilde{q}_0 \to \infty} \bar{w}(q_0) = \lim_{\tilde{q}_0 \to \infty} \bar{\bar{w}}(q_0) = w^*.$$

Condition 1 of the theorem is the union of the inequalities (2.89), (2.90), (2.91), and (2.93). The truth of condition 2 was demonstrated above and condition 3 follows from Theorem 2.5.

For a finite value of the model parameter $\tilde{q}_0$ (the penalty parameter) we can thus obtain two pairs of vectors: a feasible and a non feasible vector for the primal problem and a feasible and a non feasible vector for the dual problem. Correspondingly we obtain two-sided estimates for the optimal value of the objective function of the primal and dual problems (condition 1 of Theorem 2.6.).

It is important to note that the inner inequalities in (2.94), that is the inequalities

$$\sum_{s=1}^{m} b_s \bar{w}_s \geqslant \sum_{i=1}^{n} p_i x_i^* = \sum_{s=1}^{m} b_s w_s^* \geqslant \sum_{i=1}^{n} p_i \bar{\bar{x}}_i$$

are not strict, and that there exists a sufficiently high value of the parameter $\tilde{q}_0$ such that the vectors $\bar{x}$ and $\bar{w}$ are optimal vectors of respectively the primal and the dual problem. That this statement is correct follows by a comparison of the equilibrium conditions of the models for the linear programming problem when the containers are filled with an incompressible liquid, and when they are filled with an ideal gas. Obviously both sets of conditions define the same vector $\bar{w} = w^*$ provided

$$\begin{aligned} \operatorname{sign} x_i^* &= \operatorname{sign} \bar{x}_i, && i \in N_2 \\ \operatorname{sign}(b_s - \xi_s(x^*)) &= \operatorname{sign}(b_s - \xi_s(\bar{x})), && s \in M_1 \end{aligned} \tag{2.95}$$

where

$$\left.\begin{aligned} \xi_s^* &= b_s - \sum_{i=1}^{n} a_{si} x_i^* \\ \bar{\xi}_s &= b_s - \sum_{i=1}^{n} a_{si} \bar{x}_i . \end{aligned}\right\} \quad s \in M_1$$

Remark. In Section 1.4 it was shown that for all $\lambda \geqslant 0$ the vector λx^* is an optimal vector for problem (1.43). Essentially the restriction of the problem is the result of a similarity transformation with a similarity coefficient λ. Clearly for a fixed value of the parameter q_0 there exists a sufficiently large number L such that for all $\lambda \geqslant L$ the vectors $\bar{\bar{x}}(\tilde{q}_0)$ and $\bar{w}(\tilde{q}_0)$ are optimal vectors of the corresponding primal and dual linear programming problems. Indeed the choice of a sufficiently large value of the similarity coefficient λ is equivalent to dividing by λ the coefficients of the objective function (p,x) of the initial problem. The essential advantage of problems of type 1.43 for large values of the parameter λ lies in the possibility of speeding up the process of obtaining exact or approximate solutions, without encountering the difficulties connected with errors of calculation produced by the growing magnitude of the penalty parameter $\tilde{q}$ or the diminishing magnitude of unimportant constraints.

2.7. Models for transport type problems

It is natural to expect that the specific properties of problems of transport type will be reflected in some way in the models of these problems. Below we shall see that this is indeed the case by considering as examples the problem of maximal flow through a network and the transport linear programming problem. The models which we shall consider are so simple that to understand them really no more than a very modest knowledge of phsycis or mechanics is required, which in the main reduces to Torelli's principle and the law of equilibrium of Maupertuis. The physical interpretation of the main results and methods of solution is equally simple, which underlines the remarks of H. Poincaré: "Physics not only gives us the occasion to solve problems; it aids us to find the means thereto, ..." [6].

The problem of the maximal flow through a network. The physical and mechanical models of the well-known problem of the maximal flow through a network have very useful properties of simplicity and transparency. The problem which we shall consider below belongs to a special class of linear programming problems; here it plays the role of an example, illustrating the value of analogy in the process of cognition and reaffirming the correctness of remarks to the effect that "in mathematics itself the main means to obtain truth are induction and analogy" [7]. The simplicity of the model leads to equally simple interpretations of the results of the theory and the solution methods, sometimes known ones, sometimes new ones.

The problem of maximal flows arises in the study of transport or communication networks. Consider a network which connects two vertices (points) A_1 and A_n by means of a collection of intermediate vertices $A_2,\ldots,A_{n-1}$. See Figure 2.4.

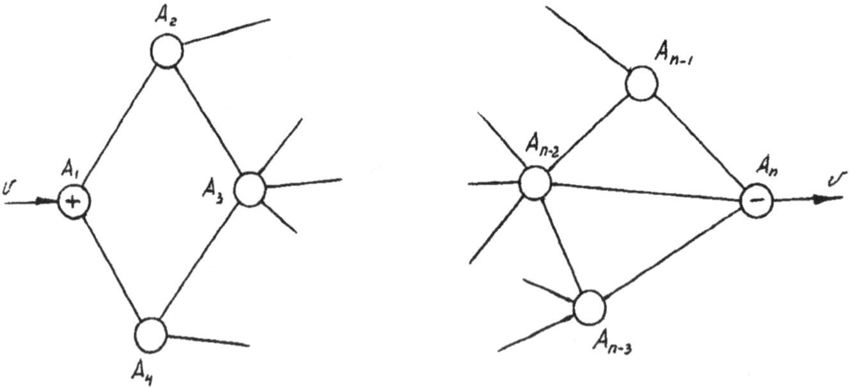

Fig. 2.4.

Each arc (edge) of the network can only let pass through a

finite amount of material to be transported in a unit time inverval. The problems consists of finding the maximal intensity by which loads can be transported from the vertex A_1 to the vertex A_n under the restrictions caused by the absorption capacities of the arcs of the network. The vertices A_1 and A_n will respectively be called the source and the target, and the vertices $A_2,\ldots,A_{n-1}$ intermediate vertices. The arc connecting A_i to A_j will be denoted (ij) and the number x_{ij} is the flow through (ij). In transport networks x_{ij} represents the amount of goods sent in the direction from point A_i to point A_j, through the arc (ij). Let us remark that the direction of movement through the arc (ij) is not prescribed and that the quantity x_{ij} can be both positive and negative. In this general setting $x_{ij} > 0$ means that the flow is directed from A_i to A_j and $x_{ij} < 0$ means that the flow is directed from A_j to A_i. Each arc (ij) of the network comes with a corresponding pair of real numbers (a_{ij},b_{ij}) which denote upper and lower bounds for the flow x_{ij}. Thus the quantity x_{ij} is required to satisfy

$$b_{ij} \leqslant x_{ij} \leqslant a_{ij}. \tag{2.96}$$

Because $x_{ij} = -x_{ji}$ and $b_{ji} \leqslant x_{ji} \leqslant a_{ji}$ it follows from (2.96) that $a_{ji} = -b_{ji}$, $b_{ji} = -a_{ji}$.

If $a_{ij} > b_{ij} \geqslant 0$ then $x_{ij} \geqslant 0$ and the admissible flows through arc (ij) are all directed from vertex A_i to vertex a_j. In case $a_{ij} > 0$ and $b_{ij} < 0$ the flow through arc (ij) may be directed both ways: $A_i \to A_j$ or $A_j \to A_i$. We denote with S_i the set of vertices of the network connected by an arc to the vertex A_i. Then the outwards flow from A_i is equal to the quantity $\sum_{j\in S_i} x_{ij}$, that is it is equal to the algebraic sum of the flows through the arcs $\{(ij), j \in S_i\}$ coming out of A_i. The subdivision of the vertices of the network in source, intermediate vertices and target is reflected in the rules

$$\sum_{j \in S_i} x_{ij} = \begin{cases} > 0 \text{ if } A_i \text{ is a source} \\ = 0 \text{ if } A_i \text{ is an intermediate vertex} \\ < 0 \text{ if } A_i \text{ is a target.} \end{cases} \tag{2.97}$$

In the problem under consideration only the vertex A_1 is a source and only A_n is a target, and the vertices $A_2,\dots,A_{n-1}$ are intermediate vertices. Let us denote with v the flow flowing out of vertex A_1. Then clearly

$$v = \sum_{j \in S_1} x_{1j} = - \sum_{j \in X_n} x_{nj} \tag{2.98}$$

and the maximal flow through a network problem consists of finding quantities x_{ij} which satisfy the conditions

$$\sum_{j \in S_1} x_{1j} \to \max \tag{2.99}$$

$$\sum_{j \in S_i} x_{ij} = 0, \quad i = 2,\dots,n-1 \tag{2.100}$$

$$\begin{gathered} b_{ij} \leqslant x_{ij} \leqslant a_{ij},\ x_{ji} = -x_{ji}, \quad i,j = 1,\dots,n,\ i \neq j \\ a_{ji} = -b_{ij},\ b_{ji} = -a_{ij},\ i,j = i,\dots,n,\ i \neq j. \end{gathered} \tag{2.101}$$

Note that conditions (2.100) are analogous to Kirchhoff's law.

The problem of finding the maximal flow through a network is equivalent to the problem of finding the equilibrium of a certain mechanical system depicted in Figure 2.5, and from this analogy there follows the duality theorem of Ford-Fulkerson [53] and also a numerical algorithm.

A model for the maximal flow problem. A mechanical model for the maximal flow through a network problem is depicted by Figure 2.5. It consists of a system of connected containers filled with an incompressible liquid. The model is a particular case of the general model for a linear programming problem (see Sdction 1.4) and its structure is similar to the structure of the network. This was first remarked by L. B. Rapoport. The model for the arc (ij) is

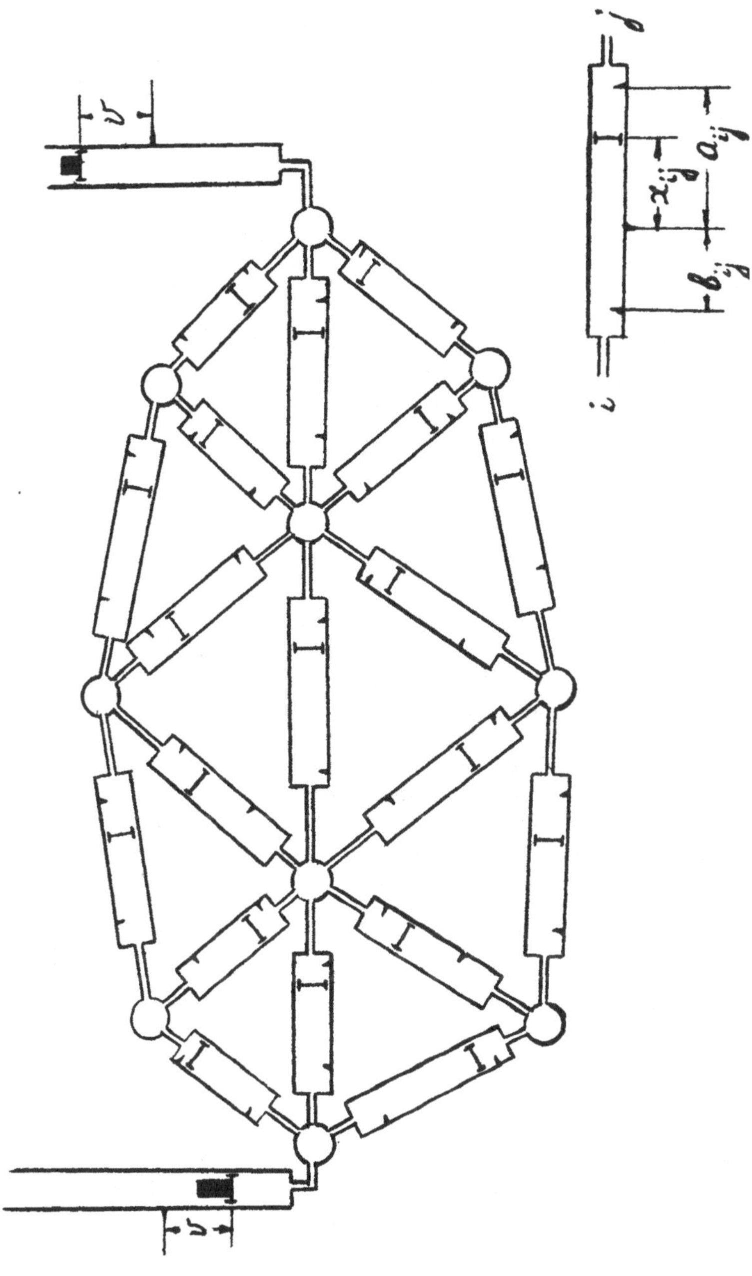

Fig. 2.5.

also depicted in Figure 2.5. It represents a cylinder with a cross-section surface of magnitude one. Inside the cylinder there is inserted a piston which divides the interior volume into two parts. The position of the piston in the cylinder determines the quantity x_{ij}, and condition (2.101) is modelled by stops restricting the possible positions of the piston. Now let us imagine that each piston of each cylinder is put in the position $x_{ij} = 0$. Denote with $V_i^{(0)}$ the sum of all the cylinder volumes connected to A_i when the pistons x_{ij}, $j \in S_i$ are all in the zero position. Let the volumes $A_i^{(0)}$ be filled with an incompressible liquid. Then for each admissible state x_{ij} of the model the following equality will hold

$$V_i = V_i^{(0)} + \sum_{j \in S_i} x_{ij}, \quad i = 2,\dots,n-1. \tag{2.102}$$

At the same time because the liquid is incompressible the condition $V_i^{(0)} = V_i$ must hold and consequently condition (2.100) is fulfilled as a simple consequence of the incompressibility property of the liquid filling the volumes of the model of Figure 2.5.

Consider now the models of the source A_1 and the target A_n of the network which are also clearly parts of the model of Figure 2.5. In contradistinction to the models of the intermediate vertices the model of the source contains an additional cylinder and the position of the piston in that cylinder x_{01} indicates the quantity v of ingoing flow. Here also, exactly as in the case of the intermediate vertices $V_1^{(0)}$ is the sum of the volumes connected to vertex $A_1 = 0$ for $v = 0$ and $x_{ij} = 0$, $j \in S_1$. The construction of the model at the target A_n is analogous to the model of the source; see Figure 2.5. To conclude the description of the model of (2.99)-(2.101) apply forces to the pistons of the ingoing and the outgoing cylinders. Let these be equal to G_1 and G_n and choose $G_1 > G_n$. Now consider the equilibrium problem for the resulting mechanical system. The first principle of statics [19, 38] and Lagrange's theorem on the stability of equilibria assert: a state of the mechanical systems which is admissible for the constraints is a stable equilibrium state if the potential energy is minimal there.

One easily convinces oneself that the potential energy of the mechanicsl system constructed above is equal to $(G_n - G_1)v$, and that $(G_n - G_1)v$ is minimal is, in virtue of the inequality $G_1 > G_n$, equivalent to the condition that v is maximal over the set of states $\{x_{ij}\}$ satisfying conditions (2.100) and (2.101).

Thus there is established the equivalence between the problem of the maximal flow through the network and the equilibrium problem of the mechanical model described above.

Equilibrium conditions. Minimal cuts. The duality theorem. An equilibrium state of the model of Figure 2.5 means that each of the movable parts of the model is in equilibrium, that is each of the pistons dividing the cylinder volumes. Let x^*_{ij} be an equilibrium state of the model. Denote with p_i the pressure of the liquid in vertex A_i at equilibrium. Because the unique active forces acting on the pistons at state x^*_{ij} are the differences of pressure in the adjacent vertices A_i and A_j the equilibrium conditions of the model obviously take the form

$$\begin{aligned} &p_1 = G_1 \\ &p_i = p_j \quad \text{for} \quad b_{ij} < x_{ij} < a_{ij} \\ &p_i \geqslant p_j \quad \text{for} \quad x_{ij} = a_{ij} \qquad\qquad i,j = 1,\dots,n,\ i \neq j \\ &p_i \leqslant p_j \quad \text{for} \quad x_{ij} = b_{ij} \\ &p_n = G_n \end{aligned} \tag{2.103}$$

The first one and the last one of these conditions are respectively the equilibrium conditions for the incoming and the outgoing cylinder pistons. From these equilibrium conditions there follows the Ford-Fulkerson duality theorem. Let us consider an arbitrary path through the network going from vertex A_1 to vertex A_n. As is well known [53] such a path can be represented by an ordered sequence of vertices connected by arcs $A_{\alpha_0}, A_{\alpha_1}, \dots, A_{\alpha_k}$ where $\alpha_0 = 1$ and $\alpha_k = n$. From conditions (2.103) it follows that in an equilibrium state such a path contains at least one arc (α_s, α_{s+1}) for which $x_{\alpha_s \alpha_{s+1}} = a_{\alpha_s \alpha_{s+1}}$ [8)]. Suppose that the assertion is not true and that the following inequalities hold

$$b_{\alpha_\nu \alpha_{\nu+1}} \leqslant x_{\alpha_\nu \alpha_{\nu+1}} < a_{\alpha_\nu \alpha_{\nu+1}}, \quad \nu = 0,1,\ldots,k-1$$

This leads innediately to a contradiction because then the conditions (2.103) imply that

$$p_{\alpha_0} = p_{\alpha_1} = \cdots = p_{\alpha_k}$$

which contradicts the condition that $p_{\alpha_0} = G_1 > G_n = p_{\alpha_k}$.

In an equilibrium state the set of vertices of the network is divided into two nonintersecting subsets $S^{(0)}$ and $S^{(1)}$ according to the rule

$$\begin{aligned} S^{(0)} &= \{A_i \mid p_i = G_1\} \\ S^{(1)} &= \{A_i \mid p_i = G_n\}. \end{aligned} \tag{2.104}$$

Obviously $A_1 \in S^{(0)}$ and $A_n \in S^{(1)}$. Thus under the conditions of equilibrium (2.103) the quantities $p_1,\ldots,p_n$ take only two values: G_1 or G_n. A division of the set of vertices of the network into two nonintersecting subsets $\bar{S}$ and $\bar{\bar{S}}$ such thath $A_1 \in \bar{S}$ and $A_n \in \bar{\bar{S}}$ is called a <u>cut</u> of the network. The set of arcs (ij) for which $i \in \bar{S}$ and $j \in \bar{\bar{S}}$ is called the set of arcs of the given cut. In virtue of the incompressibility of the liquid filling the containers of the model the maximal flow through the network is equal to the sum of the flows through the arcs in an arbitrary cut. From the equilibrium conditions (2.103) it follows that the flow through the arcs of the cut $(S^{(0)},S^{(1)})$, defined by condition (2.104) is equal to the absorption capacities a_{ij}, $i \in S^{(1)}$ of this arc. Such a cut $(S^{(0)},S^{(1)})$ is called minimal. Thus one arrives at the theorem of Ford-Fulkerson.

<u>THEOREM</u>. <u>The maximal flow through a network is equal to the sum of the absorption capacities in a minimal cut</u>

$$v_{max} = \sum_{i \in S^{(0)},\, j \in S^{(1)}} a_{ij}$$

<u>A physical model for the maximal flow problem</u>. Consider now the equilibrium problem for a physical model which differs

from the model described above for the maximal flow problem and depicted in Figure 2.5 only in the circumstances that it contains instead of incompressible liquids ideal gases in such quantities that the pressures in the volumes V_i for the state $x_{ij} = 0$, $i,j = 1,\dots,n$ would all be equal to p_0, where p_0 is a positive number which can be chosen arbitrarily. Assume that the model is placed in a thermostat, then the dependence of the pressures in its volumes for arbitrary positions of the coordinates is given by the law of Boyle-Mariotte

$$p_i V_i = p_0 V_i^{(0)} \tag{2.105}$$

where V_i is the sum of all the volumes connected to vertex A_i and is determined by the equality (2.102), and p_i is the pressure at vertex A_i. From (2.102) and (2.105) it follows that

$$p_i = (V_i^{(0)} + \sum_{j\in S_i} x_{ij})^{-1} p_0 V_i^{(0)}.$$

Further we shall see that it is convenient to choose the quantity G_n equal to p_0. For the physical system under consideration the potential energy consists of the potential energy of the weights G_1 and $G_n = p_0$ and the Helmholtz free energies of the gases in the volumes of the model. Using the fact that we can freely set the potential energy to be zero for one arbitrarily chosen state we take the potential energy to be zero in the state

$$x_{ij} = 0, \ i \neq j, \ i,j = 1,2,\dots,n. \tag{2.106}$$

Then in an arbitrary state the potential energy is equal to

$$\Pi = (p_0 - G_1)\, v - G_1\Delta + \sum_{i=1} \int_{V_i(x_{ij})}^{V_i^{(0)}} p_i dV_i . \tag{2.107}$$

In the expression (2.107) the number Δ is the sum of all the changes in volume of the gases in all volumes of the model and the quantities p_i and V_i are connected by relation (2.105). Obviously in an equilibrium state of the model filled with an ideal gas the conditions (2.103) still hold, as the mechanical meaning of these conditions does not depend on the properties of incompressibility or compressibility. The difference only consists in that in an equilibrium state of the model under consideration the condition (2.100) may not anymore be fulfilled. The value of that fact lies in that the arguments are still valid which led to the Ford-Fulkerson theorem on the equality of the maximal flow and the absorption capacity of a minimal cut. Condition (2.104) which determines a minimal cut of the network must in the case of a model with its volumes filled with an ideas gas be replaced by the conditions

$$S^{(0)} = \{A_i | p_i > p_0\}$$
$$S^{(1)} = \{A_i | p_i = p_0 = G_n\}. \tag{2.108}$$

From the fact that the set $S^{(1)}$ defined by conditions (2.104) and (2.108) coincides in two cases, it follows that for each $G_1 > p_0$ an equilibrium state of the model filled with ideal gases determines a minimal cut of the network and consequently determines the maximal flow. Moreover from the condition $p_i = p_0$, $i \in S^{(1)}$ and the state equation (2.105) it follows that

$$\sum_{j \in S_i} \bar{x}_{ij} = 0 \quad \text{for} \quad A_i \in S^{(1)} \tag{2.109}$$

where $(\bar{x}_{ij})$ is the equilibrium state of the model. Thus a solution of the equilibrium problem for arbitrary $G_1 > p_0$ not only determines the maximal flow v^* and a minimal cut but it also determines the flows through the arcs connecting all vertices of the set $S^{(1)}$ with vertices of the set $S^{(0)} \cup S^{(1)} = S$ and consequently the following inequality holds

$$\bar{x}_{ij} = x^*_{ij}, \; i \in S^{(1)}, \; j \in S^{(0)} \cup S^{(1)}. \tag{2.110}$$

The properties indicated above play a decisive role in the construction of algorithms to calculate solutions of maximal flow in network problems.

To conclude let us turn our attention to an expression for the potential energy of the physical model which in an equilibrium state must assume its minimum value on the set of states which satisfy only condition (2.110). Let us calculate that part which expresses the potential energy of an ideal gas 9)

$$\pi = \sum_{i=1}^{n} \int_{V_i}^{V_i^{(0)}} p_i dV_i .$$

Using the state equation (2.105) one obtains

$$\pi = p_0 \sum_{i=1}^{n} V_i^{(0)} \ln \frac{V_i^{(0)}}{V_i^{(0)}+X_i} \tag{2.111}$$

where X_i is introduced to stand for

$$X_i = \sum_{j \in S_i} x_{ij}, \quad i = 1,\dots,n \tag{2.112}$$

Consider the problem of finding the minimum Helmholtz free energy for a fixed value of the quantity Δ. From $G_1 > 0$ it follows that $\Delta > 0$. Further the following remark is very important. It was shown above that in an equilibrium state $\{\bar{x}_{ij}\}$ for all $G_1 > p_0$, $v = v^*$ the value of the maximal flow. It is therefore always possible to give an arbitrary positive number $\bar{\Delta} > 0$ and then choose $\bar{G}_1 > p_0$ such that for $G_1 = \bar{G}_1$ the minimum of the function π within the set of admissible states is assumed in an admissible equilibrium state $\{\bar{x}_{ij}\}$ such that moreover $\Delta = \bar{\Delta}$.

This means that the problem of finding an equilibrium state $\{\bar{x}_{ij}\}$ reduces to the problem

$$\min\{ \sum_{i=1}^{n} V_i^{(0)} \ln \frac{V_i^{(0)}}{V_i^{(0)}+X_i} \mid \sum_{i=1}^{n} X_i + \bar{\Delta} = 0,\ X_i = \sum_{j \in S_i} x_{ij},$$

$$i = 1,\dots,n;\ a_{ij} \leqslant x_{ij} \leqslant b_{ij},\ i,j = 1,\dots,n\} \tag{2.113}$$

where $\bar{\Delta}$ is an arbitrary given positive number. The problem (2.113) is equivalent to the problem of minimizing the function π under conditions (2.101) in the sense that its solution for each $\Delta > 0$ defines a minimal cut $(S^{(0)}, S^{(1)})$, the maximal flow v^* and the optimal flows through the arcs x_{ij}, $i \in S^{(1)}$, $j \in S^{(1)} \cup S^{(2)} = S$. The flows through the remaining arcs of the network are functions of $\bar{\Delta}$, $\bar{x}_{ij} = \bar{x}_{ij}(\bar{\Delta})$ and there is the limit equality

$$\lim_{\Delta \to 0} \bar{x}_{ij}(\bar{\Delta}) = x^*_{ij}.$$

The flows through the arcs x_{ij}, $i \in S^{(1)}$, $j \in S$ are equal to x^*_{ij} and do not depend on the quantity $\bar{\Delta} > 0$. Because in the equilibrium state to be determined $\{\bar{x}_{ij}(\bar{\Delta})\}$ the conditions

$$X_i = 0 \quad \text{for} \quad i \in S^{(1)} \tag{2.114}$$

hold and because the dual values of the restrictions (2.101) are different from zero only for the arcs (ij) with $i \in S^{(0)}$ and $j \in S^{(1)}$, problem (2.113) is equivalent to the problem

$$\min\left(\sum_{i \in S^{(0)}} V_i^{(0)} \ln \frac{V_i^{(0)}}{V_i^{(0)} + X_i} \;\middle|\; \sum_{i \in S^{(0)}} X_i + \bar{\Delta} = 0 \right) \tag{2.115}$$

A solution of the last problem can be easily obtained by using the method of multipliers which means to solve the unconstrained minimization problem

$$\min_{X_i, i \in S^{(0)}} \left[\sum_{i \in S^{(0)}} V_i^{(0)} \ln \frac{V_i^{(0)}}{V_i^{(0)} + X_i} + \lambda\left(\sum_{i \in S^{(0)}} X_i + \bar{\Delta}\right)\right]. \tag{2.116}$$

The conditions for minimality of this problem have the form

$$\frac{V_i^{(0)}}{V_i^{(0)} + X_i} = \lambda, \quad i \in S^{(0)} \tag{2.117}$$

from which it follows that

$$X_i = -\bar{\Delta}\, V_i^{(0)} \Big(\sum_{i \in S^{(0)}} V_i^{(0)} \Big), \quad i \in S^{(0)} \tag{2.118}$$

and after a simple calculation one obtains the minimal value π_{min} of the free energy

$$\pi_{min} = V^{(0)}(S^{(0)}) \ \ell n \frac{V^{(0)}(S^{(0)})}{V^{(0)}(S^{(0)})-\bar{\Delta}} \tag{2.119}$$

where

$$V^{(0)}(S^{(0)}) = \sum_{i \in S^{(0)}} V_i^{(0)}.$$

It follows from relations (2.118) and (2.119) that

$$\lim_{\bar{\Delta}\to 0} X_i = 0, \quad i \in S^{(0)}$$

$$\lim_{\bar{\Delta}\to 0} \pi_{min} = 0$$

that is, it follows that the limit equality (2.100) is correct. The inequality $\pi_{min} > 0$ is clear.

Methods for numerical solution. Of course all known methods to calculate the solution to maximal flow problems in networks are directly or indirectly contained in the models for these problems, and it is perhaps easy to give them explicit physical interpretations. Moreover an attentive consideration of these models should stimulate inventiveness as work in the area of creating algorithms for the numerical solution of problems in an inventive activity. Here the method of redundant constraints exposed in Chapter III turns out to be very fruitful. Algorithms for the numerical solution of these problems following from these models and the method of redundant constraints are described in [54] and it clearly suffices to restrict ourselves here to a reference.

There may also be interest in another purely mechanical model for the flow through networks problem. This model is sketched in Figure 2.6. The construction elements of this model are inextensible threads and movable and immovable blocks and doubtless it will be useful for the reader to investigate its structure and convince himself that the maximal flow through a network problem is equivalent to an equilibrium problem for the mechanical one depicted in Figure 2.6.

The transport problem of linear programming. To conclude this section let us consider very briefly the constrained minimization problem

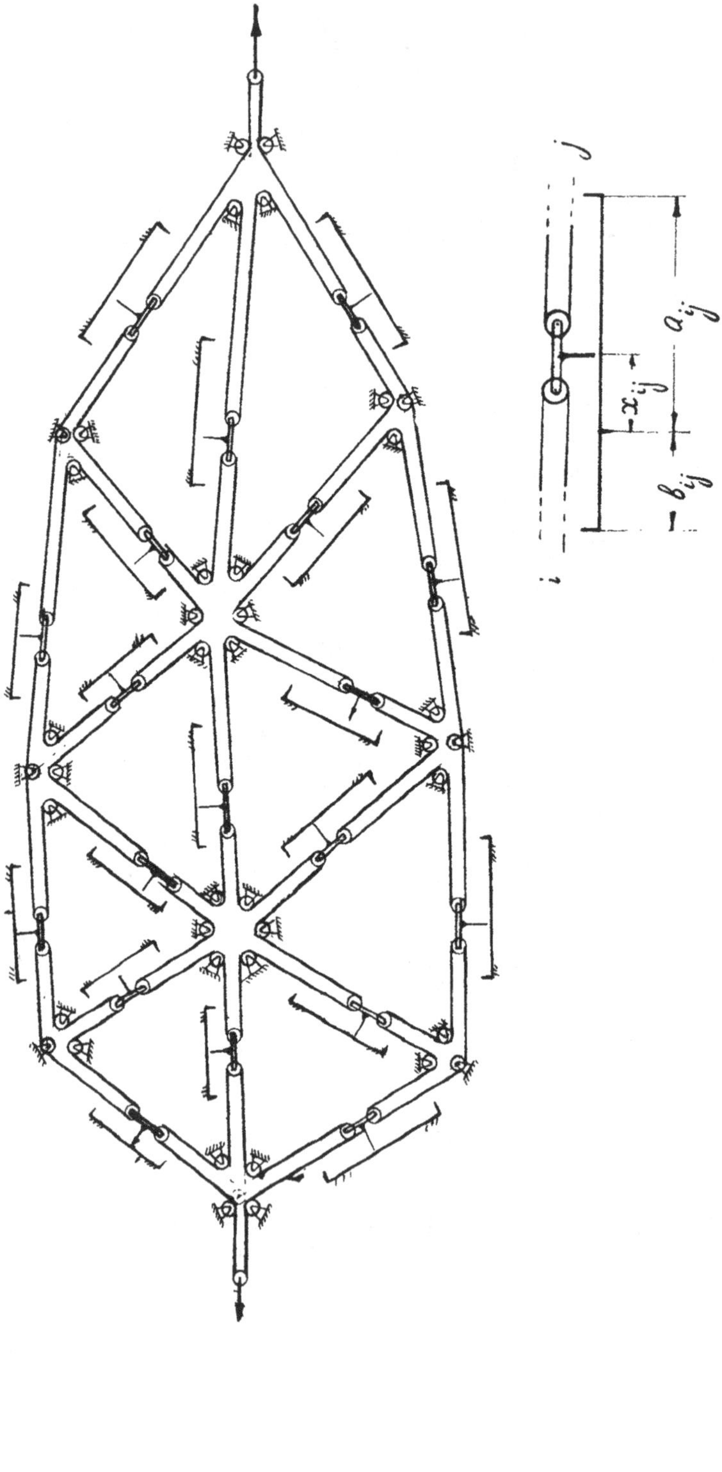

Fig. 2.6.

$$\min\{\sum_{i=1}^{n}\sum_{j=1}^{m} c_{ij}x_{ij} \mid \sum_{j=1}^{m} x_{ij} = a_i,\ i = 1,\ldots,n;$$
$$\sum_{i=1}^{n} x_{ij} = b_j,\ j = 1,\ldots,m;\ x_{ij} \geqslant 0,\ i = 1,\ldots,n,$$
$$j = 1,\ldots,m\} \qquad (2.120)$$

which is known as the transport problem. Let $A_1,\ldots,A_n$ be points of supply for goods containing respectively $a_1,\ldots,a_n$ units of goods. Let $B_1,\ldots,B_m$ be points of demand of goods and let $b_1,\ldots,b_m$ be the amounts demanded. The elements of the matrix $C = (c_{ij})$ are numbers c_{ij} which determine the costs of transporting one unit of goods from point A_i to point B_j. The matrix $X = (x_{ij})$ is called a transport plan because x_{ij} represents the amount of goods despatched from point A_i to point B_j. To make the transport problem well defined the following condition must hold:

$$\sum_{i=1}^{n} a_i = \sum_{j=1}^{m} b_j.$$

Because the problem (2.120) is a particular case of the linear programming problem it is easy to construct also the corresponding particular case of the model of that problem consisting of containers filled either with an incompressible liquid or with an ideal gas.

In Figure 2.7 there is shown another purely mechanical model for the transport problem (2.120) and in Figure 2.8 a fragment of this model. Doubtless the reader will be able to convince himself without too much trouble that the equilibrium problem for the mechanical system of Figure 2.7 consisting of inextensible threads, pulleys and loads is equivalent to problem (2.120).

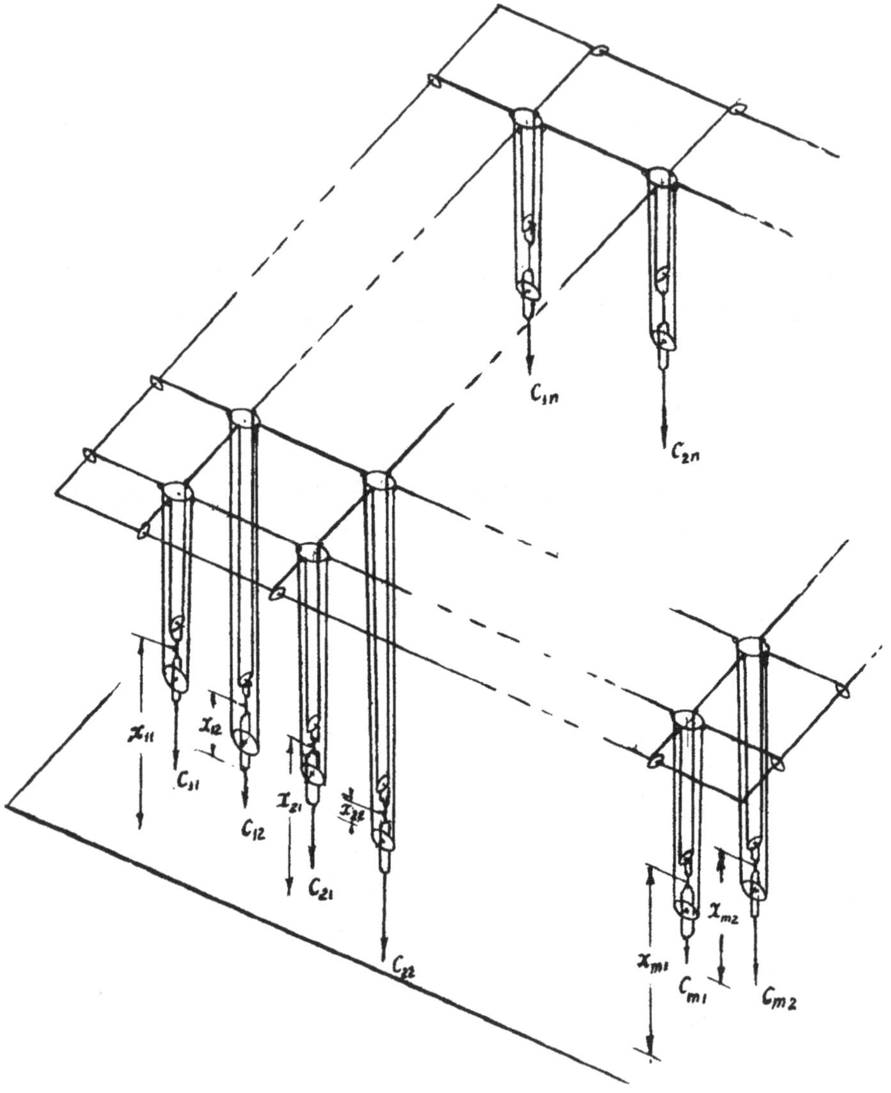

Fig. 2.7.

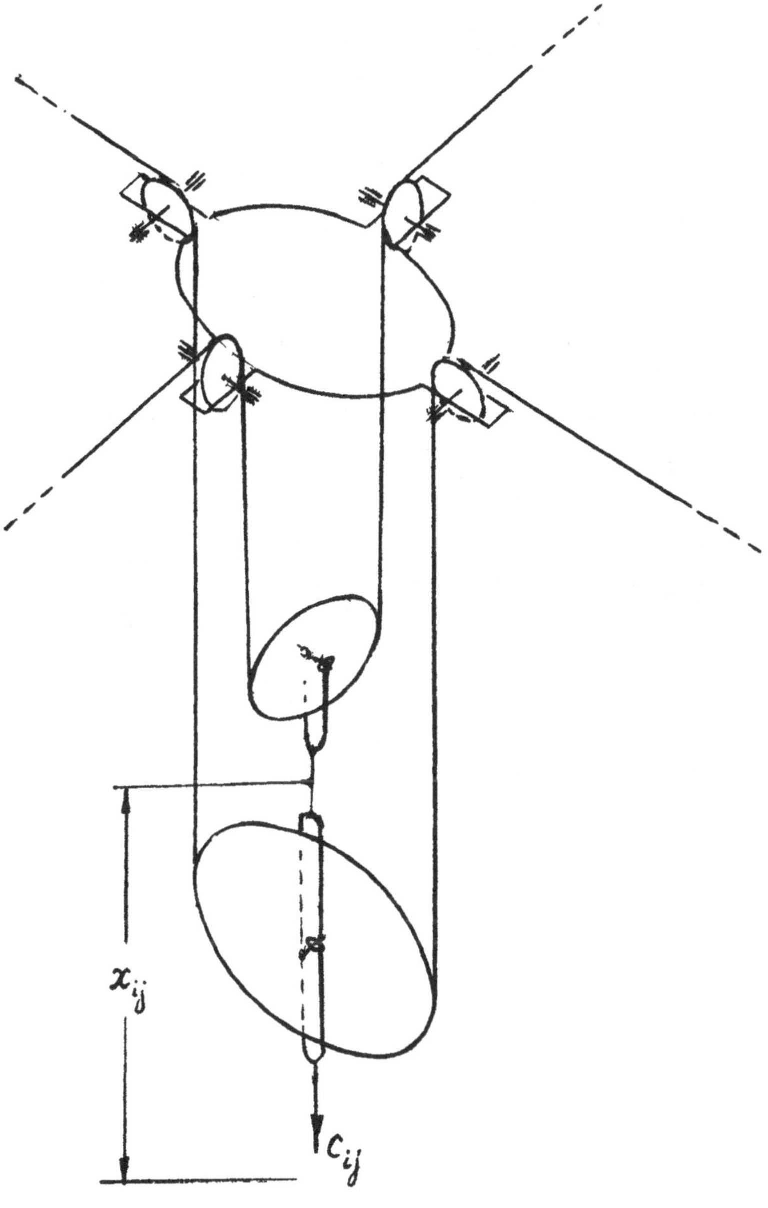

Fig. 2.8.

NOTES

1. Recall that one mol of a gas has a mass in grams which is numerically equal to its molecular weight (so that every mol of every (ideal) gas has the same number of molecules in it).
2. The conditions $x_i \geqslant 0$ for $i \in N_2$ are modelled by means of stops. This is shown in Figure 1.8 which we used as a model for linear programming problems. For $p_1 = \ldots = p_n = 0$ Figure 1.8 is a model for the system.
3. This means that the property $g(x^{(k)}) \to 0$ for $k \to \infty$ implies $\rho(x^{(k)},Q) \to 0$ where $\rho(x^{(k)},Q) = \inf_{x \in Q} \|x^{(k)}-x\|$.
4. On comprend en effet, qu'une fois l'équilibre établi, le fil ayant pris une certaine longueur qui ne varie plus, peu importe que cette longueur soit ou ne soit pas asujettie à demeurer constante.
5. Inequality (2.89), clearly, is proved just as inequality (2.69), see Section 2.4.
6. H. Poincaré, The value of science, p. 82 of the Dover reprint, 1958.
7. Laplace, Essay on the philosophy of probability theory.
8. Such an arc is called saturated.
9. The quantity π in thermodynamics bears the name of the Helmholtz free energy, see also Section 2.2.

Chapter III

THE METHOD OF REDUNDANT CONSTRAINTS AND ITERATIVE ALGORITHMS

3.1. Introduction

In this chapter more general methods will be described for obtaining iterative (recursive) algorithms for the numerical solution of mathematical programming problems and problems from mathematical economics. And these will also be applied to linear programming problems.

The method is essentially based on physics and we shall use the physical properties of bodies subject to constraints. In this way we shall again study equilibrium states of physical models and transition processes between such states considering the constraining equalities or inequalities as ideal constraints, that is as a limiting case of elastic constraints. Those who have attentively perused the introduction to this book will understand that it is necessary to be careful concerning the quality of the approximate solutions which can be obtained quite simply by the formulas proposed. Certainly one can to this end use this or that norm of the errors made. However, one should not forget that in general we do not have a norm by means of which one can judge the quality of a mathematical model. It seems that the numerical difficulties connected with a passage to the limit are the price we pay for the coarseness of (idealized) mathematical models, that is for the violations of that law which Johann Bernouilli called the immutable and eternal law of continuity [22].

3.2. The method of redundant constraints

An important class of algorithms for finding a minimum goes under the name of descent methods; that is they are methods for constructing a sequence of points such that its limit is a minimum point or belongs to a given neighbourhood of a minimum point. The algorithms which are contained in this chapter and in the following chapters of this book also belong to the class of descent algorithms. They are rather special in that they follow immediately from the physical properties of the models of the corresponding problems. In physical models the process of descent obviously manifests itself as a transition process from an arbitrary given initial state to an equilibrium state. These processes can be controlled by changing, according to the control aims, either the exterior conditions or the constraints to which the physical model is subject. The various ways of bringing about quasistatic processes in thermodynamics are well known examples of such controlled physical systems. Changing the exterior conditions or the constraints it is possible to realize different trajectories in state space for bringing the system to the desired equilibrium state. Below we shall consider quasi static and isothermic processes in the physical models for extremum problems and the control of these processes will be realized by imposing additional or superfluous stationary constraints. Thus a set of constraints will be said to be redundant if they are absent from the conditions modelling the problem and are imposed temporarily on the system in order to realize a transition to a required equilibrium state. Of course the redundant constraints imposed on the system in a given state must be compatible with that state. By means of these redundant constraints one diminishes the dimension (i.e. the number of variables) of the initial problem to an acceptable number and thus one reduces a complicated high dimensional system to a sequence of simpler problems of small dimension. The method of redundant constraints consists of decomposing complicated transition processes to equilibrium state of physical systems into a series of elementary processes by means of imposing a sequence of redundant constraints. Obviously a gradient method with small steps and a method of coordinate-wise descent and many others can be considered as instances of the method of redundant constraints. In the following we shall use redundant constraints which naturally fall into one of two types:

(1) Holonomic (geometric) constraints which express relations connecting the coordinates of the bodies making up the physical system. For example constraints which fix the positions or all or some of these bodies.

(2) Constraints of partitioning type which divide the physical system into some collection of isolated subsystems.

It is well known that the imposing of additional

stationary constraints compatible with a given stable equilibrium state x of the system preserves that stability property and can only reinforce the equilibrium. If the state x is a unstable equilibrium state then an additional stationary constraint can either stabilize that equilibrium or diminish the degree of instability. If $\bar{x}$ is an equilibrium of the physical system without additional redundant constraints and $\hat{x}$ is an equilibrium state of the same system but subjected to additional redundant constraints, then clearly the values of the general thermodynamic potential (2.14) for those states satisfy the inequality

$$\phi(\bar{x}) \leqslant \phi(\hat{x}).$$

In the present chapter we shall describe algorithms for the numerical solution of linear programming problems which are based on the idea of controlling by means of redundant constraints the transition processes to equilibrium states of the physical model of these problems. We shall assume that the conditions are satisfied which guarantee that the processes which take place in the systems under consideration will be isothermic ones. In subsequent chapters we shall show the efficacy of the method of redundant constraints for nonlinear programming problems and problems from mathematical economics.

The most important property of all the algorithms which come out of the method of physical modelling and redundant constraints is the monotone convergence of the corresponding iterative calculation processes. This follows from the fundamental laws of nature, that is from the first and second law of thermodynamics, see also Section 2.2. The convergence of the numerical processes, which are merely the mathematical descriptions of the physical processes which take place in the system, is guaranteed by the basic principles of thermodynamics if the following two conditions are satisfied:

(1) The processes which take place in the physical system which models the initial problem are spontaneous processes.

(2) The sequence of redundant constraints must be cyclic in nature and during every cycle all degrees of freedom of the physical system must be tried out.

The first condition means that the processes which take

place in the system must be self generated; that is they must be produced under the action of only interior forces or exterior forces which naturally result from the replacement of the initial problem by its physical model. This means that the processes in the system must take place only within the field of forces defined by the objective function which, naturally, can be considered as a potential or (resultant) force-function. Consequently the control of the processes in question can only be done by the means of imposing stationary redundant constraints which produce passive reaction forces. Under such interventions in the physical process its evolution remains spontaneous and its development satisfies the first and second principles of thermodynamics. By satisfying the first condition we acknowledge in fact the wisdom of nature which in the words of J. Bernouilli always acts in best possible way [1].

The only result of imposing stationary redundant constraints is to break up the natural process in elements which admit a simple mathematical description. The second condition guarantees that the limit of the sequence of equilibrium states of the physical model subject to redundant constraints constitutes a solution for the problem being modelled.

Let us now turn to the physical model of a linear programming problem as depicted by Figure 1.8. The state variables or indicators of the model are the coordinates $x_1,\dots,x_n$ of the bars and the 2m quantities $q_s^{(+)}$ and $q_s^{(-)}$ giving the pressures in the volumes $V_s^{(+)}$ and $V_s^{(-)}$, $s = 1,\dots,m$. These quantities are connected by means of the state equation for an ideal gas (the law of Boyle-Mariott) which says that

$$q_s^{(+)}V_s^{(+)} = q_0\bar{V}_s^{(+)}, \quad q_s^{(-)}V_s^{(-)} = q_0\bar{V}_s^{(-)}, \quad s = 1,\dots,m \quad (3.1)$$

and the isochoricity conditions

$$V_s^{(+)} + V_s^{(-)} = \bar{V}_s^{(+)} + \bar{V}_s^{(-)}, \quad s = 1,\dots,m.$$

In this manner, in virtue of relation (1.16) of Chapter I and the state equations (3.1), the 2m quantities $q_s^{(+)}$ and $q_s^{(-)}$ are functions of the coordinates $x_1,\dots,x_n$ which can be taken to be independent generalized coordinates for the physical model. Recall that in an equilibrium state the quantities $\xi_1,\ldots,\xi_n$ which occur in the formulas (1.16) must be considered dependent on $x_1,\ldots,x_n$ and that dependence is given by the formulas (2.33).

3.3. The first iterative algorithm for solving linear programming problems and for solving systems of linear equations and inequalities

In Chapter II it was shown that the equilibrium state vector $\bar{x}(\tilde{q}_0)$ of the physical model depicted in Figure 1.8 is an approximately optimal vector of the corresponding linear programming problem. Hereby the independent state parameters of the physical model are only subject to the restrictions

$$x_i \geqslant 0 \quad \text{for} \quad i \in N_2. \tag{3.2}$$

Let $Q_1,\ldots,Q_n$ be the generalized forces acting respectively on the generalized coordinates $x_1,x_2,\ldots,x_n$ of the state of the model and let $\bar{x}_1,\ldots,\bar{x}_n$ be the resulting equilibrium state coordinates whereby part of the constraints (3.2) are active. Thus

$$\bar{x}_i = 0 \quad \text{for} \quad i \in N_2^{(1)} \subset N_2.$$

Because all problems which will be considered in this book are equilibrium problems or can be reduced to such problems it is natural to base the study of these problems on the principle of virtual displacements (Theorem 1.5). According to Theorem 1.5 the state $\bar{x}$ is an equilibrium state if and only if the following condition holds

$$\sum_{i=1}^{n} Q_i \delta x_i \leqslant 0$$

for all $\delta x_1,\ldots,\delta x_n$ satisfying the conditions

$$\delta x_i \geqslant 0 \quad \text{for} \quad i \in N_2^{(1)} \subset N_2. \tag{3.3}$$

Therefore

$$\sum_{i=1}^{n} Q_i \delta x_i = 0 \quad \text{if} \ \delta x_i = 0, \quad i \in N_2^{(1)}$$

and $\sum_{i=1}^{n} Q_i \delta x_i \leqslant 0$ if for some of the indexes $i \in N_2^{(1)}$ the inequality $\delta x_i > 0$ holds.

Because the quantities δx_i are only subject to the condition (3.3) the principle of virtual displacement yields the following equilibrium conditions

$$Q_1 = 0 \quad \text{for} \quad i \in N_1$$
$$Q_i \begin{cases} = 0 & \text{for } \bar{x}_i > 0,\ i \in N_2 \\ \leqslant 0 & \text{for } \bar{x}_i = 0,\ i \in N_2. \end{cases} \tag{3.4}$$

These conditions coincide with the conditions (1.31)-(1.33) because clearly

$$Q_i = p_i - \sum_{s=1}^{m} \bar{w}_s a_{si}, \quad i = 1,\ldots,n. \tag{3.5}$$

The first algorithm now consists of the following. Let $x^{(0)} = (x_1^{(0)},\ldots,x_n^{(0)})$ be an arbitrary initial state of the physical model which satisfies conditions (3.2). Consider the equilibrium problem for the physical model under the extra redundant constraints

$$x_i = x_i^{(0)}, \quad i = 2,\ldots,n$$
$$\xi_s = \begin{cases} \min(\sum_{i=1}^{n} a_{si} x_i^{(0)}, b_s) & \text{for} \quad s \in M_1 \\ b_s & \text{for} \quad s \in M_2 \end{cases} \tag{3.6}$$

Under these conditions the state of the model is determined solely by the coordinate x_1 and the solution of the corresponding equilibrium problem is not difficult. Let $x_1^{(1)}$ be the coordinate of the equilibrium state of the model under the redundant constraints (3.6). The next problem is equally simple and asks for the equilibrium of the model under the redundant constraints

$$x_1 = x_1^{(1)},\ x_3 = x_3^{(0)},\ldots,x_n = x_n^{(0)}$$
$$\xi_s = \begin{cases} \min(a_{11}x_1^{(1)} + \sum_{i=2}^{n} a_{si} x_i^{(0)}, b_s) & s \in M_1 \\ b_s & s \in M_2 \end{cases}$$

It is obvious how to continue and we obtain a cyclic process: each cycle consists of n one dimensional equilibrium problems, and concludes with the construction of a vector $x^{(\nu)} = (x_1^{(\nu)},\ldots,x_n^{(\nu)})$ where ν is the index of the cycle.

Let us consider the problem of determining the coordinate $x_\alpha^{(\nu+1)}$ of the equilibrium state of the model under the redundant constraints

$$x_i = \begin{cases} x_i^{(\nu+1)} & \text{for } i = 1,\ldots,\alpha-1 \\ x_i^{(\nu)} & \text{for } i = \alpha+1,\ldots,n \end{cases}$$

$$\xi_s = \begin{cases} \min\left(\sum_{i=1}^{\alpha-1} a_{si} x_i^{(\nu+1)} + \sum_{i=\alpha}^{n} a_{si} x_i^{(\nu)}, b_s\right), & s \in M_1 \\ b_s, & s \in M_2 \end{cases}$$

The conditions for equilibrium of the model under the redundant constraints listed above follow from (3.4)

$$Q_\alpha(x_1^{(\nu+1)},\ldots,x_\alpha^{(\nu+1)}, x_{\alpha+1}^{(\nu)},\ldots,x_n^{(\nu)})$$

$$\begin{cases} = 0 \text{ for } \alpha \in N_1 \\ = 0 \text{ for } \alpha \in N_2 \text{ and } x_\alpha^{(\nu+1)} > 0 \\ \leq 0 \text{ for } \alpha \in N_2 \text{ and } x_\alpha^{(\nu+1)} = 0. \end{cases} \quad (3.7)$$

An expression for Q_α in terms of coordinates is obtained by substituting in (3.5) the expressions for $w_1,\ldots,w_m$ (2.32) in terms of the errors in the constraints

$$w_s(x_1^{(\nu+1)},\ldots,x_\alpha^{(\nu+1)}, x_{\alpha+1}^{(\nu)},\ldots,x_n^{(\nu)}) =$$
$$= \|a_s\|^{-1}\, \tilde{q}_0 y_s(x_1^{(\nu+1)},\ldots,x_\alpha^{(\nu+1)}, x_{\alpha+1}^{(\nu)},\ldots,x_n^{(\nu)}). \quad (3.8)$$

Let us introduce the n-dimensional vector $x^{(\nu,\alpha)} = (x_1^{(\nu,\alpha)},\ldots,x_n^{(\nu,\alpha)})$ with components

$$x_i^{(\nu,\alpha)} = \begin{cases} x_i^{(\nu+1)} & \text{for } i = 1,\ldots,\alpha-1 \\ x_i^{(\nu)} & \text{for } i = \alpha,\ldots,n \end{cases} \quad (3.9)$$

and the m-dimensional vector $\xi^{(\nu,\alpha)} = (\xi_1^{(\nu,\alpha)},\ldots,\xi_m^{(\nu,\alpha)})$

with the components

$$\xi_s^{(\nu,\alpha)} = \begin{cases} \min(\sum_{i=1}^{n} a_{si} x_i^{(\nu,\alpha)}, b_s) & \text{for} \quad s \in M_1 \\ b_s & \text{for} \quad s \in M_2. \end{cases} \tag{3.10}$$

Then the problem under consideration becomes the problem of finding the equilibrium vector $x^{(\nu,\alpha+1)}$ with coordinates

$$x_i^{(\nu,\alpha+1)} = \begin{cases} x_i^{(\nu+1)} & \text{for} \quad i = 1,\dots, \\ x_i^{(\nu)} & \text{for} \quad i = \alpha+1,\dots,n \end{cases}$$

for fixed values of the components of the vector $\xi = \xi^{(\nu,\alpha)}$. In the equilibrium state of the model which is defined by the bivector $(x^{(\nu,\alpha+1)}, \xi^{(\nu,\alpha)})$ which we are seeking, the errors are therefore given by the expressions

$$y_s(x^{(\nu,\alpha+1)}, \xi_s^{(\nu,\alpha)}) = \sum_{i=1}^{n} a_{si} x_i^{(\nu,\alpha+1)} - \xi_s^{(\nu,\alpha)}, \quad s = 1,\dots,m$$

and in virtue of the definition of the vectors $x^{(\nu,\alpha)}$ and $x^{(\nu,\alpha+1)}$ the following equality holds

$$y_s(x^{(\nu,\alpha+1)}, \xi_s^{(\nu,\alpha)}) = y_s(x^{(\nu,\alpha)}, \xi_s^{(\nu,\alpha)}) + \\ + a_{s\alpha}(x_\alpha^{(\nu+1)} - x_\alpha^{(\nu)}), \quad s = 1,\dots,m \tag{3.11}$$

where $y_s(x^{(\nu,\alpha)}, \xi_s^{(\nu,\alpha)})$, $s = 1,\dots,m$ are known components of the errors at the starting state for the problem under consideration of finding coordinate $x_\alpha^{(\nu+1)}$. Substituting (3.11) in (3.8) one finds

$$w_s(x^{(\nu,\alpha+1)}, \xi_s^{(\nu,\alpha)}) = \|a_s\|^{-1} \cdot \tilde{q}_0 \{y_s(x^{(\nu,\alpha)}, \xi_s^{(\nu,\alpha)}) + \\ + a_{s\alpha}(x_\alpha^{(\nu+1)} - x_\alpha^{(\nu)})\} \tag{3.12}$$

By virtue of (3.5), (3.12), and (3.7) the equilibrium conditions now take the form

$$p_\alpha - \tilde{q}_0 \sum_{s=1}^{m} \|a_s\|^{-1} a_{s\alpha} y_s(x^{(\nu,\alpha)}, \xi_s^{(\nu,\alpha)}) -$$

$$- (x_\alpha^{(\nu+1)} - x_\alpha^{(\nu)}) \tilde{q}_0 \sum_{s=1}^{m} \|a_s\|^{-1} a_{s\alpha}^2 =$$

$$= \begin{cases} = 0 & \text{for } \alpha \in N_1 \\ = 0 & \text{for } \alpha \in N_2,\ x_\alpha^{(\nu+1)} > 0 \\ \leqslant 0 & \text{for } \alpha \in N_2,\ x_\alpha^{(\nu+1)} = 0 \end{cases} \tag{3.13}$$

Let us introduce the following notation

$$\Psi_\alpha(x^{(\nu,\alpha)}, \xi^{(\nu,\alpha)}) =$$

$$= x_\alpha^{(\nu)} + (\tilde{q}_0 \sum_{s=1}^{m} \|a_s\|^{-1} a_{s\alpha}^2)^{-1} (p_\alpha - \tilde{q}_0 \sum_{s=1}^{m} \|a_s\|^{-1} a_{s\alpha} y_s(x^{(\nu,\alpha)}, \xi_s^{(\nu,\alpha)}) \tag{3.14}$$

Starting from the equilibrium conditions (3.13) one now obtains a formula for the quantity to be found

$$x_\alpha^{(\nu+1)} = \begin{cases} \Psi_\alpha(x^{(\nu,\alpha)}, \xi^{(\nu,\alpha)}), & \alpha \in N_1 \\ \Psi_\alpha(x^{(\nu,\alpha)}, \xi^{(\nu,\alpha)} \ \underline{\underline{1}}\ [\Psi_\alpha(x^{(\nu,\alpha)}, \xi^{(\nu,\alpha)}], & \alpha \in N_2 \end{cases} \tag{3.15}$$

<u>Remark</u> 1. Denote with $z_s^{(\nu,\alpha)}$ the quantity

$$z_s^{(\nu,\alpha)} = z_s(x^{(\nu,\alpha)}) = \sum_{i=1}^{n} a_{si} x_i^{(\nu,\alpha)} - b_s .$$

From the definitions

$$y_s(x^{(\nu,\alpha)}, \xi_s^{(\nu,\alpha)}) = \sum_{i=1}^{n} a_{si} x_i^{(\nu,\alpha)} - \xi_s^{(\nu,\alpha)}$$

and formulas (3.9) and (3.10) one obtains

$$y_s(x^{(\nu,\alpha)}, \xi_s^{(\nu,\alpha)}) = \begin{cases} z_s(x^{(\nu,\alpha)}) \ \underline{\underline{1}}\ [z_s(x^{(\nu,\alpha)})], & s \in M_1 \\ z_s(x^{(\nu,\alpha)}), & s \in M_2 . \end{cases}$$

Consequently the formula (3.14) for Ψ_α may be rewritten in the form

$$\Psi_\alpha(x^{(\nu,\alpha)}) = x_\alpha^{(\nu)} + $$

$$+ \frac{p_\alpha - \tilde{q}_0 \left[\sum_{s=1}^{m_1} \frac{a_{s\alpha}}{\|a_s\|} z_s^{(\nu,\alpha)} \underline{\underline{1}} [z_s^{(\nu,\alpha)}]+ \sum_{s=m_1+1}^{m} \frac{a_{s\alpha}}{\|a_s\|} z_s^{(\nu,\alpha)} \right]}{\tilde{q}_0 \sum_{s=1}^{m} \frac{a_{s\alpha}^2}{\|a_s\|}},$$

(3.16)

which shows the dependence of the function $\Psi_\alpha(x^{(\nu,\alpha)})$ on the components of the vector $z^{(\nu,\alpha)}$.

Remark 2. It must be remembered that it follows from formula (3.9) that

$$x^{(\nu,n+1)} = x^{(\nu+1,1)}.$$

This equality assures the cyclic nature of the numerical process.

Formulas (3.15) and (3.16) express the first iterative algorithm for calculating a solution to a general linear programming problem. The matrix (a_{si}) can always be normalized in such a way that the conditions

$$\|a_s\| = 1 + \sum_{i=1}^{n} |a_{si}| = a > 1, \quad s = 1,\dots,m \qquad (3.17)$$

hold. Then formula (3.16) takes the following simpler form

$$\Psi_\alpha(x^{(\nu,\alpha)}) = x_\alpha^{(\nu)} + $$

$$+ \frac{p_a - \hat{q}_0 \left[\sum_{s=1}^{m_1} a_{s\alpha} z_s^{(\nu,\alpha)} \underline{\underline{1}} [z_s^{(\nu,\alpha)}]+ \sum_{s=m_1+1}^{m} a_{s\alpha} z_s^{(\alpha,\nu)} \right]}{\hat{q}_0 \sum_{s=1}^{m} a_{s\alpha}^2}$$

where $\hat{q} = \tilde{q}_0 a^{-1}$.

In the case $p_1 = p_2 = \dots = p_n = 0$ formulas (3.15) and (3.16) define an algorithm for solving a system of linear equations and inequalities. Let us write down these iterative formulas for particular linear algebra problems. In this case, as can be seen from (3.16), this algorithm makes it possible to obtain a solution of the problem, the precision of which only depends on the number of iterations and does not depend on the magnitude of the parameter $\tilde{q}_0$.

1. An algorithm for solving a system of linear equations and inequalities

$$\sum_{i=1}^{n} a_{si}x_i - \xi_s = 0, \quad s = 1,\dots,m$$

$$x_i \geqslant 0 \text{ for } i \in N_2, \ \xi_s = \begin{cases} \leqslant b_s & \text{for } s \in M_1 \\ = b_s & \text{for } s \in M_2. \end{cases} \tag{3.18}$$

The algorithm has the same form as (31.5), where

$$\Psi_\alpha(x^{(\nu,\alpha)}) = x_\alpha^{(\nu)} - \frac{\sum_{s=1}^{m_1} \frac{a_{s\alpha}}{\|a_s\|} z_s^{(\nu,\alpha)} \underset{=}{1} [z_s^{(\nu,\alpha)}]_+ \sum_{s=m_1+1}^{m} \frac{a_{s\alpha}}{\|a_s\|} z_s^{(\nu,\alpha)}}{\sum_{s=1}^{m} \frac{a_{s\alpha}^2}{\|a_s\|}}, \tag{3.19}$$

while here it is also always possible to normalize the matrix (a_{si}) according to (3.17)

2. Nonnegative solutions of a system of linear equations

$$\sum_{i=1}^{n} a_{si}x_i = b_s, \quad s = 1,\dots,m$$

$$x_i \geqslant 0, \qquad i = 1,\dots,n \tag{3.20}$$

In this case the algorithm under consideration takes the form

$$x_\alpha^{(\alpha+1)} = \Psi_\alpha(x_\alpha^{(\nu,\alpha)}) \underset{=}{1} [\Psi_\alpha(x^{(\nu,\alpha)})]$$

where

$$\Psi_\alpha(x^{(\nu,\alpha)}) = x_\alpha^{(\nu)} - \frac{\sum_{s=1}^{m} \frac{a_{s\alpha}}{\|a_s\|} z_s(x^{(\nu,\alpha)})}{\sum_{s=1}^{m} \frac{a_{s\alpha}^2}{\|a_s\|}},$$

$$z_s(x^{(\nu,\alpha)}) = \sum_{i=1}^{\alpha-1} a_{si}x_i^{(\nu+1)} + \sum_{i=\alpha}^{n} a_{si}x_i^{(\nu)} - b_s.$$

3. <u>An iterative algorithm for solving systems of linear equations</u>

$$\sum_{i=1}^{n} a_{si}x_i = b_s, \quad s = 1,\ldots,n.$$

The algorithm takes the form

$$x_\alpha^{(\nu+1)} = \Psi_\alpha(x^{(\nu,\alpha)}), \quad \alpha = 1,\ldots,n, \quad \nu = 0,1,\ldots$$

where

$$\Psi_\alpha(x^{(\nu,\alpha)}) = x_\alpha^{(\nu)} - \frac{\sum_{s=1}^{n} \frac{a_{s\alpha}}{\|a_s\|} z_s(x^{(\nu,\alpha)})}{\sum_{s=1}^{n} \frac{a_{s\alpha}^2}{\|a_s\|}}.$$

To conclude we remark that the components of the feasible approximately optimal vector $w(\tilde{q}_0, x^{(\nu)})$ can be calculated by means of formula (3.8).

3.4. <u>The second algorithm</u>

The use of redundant constraints in the form of partitions (impermeable membranes) leads to a second iterative algorithm. The essential difference in using such redundant constraints compared with geometric (holonomic) constraints is that imposing such constraints on the physical system amounts to dividing it into a corresponding number of isolated subsystems. Also for a large class of systems, which contains all the physical models considered in this book, imposing redundant constraints of the impermeable membrane type does not change the total number of degrees of freedom. It just leads to a partition of these degrees of freedom among the subsystems and the equilibrium problem of the physical system thus splits into a number of independent low dimensional equilibrium problems by imposing redundant constraints of the impermeable membrane type. The essentials of the method are presented below.

Suppose we are considering a physical system D whose state is described by an n-dimensional vector $x = (x_1,\dots,x_n)$ of extensive variables and an m-dimensional vector $w = (w_1,\dots,w_m)$ of intensive variables [2]. We shall assume that it is possible to divide the system D into k isolated subsystems by means of impermeable partitions and that this can be done in such a way that the subsystem D is determined by the vector [3] $x_{(\alpha)}$ of extensive variables and the vector $w_{(\alpha)}$ of intensive variables. Thus (loosely speaking) $x = x_{(1)} \cup x_{(2)} \cup \dots \cup x_{(k)}$. If the system to be divided is in a state $(x^{(0)},w^{(0)})$ which is a thermal state [4] but not a mechanical equilibrium, then after the division into subsystems $D_1,\dots,D_k$ each part remains in a thermic state and the values of the intensive variables remain unchanged (in virtue of their definition). In the same sense also the (total) thermodynamic potential of the system D remains unchanged. Thus in the first stage we have k problems of equilibrium of isolated subsystems $D_1,\dots,D_k$.
Let $(x^{(1)}_{(1)},w^{(1)}_{(1)}),\dots,(x^{(1)}_k,w^{(1)}_k)$ be the extensive and intensive variables (indicators) of the isolated subsystems $D_1,\dots,D_k$ in a state of mechanical equilibrium. Because the general thermodynamic potential is an extensive variable [36] it follows that the following conditions hold

$$\phi(x^{(0)},w^{(0)}) = \sum_{\alpha=1}^{k} \phi_\alpha(x^{(0)}_{(\alpha)},w^{(0)}_{(\alpha)})$$

$$\phi_\alpha(x^{(0)}_\alpha,w^{(0)}_{(\alpha)}) \geqslant \phi_\alpha(x^{(1)}_{(\alpha)},w^{(1)}_{(\alpha)}),\ \alpha = 1,\dots,k,$$

$$\phi(x^{(1)}_{(1)},w^{(1)}_{(1)},x^{(1)}_{(2)},w^{(1)}_{(2)},\dots,x^{(1)}_{(k)},w^{(1)}_{(k)}) =$$

$$= \sum_{\alpha=1}^{k} \phi_\alpha(x^{(1)}_{(\alpha)},w^{(1)}_{(\alpha)}) \leqslant \phi(x^{(0)},w^{(0)}), \tag{3.21}$$

where $\phi_\alpha(x_{(\alpha)},w_{(\alpha)})$ is the thermodynamic potential of the the subsystem

D_α and Φ is the general thermodynamic potential of the system D^α subject to the redundant impermeable partition constraints. There is equality in conditions (3.21) only if $(x^{(0)},w^{(0)})$ is a state which is both a thermal and mechanical equilibrium.

The second stage consists of the problem of calculating the quantities $w^{(1)} = (w_1^{(1)},\dots,w_m^{(1)})$ which are the components of the vector w of intensive variables of the physical system D freed from the redundant constraints. The vector $w^{(1)}$ is obtained from the thermal state equations by fixing the components of the extensive variables $x^{(1)} = x_{(1)}^{(1)} \cup \dots \cup x_{(k)}^{(1)}$. Here we are concerned with the law of maximal entropy of a physical system under fixed values of the extensive parameters $x_1,\dots,x_n$. Therefore we obviously have the following inequality

$$\phi(x^{(1)},w^{(1)}) \leqslant \phi(x_{(1)}^{(1)},w_{(1)}^{(1)},x_{(2)}^{(1)},w_{(2)}^{(1)},\dots,x_{(k)}^{(1)},w_{(k)}^{(1)} \leqslant$$
$$\leqslant \phi(x^{(0)},w^{(0)}). \tag{3.22}$$

In condition (3.22) equality (between the extreme terms) is possible only in the case where both $(x^{(0)},w^{(0)})$, and $(x^{(1)},w^{(1)})$ are thermal and mechanical equilibria.

In this manner this particular variant of the method of redundant constraints makes it possible to construct a sequence of states of the physical system $(x^{(0)},w^{(0)})$, $(x^{(1)},w^{(1)}),\dots$ such that the general thermodynamic potential (2.14) diminishes monotonically and the limit (x,w) of this sequence of states is a thermal and mechanical equilibrium. Consequently

$$\phi(\bar{x},\bar{w}) = \min_{x,w} \phi(x,w) \leqslant \phi(x^{(\nu)},w^{(\nu)}) \leqslant \phi(x^{(\nu-1)},w^{(\nu-1)}),$$
$$\phi(\bar{x},\bar{w}) = \lim_{\nu\to\infty} \phi(x^\nu,w^\nu).$$

Consider again the physical model of a general linear programming problem. In this model the extensive variables are the coordinates $x_1,\dots,x_n$ (the positions of the bars) and the intensive variables are pressures $q_s^{(+)}$ and $q_s^{(-)}$

in the volumes $V_s^{(+)}$ and $V_s^{(-)}$, $s = 1,\dots,m$. Let $x^{(0)}$ be an arbitrary initial state vector of the model of Figure 1.8 which is only required to satisfy the conditions

$$x_i^{(0)} \geqslant 0, \quad i \in N_2$$

The quantities $\xi_1^{(0)},\dots,\xi_m^{(0)}$ are defined by the formulas

$$\xi_s^{(0)} = \begin{cases} \min\left(\sum_{i=1}^{n} a_{si}x_i^{(0)}, b_s\right) & \text{for } s \in M_1 \\ b_s & \text{for } s \in M_2, \end{cases}$$

and the pressure variables $q_s^{(+)}(x^{(0)})$ and $q_s^{(-)}(x^{(0)})$ are connected to the magnitudes of the volumes $V_s^{(+)}$, $V_s^{(-)}$ by the state equation of Boyle-Mariott

$$q_s^{(+)}(x^{(0)}) = q_0 \frac{\bar{V}_s^{(+)}}{V_s^{(+)}(x^{(0)})}, \quad q_s^{(-)}(x^{(0)}) = q_0 \frac{\bar{V}_s^{(-)}}{V_s^{(-)}(x^{(0)})},$$

$$s = 1,\dots,m. \tag{3.23}$$

where according to (1.16)-(1.17)

$$\bar{V}_s^{(+)} = \bar{V}_s^{(-)} = \ell\|a_s\| \tag{3.24}$$

$$V_s^{(+)}(x^{(0)}) = \ell\|a_s\| + \sum_{i=1}^{n} a_{si}x_i^{(0)} - \xi_s^{(0)} \tag{3.25}$$

$$V_s^{(-)}(x^{(0)}) = \ell\|a_s\| - \sum_{i=1}^{n} a_{si}x_i^{(0)} + \xi_s^{(0)}. \tag{3.26}$$

Hence, using the notations of Section 2.3

$$V_s^{(+)}(x^{(0)}) = \begin{cases} \ell\|a_s\| + z_s^{(0)}\ \underline{\underline{1}}\,[z_s^{(0)}], & s \in M_1 \\ \ell\|a_s\| + z_s^{(0)}, & s \in M_2 \end{cases}$$

$$V_s^{(-)}(x^{(0)}) = \begin{cases} \ell\|a_s\| - z_s^{(0)}\ \underline{\underline{1}}\,[z_s^{(0)}], & s \in M_1 \\ \ell\|a_s\| - z_s^{(0)}, & s \in M_2, \end{cases}$$

with

$$z_s^{(0)} = \sum_{i=1}^{n} a_{si} x_i^{(0)} - b_s, \quad s = 1,\ldots,m.$$

Introduce in a state of thermal (isothermal) equilibrium $x^{(0)}$ redundant constraints in the form of impermeable partitions which cut all the channels connecting the volumes $V_{si}^{(+)}$ and $V_{si}^{(-)}$ and consider the n one-dimensional equilibrium problems of the isolated subsystems. The state of the i-th isolated subsystem, sketched in Figure 1.9, is clearly defined by the single coordinate x_i subject to the condition $x_i \geqslant 0$ in the case $i \in N_2$. Let $x_i^{(1)}$, $i = 1,\ldots,n$ be the coordinate of the equilibrium state of the n subsystems (obtained after the introduction of the redundant constraints).

The conditions for equilibrium of the i-th subsystem are analogous to (1.31)-(1.33) with the difference that the pressures $q_{si}^{(+)}$ and $q_{si}^{(-)}$ in the isolated volumes $V_{si}^{(+)}$ and $V_{si}^{(-)}$ are different for different values of the index i. In the two possible cases $i \in N_1$ and $i \in N_2$ the equilibrium conditions for the i-th column (the i-th subsystem) have the form

$$\sum_{s=1}^{m} a_{si}(q_{si}^{(-)} - q_{si}^{(+)}) - p_i = 0, \quad i \in N_1, \tag{3.27}$$

$$\sum_{s=1}^{m} a_{si}(q_{si}^{(-)} - q_{si}^{(+)}) - p_i \begin{cases} = 0, & x_i^{(1)} > 0 \\ \geqslant 0, & x_i^{(1)} = 0 \end{cases}, i \in N_2. \tag{3.28}$$

The pressure quantities in the isolated volumes $V_{si}^{(-)}$ and $V_{si}^{(+)}$ are in an equilibrium state connected with the coordinates of the vector $x^{(1)}$ by means of the equation of Boyle-Mariott

$$V_{si}^{(+)}(x^{(1)})q_{si}^{(+)}(x^{(1)}) = V_{si}^{(+)}(x^{(0)})q_s^{(+)}(x^{(0)}),$$
$$V_{si}^{(-)}(x^{(1)})q_{si}^{(-)}(x^{(1)}) = V_{si}^{(-)}(x^{(0)})q_s^{(-)}(x^{(0)}).$$

From these equations we find

$$q_{si}^{(-)}(x^{(1)}) - q_{si}^{(+)}(x^{(1)}) = \frac{V_{si}^{(-)}(x^{(0)})q_s^{(-)}(x^{(0)})}{V_{si}^{(-)}(x^{(1)})} - \frac{V_{si}^{(+)}(x^{(0)})q_s^{(+)}(x^{(0)})}{V_{si}^{(+)}(x^{(1)})}.$$

Using formulas (3.23)-(3.26) and introducing the notation

$$w_{si}(x^{(1)}) = q_{si}^{(-)}(x^{(1)}) - q_{si}^{(+)}(x^{(1)})$$

we obtain after some calculations

$$w_{si}(x^{(1)}) = \begin{cases} 2q_0\|a_s\|\dfrac{\ell^3[y_s(x^{(0)})+\|a_s\|(x_i^{(1)}-x_i^{(0)})]-\ell y_s(x^{(0)})x_i^{(1)}x_i^{(0)}}{[\ell^2-(x_i^{(1)})^2][\ell^2\|a_s\|^2-y_s^2(x^0)]}, & a_{si} \geqslant 0, \\ 2q_0\|a_s\|\dfrac{\ell^3[y_s(x^{(0)})-\|a_s\|(x_i^{(1)}-x_i^{(0)})]-\ell y_s(x^{(0)})x_i^{(1)}x_i^{(0)}}{[\ell^2-(x_i^{(1)})^2][\ell^2\|a_s\|^2-y_s^2(x^{(0)})]}, & a_{si} < 0. \end{cases} \quad (3.29)$$

Now as in Section 2.3 of Chapter II, let

$$q_0 = \frac{1}{2}\ell\tilde{q}_0$$

substitute this for q_0 in (3.29) and take the limit as $\ell \to +\infty$. The result is

$$w_{si}(x^{(1)}) = \begin{cases} \tilde{q}_0\left[\dfrac{y_s(x^{(0)})}{\|a_s\|} + (x_i^{(1)}-x_i^{(0)})\right], & a_{si} \geqslant 0, \\ \tilde{q}_0\left[\dfrac{y_s(x^{(0)})}{\|a_s\|} - (x_i^{(1)}-x_i^{(0)})\right], & a_{si} < 0. \end{cases}$$

The equilibrium conditions (3.27)-(3.28) therefore take the

form

$$\widetilde{q}_0 \sum_{s=1}^{m} \frac{a_{si}}{\|a_s\|} y_s(x^{(0)}) + $$
$$+ \widetilde{q}_0(x_i^{(1)}-x_i^{(0)}) \sum_{s=1}^{m} |a_{si}|-p_i \begin{cases} = 0, & i \in N_1, \\ = 0,\ x_i^{(1)} > 0 & i \in N_2. \\ \geqslant 0,\ x_i^{(1)} = 0 & \end{cases} \tag{3.30}$$

Now solve (3.30) with respect to the quantities to be determined $x_1^{(1)},\ldots,x_n^{(1)}$. The result is

$$x_i^{(1)} = \begin{cases} = x_i^{(0)} + \dfrac{p_i-\widetilde{q}_0 \sum_{s=1}^{m} \frac{a_{si}}{\|a_s\|} y_s(x^{(0)})}{\widetilde{q}_0 \sum_{s=1}^{m} |a_{si}|}, & \\ \qquad i \in N_1,\ i \in N_2,\ x_i^{(1)} > 0, & \\ \leqslant x_i^{(0)} + \dfrac{p_i-\widetilde{q}_0 \sum_{s=1}^{m} \frac{a_{si}}{\|a_s\|} y_s(x^{(0)})}{\widetilde{q}_0 \sum_{s=1}^{m} |a_{si}|}, & \\ \qquad i \in N_2,\ x_i^{(1)} = 0. & \end{cases}$$

Denoting with $\Omega_i(x^{(0)})$ the following function of the initial state

$$\Omega_i(x^{(0)}) = x^{(0)} + \frac{p_i-\widetilde{q}_0 \sum_{s=1}^{m} \frac{a_{si}}{\|a_s\|} y_s(x^{(0)})}{\widetilde{q}_0 \sum_{s=1}^{m} |a_{si}|},$$

we obtain a formula for calculating the coordinates $x^{(1)}(\widetilde{q}_0)$ of the equilibrium state of the physical model of the linear programming problem (1.20)-(1.21) when it is additionally and redundantly constrained by impermeable partitions:

$$x_i^{(1)} = \begin{cases} \Omega_i(x^{(0)}), & i \in N_1, \\ \Omega_i(x^{(0)}) \stackrel{1}{=} [\Omega_i(x^{(0)})], & i \in N_2. \end{cases}$$

Now use the relations

$$y_s = \begin{cases} z_s \stackrel{1}{=} [z_s], & s \in M_1 \\ z_s, & s \in M_2 \end{cases}$$

$$z_s = \sum_{i=1}^{n} a_{si} x_i - b_s, \quad s = 1,2,\ldots,m$$

to arrive at the following expression

$$\Omega_i(x^{(0)}) = x_i^{(0)} + \\ + \frac{p_i - \tilde{q}_0 \left[\sum_{s=1}^{m_1} \frac{a_{si}}{\|a_s\|} z_s(x^{(0)}) \stackrel{1}{=} [z_s(x^{(0)})] + \sum_{s=m_1+1}^{m} \frac{a_{si}}{\|a_s\|} z_s(x^{(0)}) \right]}{\tilde{q}_0 \sum_{s=1}^{m} |a_{si}|}$$

To proceed one takes the state obtained in this way as a new initial state for the physical model without (i.e. freed from) the redundant constraints. Repeating all arguments permits one to arrive at the formulas defining the coordinates of the next approximation $x^{(2)}$ to the equilibrium state to be determined $\bar{x}(\tilde{q}_0)$

$$x_i^{(2)} = \left\{ \begin{array}{ll} \Omega_i(x^{(1)}), & i \in N_1, \\ \Omega_i(x^{(1)}) \stackrel{1}{=} [\Omega_i(x^{(1)})], & i \in N_2 \end{array} \right\}$$

In this manner the iterative formulas for the second algorithm are found to take the form

$$x_i^{(\nu+1)} = \begin{cases} \Omega_i(x^{(\nu)}), & i \in N_1, \\ \Omega_i(x^{(\nu)}) \stackrel{1}{=} [\Omega_i(x^{(\nu)})], & i \in N_2. \end{cases} \tag{3.31}$$

$$\nu = 0,1,2,\ldots ,$$

where

$$\Omega_i(x^{(\nu)}) = x_i^{(\nu)} + $$

$$+ \frac{p_i - \tilde{q}_0\left[\sum_{s=1}^{m_1} \frac{a_{si}}{\|a_s\|} z_s(x^{(\nu)}) \ \underline{1}\ [z_s(x^{(\nu)})] + \sum_{s=m_1+1}^{m} \frac{a_{si}}{\|a_s\|} z_s(x^{(\nu)})\right]}{\tilde{q}_0 \sum_{s=1}^{m} |a_{si}|}$$

The sequence of (corresponding) approximations to the vector $\bar{w}(\tilde{q}_0)$ of intensive variables of the model at equilibrium are found by means of the usual formulas (see Section 2.3 of Chapter II)

$$w_s^{(\nu)}(\tilde{q}_0) = \begin{cases} \frac{\tilde{q}_0}{\|a_s\|} z_s(x^{(\nu)}) \ \underline{1}\ [z_s(x^{(\nu)})], & s \in M_1, \\ \frac{\tilde{q}_0}{\|a_s\|} z_s(x^{(\nu)}), & s \in M_2. \end{cases} \tag{3.32}$$

$$\nu = 0,1,2,\ldots$$

Recall (see Section 2.6) of Chapter II) that the vector $\bar{x}(\tilde{q}_0)$ of extensive variables of the physical model at equilibrium satisfies

$$\bar{x}(\tilde{q}_0) = \lim_{\nu\to\infty} x^{(\nu)}(\tilde{q}_0),$$

and is a non feasible approximately optimal vector for the primal problem and that the vector $\bar{w}(\tilde{q}_0)$ of intensive variables of the same physical model at mechanical equilibrium

$$\bar{w}(\tilde{q}_0) = \lim_{\nu\to\infty} w^{(\nu)}(q_0).$$

is a feasible approximately optimal solution of the dual problem.

Consider now the dual problem (1.22)-(1.23) as the primal problem. Repeating the arguments of this section one easily obtains iterative formulas which define a sequence of vectors $\hat{w}^{(0)}, \hat{w}^{(1)}, \ldots$

$$\hat{w}_s^{(\nu+1)} = \begin{cases} \theta_s(\hat{w}^{(\nu)}) & s \in M_2, \\ \theta_s(\hat{w}^{(\nu)}) \ \underline{1}\ [\theta_s(\hat{w}^{(\nu)})], & s \in M_1, \end{cases} \tag{3.33}$$

where

$$\theta_s(\hat{w}^{(\zeta)}) = \hat{w}_s^{(\nu)} +$$

$$+ \frac{-b_s+\tilde{q}_0\left[\sum_{i=1}^{n_1} \frac{a_{si}}{\|a_i\|}\zeta_i(\hat{w}^{(\nu)})+ \sum_{i=n_1+1}^{n} \frac{a_{si}}{\|a_i\|}\zeta_i(\hat{w}^{(\nu)})\underline{\underline{1}}[\zeta_i(\hat{w}^{(\nu)})]\right]}{\tilde{q}_0 \sum_{i=1}^{n} |a_{si}|}, \quad (3.34)$$

$$\|a_i\| = 1 + \sum_{s=1}^{m} |a_{si}|,$$

$$\zeta_i(w^{(\nu)}) = p_i - \sum_{s=1}^{m} a_{si} w_s^{(\nu)},$$

Here $\hat{w}^{(0)}$ is an arbitrary m-dimensional (starting) vector which is only required to satisfy the conditions

$$\hat{w}_s^{(0)} \geqslant 0, \quad s \in M_1,$$

The limit of the sequence $\hat{w}^{(0)}, \hat{w}^{(1)}, \ldots$ is clearly a non feasible approximately optimal vector $\bar{\bar{w}}(\tilde{q}_0) = \lim_{\nu\to\infty} \hat{w}^{(\nu)}$ for the problem (1.22)-(1.23) and it follows from the state equations for an ideal gas that the formulas

$$\bar{\bar{x}}_i(\tilde{q}_0) = \begin{cases} \frac{\tilde{q}_0}{\|a_i\|} \zeta_i(\bar{\bar{w}}(q_0)), & i \in N_1, \\ \frac{\tilde{q}_0}{\|a_i\|} \zeta_i(\bar{\bar{w}}(\tilde{q}_0))\, \underline{\underline{1}}\, [\zeta_i(\bar{\bar{w}}(\tilde{q}_0))], & i \in N_2, \end{cases} \quad (3.35)$$

then determine a feasible approximately optimal vector for the problem (1.20)-(1.21).

The formulas (3.31)-(3.35) make it possible to calculate the components of two approximately optimal bivectors $(\bar{x},\bar{w})$ and $(\bar{\bar{x}},\bar{\bar{w}})$ for the dual pair of linear programming problems. These bivectors satisfy the conditions of Theorem 2.7 (in Section 2.6 of Chapter II) and yield two sided estimates for the optimal value of the objective function for the primal and dual problem.

To conclude let us apply the formulas of the second algorithm to the problems of linear algebra considered in the previous section (Section 3.3).

1. <u>An algorithm for solving systems of linear equations and inequalities</u> (3.18). In this problem $p_1 = \ldots = p_n = 0$ and an exact solution may be obtained by means of the formula (3.31) for any arbitrary positive value of the parameter $\tilde{q}_0$, where

$$\Omega_i(x^{(\nu)}) = x_i^{(\nu)} - \frac{\sum_{s=1}^{m_1} \frac{a_{si}}{\|a_s\|} z_s(x^{(\nu)}) \underset{=}{1} [z_s(x^{(\nu)})] + \sum_{s=m_1+1}^{m} \frac{a_{si}}{\|a_s\|} z_s(x^{(\nu)})}{\sum_{s=1}^{m} |a_{si}|},$$

$$i = 1,2,\ldots,n; \ \nu = 0,1,\ldots \ . \qquad (3.36)$$

Here $x^{(0)}$ is an arbitrary vector satisfying the conditions $x_i^{(0)} \geqslant 0$ for $i \in N_2$.

2. <u>Nonnegative solutions of a system of linear equations</u>. The algorithm results from (3.31) with $m_1 = 0$, $n_1 = 0$ and $p_1 = \ldots = p_n = 0$. Thus

$$x_i^{(\nu+1)} = \Omega_i(x^{(\nu)}) \underset{=}{1} [\Omega_i(x^{\nu})], \ i = 1,\ldots,n; \ \nu = 0,1,2,\ldots,$$

where

$$\Omega_i(x^{(\nu)}) = x_i^{(\nu)} - \frac{\sum_{s=1}^{m} \frac{a_{si}}{\|a_s\|} z_s(x^{(\nu)})}{\sum_{s=1}^{m} |a_{si}|},$$

and $x^{(0)}$ is an arbitrary nonnegative vector.

3. <u>An algorithm for solving systems of linear equations</u>.

$$\sum_{i=1}^{n} a_{si} x_i = b_s, \quad s = 1,\ldots,n.$$

The algorithm is a particular case of the algorithm (3.31) with $m = n$, $N_2 = \emptyset$ and $M_1 = \emptyset$ and it has the form

$$x_i^{(\nu+1)} = x_i^{(\nu)} - \frac{\sum_{s=1}^{m} \frac{a_{si}}{\|a_s\|} z_s(x^{(\nu)})}{\sum_{s=1}^{n} |a_{si}|},$$

$$i = 1,2,\ldots,n; \quad \nu = 0,1,\ldots,$$

where $x^{(0)}$ is an arbitrary vector.

Some remarks. It is important to notice the essential differences between the first and the second algorithm. The first algorithm is an algorithm of coordinate-wise descent and consists of a sequence of numerical iterative calculations for the coordinates of equilibrium states. In the second algorithm the equilibrium problems obtained after imposing the redundant constraints are independent and because of that the equilibrium state vector can be found by parallel iterations on the coordinates of state of the physical model. Therefore the second algorithm is of a decomposition nature which follows from the physical essence of the redundant constraints employed. In Chapter VII this property of the redundant constraint of the impermeable membrane type will turn out to be a basic method for decomposing large dimensional problems.

There is also interest in the modification of the first and second algorithm for a general linear programming problem written in the form

$$x_0 \to \max$$

under the conditions

$$\sum_{i=1}^{n} a_{si}x_i \leqslant b_s, \qquad s \in M_1,$$

$$\sum_{i=1}^{n} a_{si}x_i = b_s, \qquad s \in M_2,$$

$$\sum_{i=1}^{n} p_i x_i - x_0, \; x_i \geqslant 0, \quad i \in N_2.$$

It is easy to construct the physical model of this problem and obtain from it the formulas of the first and the second algorithm

There are also possible variants of the algorithms described in which one uses the idea of ordering the coordinates of the state vector in the order of decreasing error in the conditions for equilibrium in the state under consideration. Let, for example, $x^{(\nu)}$ be some given state of the model. It is natural to look for an equilibrium state by allowing to change (that is releasing, in the case of the first algorithm) only that one of the coordinates for which the quantity

$$\left| \sum_{s=1}^{m} a_{si} w_s(x^{(\nu)}) - p_i \right|, \quad i \in N_1,$$

or

$$\left(\sum_{s=1}^{m} a_{si} w_s(x^{(\nu)}) - p_i \right) \underline{1} \left[p_i - \sum_{s=1}^{m} a_{si} w_s(x^{(\nu)})\right], \quad i \in N_2,$$

is maximal. Thus one can obtain substantially more economical use of machine time.

To conclude we remark that both the first and the second algorithm yield maximizing sequences which always have a well defined limit because the imposing of redundant constraints leads to a series of problem which have unique solutions. In the case where the initial linear programming problem has non unique solutions, the resulting limit depends on the initial vector. In such problems one can use with great succes the regularization method of A.N. Tihonov to construct a convex set of optimal solutions.

3.5. Reduction of the general linear programming problem to a sequence of inconsistent systems. The third algorithm.

A most important property of the physical model of a system of linear equations and linear inequalities is that it is a real physical system independently of the fact whether there exists or does not exist a solution of the original system. The difference lies solely in the fact that in the case of consistent system the Helmholtz free energy is zero in every state corresponding to a solution of the system and in the case of an inconsistent system the Helmholtz free energy of its physical model is strictly positive in an equilibrium state. The value of these properties of the physical model became clear in Section 2.3 of Chapter II where they allowed us to prove some theorems of linear algebra.

In this section we present yet another algorithm for the numerical solution of linear programming problems. It consists in the construction of a maximizing sequence of which the elements are equilibrium state vectors of physical models of inconsistent systems of linear equations and inequalities. It will turn out to be very useful that these vectors do not depend on the magnitude of the model parameter $\tilde{q}_0$, so that the problem of choosing a growing sequence of values of model parameters is avoided.

Let $\beta^{(0)}$ be a number which satisfies the condition

$$\beta^{(0)} \geqslant \max \sum_{i=1}^{n} p_i x_i ,$$

where the maximum is taken over the region defined by the constraints of the linear programming problem. In general it is not difficult to find such a $\beta^{(0)}$, and in any case one can set

$$\beta^{(0)} = \sum_{i=1}^{n} p_i \bar{x}_i(\tilde{q}_0),$$

where $\bar{x}(\tilde{q}_0)$ is an equilibrium state vector of the physical linear programming problem, which is found by means of the first or second algorithm for some value of the parameter $\tilde{q}_0$ [5].

We now consider the equilibrium problem of the physical model of the following system of linear equations and inequalities

$$\left.\begin{aligned} &\sum_{i=1}^{n} a_{si} x_i \leqslant b_s^{(0)}, \quad s = 1,\dots,m_1, \\ &\sum_{i=1}^{n} a_{si} x_i = b_s^{(0)}, \quad s = m_1+1,\dots,m+1 \\ &\qquad\qquad x_i \geqslant 0, \quad i \in N_2, \end{aligned}\right\} \qquad (3.37)$$

where

$$a_{m+1,i} = p_i, \quad i = 1,2,\dots,n,$$

$$b_s^{(0)} = b_s, \quad s = 1,\dots,m, \quad b_{m+1}^{(0)} = \beta^{(0)}.$$

In case $\beta^{(0)} > \max(\sum_{i=1}^{n} p_i x_i) = \sum_{i=1}^{n} p_i x_i^*$ the system (3.37) is inconsistent and the Helmholtz free energy of its model at equilibrium is positive. For the determination of the vector $\bar{x}(b_{m+1}^{(0)})$ [6] describing the equilibrium state of the model of systems (3.37) one can use the first or the second algorithm. The first algorithm for the problem under consideration has the form (analogous to (3.15))

$$x_\alpha^{(\nu+1)} = \begin{cases} \Psi_\alpha^{(0)}(x^{(\nu,\alpha)}), & \alpha \in N_1 \\ \Psi_\alpha^{(0)}(x^{(\nu,\alpha)}) \,\underline{\underline{1}}\, [\Psi_\alpha^{(0)}(x^{(\nu,\alpha)})], & \alpha \in N_2 \end{cases} \qquad (3.38)$$

where, according to (3.19)

$$\Psi_\alpha^{(0)}(x^{(\nu,\alpha)}) = x_\alpha^{(\nu)} - \frac{\sum_{s=1}^{m_1} \frac{a_{s\alpha}}{\|a_s\|} z_s^{(0)}(x^{(\nu,\alpha)}) \,\underline{\underline{1}}\, [z_s^{(0)}(x^{(\nu,\alpha)})] + \sum_{s=m_1+1}^{m+1} \frac{a_{s\alpha}}{\|a_s\|} z_s^{(0)}(x^{(\nu,\alpha)})}{\sum_{s=1}^{m+1} \frac{a_{s\alpha}^2}{\|a_s\|}},$$

$$z^{(0)}(x^{(\nu,\alpha)}) = \sum_{i=1}^{\alpha-1} a_{si} x_i^{(\nu+1)} + \sum_{i=\alpha}^{n} a_{si} x_i^{(\nu)} - b_s^{(0)},$$

$$s = 1,\dots,m+1.$$

The second algorithm is determined by formulas analogous to (3.31), that is:

$$x_i^{(\nu+1)} = \begin{cases} \Omega_i^{(0)}(x^{(\nu)}), & i \in N_1, \\ \Omega_i^{(0)}(x^{(\nu)}) \,\underline{\underline{1}}\, [\Omega_i^{(0)}(x^{(\nu)})], & i \in N_2. \end{cases}$$

where by analogy with (3.36)

$$\Omega_i^{(0)}(x^{(\nu)}) = x_i^{(\nu)} - \frac{\sum_{s=1}^{m_1} \frac{a_{si}}{\|a_s\|} z_s^{(0)}(x^{(\nu)}) \,\underline{\underline{1}}\, [z_s^{(0)}(x^{(\nu)})] + \sum_{s=m_1+1}^{m+1} \frac{a_{si}}{\|a_s\|} z_s^{(0)}(x^{(\nu)})}{\sum_{s=1}^{m+1} |a_{si}|}$$

$$z_s^{(0)}(x^{(\nu)}) = \sum_{i=1}^{n} a_s(x_i^{(\nu)} - b_s^{(0)}.$$

Now let us verify that under the condition $b_{m+1}^{(0)} > \sum_{i=1}^{m} p_i x_i^*$ the vector $\bar{x}(b_{m+1}^{(0)})$ is not a feasible vector for the linear programming problem and that moreover the following condition holds

$$b_{m+1}^{(0)} > \sum_{i=1}^{n} p_i \bar{x}_i(b_{m+1}^{(0)}) > \sum_{i=1}^{n} p_i x_i^*. \tag{3.39}$$

That this assertion is correct follows from the fact that the Helmholtz free energy of the physical model of system (3.37) is a convex continuously differentiable function equal to the sum of the Helmholtz free energy of the model of the consistent system of equations and inequalities (2.23) and the free energy of the model of the equation

$$\sum_{i=1}^{n} a_{m+1,i}\, x_i = b_{m+1}^{(0)}. \tag{3.40}$$

Because the intersection of the set of feasible solutions of the linear programming problem and the solutions of Equations (3.40) is empty, it follows that the Helmholtz free energy of the model of system (3.37) is everywhere strictly positive and its minimum is reached in the point $\bar{x}(b_{m+1}^{(0)})$ and in this point the gradient vectors of the two parts of the Helmholtz free energy mentioned above are equal in absolute magnitude and opposite in sign. From this there follows assertion (3.39).

After the calculation of the components of vector $\bar{x}(b_{m+1}^{(0)})$ one proceeds to the equilibrium problem of the physical model of the following equally inconsistent system of equations and inequalities

$$\left.\begin{aligned} &\sum_{i=1}^{n} a_{si} x_i \leqslant b_s^{(1)}, \quad s = 1,\ldots,m_1 \\ &\sum_{i=1}^{n} a_{si} x_i = b_s^{(1)}, \quad s = m_1+1,\ldots,m+1, \\ &\qquad\qquad x_i \geqslant 0, \quad i \in N_2, \end{aligned}\right| \tag{3.41}$$

where

$$a_{m+1,i} = p_i, \quad i = 1,2,\ldots,n;$$

$$b_s^{(1)} = b_s, \; s = 1,\ldots,m; \; b_{m+1}^{(1)} = \sum_{i=1}^{n} p_i \bar{x}_i(b_{m+1}^{(0)}).$$

If $\bar{x}(b_{m+1}^{(1)})$ is the equilibrium state of the physical model of system (3.41) then the next problem will be an equilibrium problem for the model of a system which differs from (3.41) only in the second half of the (m+1)-th equation

$$b_{m+1}^{(2)} = \sum_{i=1}^{n} p_i \bar{x}_i(b_{m+1}^{(1)}).$$

It is clear how to continue and the sequence of vectors $\bar{x}(b_{m+1}^{(0)}), \bar{x}(b_{m+1}^{(1)}), \ldots$ converges monotonically in terms of a positive definite function of the errors which has the interpretation of the Helmholtz free energy of the physical model. Moreover obviously

$$b_{m+1}^{(0)} > b_{m+1}^{(1)} > b_{m+1}^{(2)} > \ldots .$$

Arguing by assuming the opposite it is also easy to show that

$$\lim_{\alpha\to\infty} \bar{x}(b_{m+1}^{(\alpha)}) = x^*, \quad \lim_{\alpha\to\infty} b_{m+1}^{(\alpha)} = \sum_{i=1}^{n} p_i x_i^* .$$

Figure 3.1 illustrates the algorithm just described.

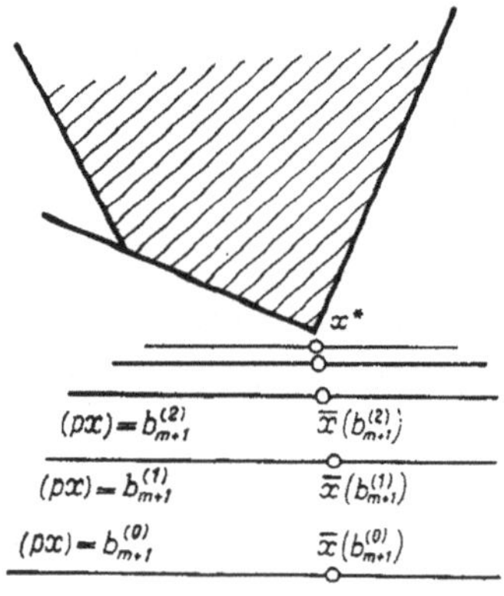

Fig. 3.1.

The method of reducing a linear programming problem to a sequence of inconsistent system of linear equations and inequalities is also a method to solve simultaneously the primal and the dual problem. In fact the conditions for equilibrium for the physical model of the inconsistent system

$$\left.\begin{aligned} &\sum_{i=1}^{n} a_{si}x_i \leqslant b_s^{(\alpha)}, \quad s = 1,\ldots,m_1, \\ &\sum_{i=1}^{n} a_{si}x_i = b_s^{(\alpha)}, \quad s = m_1+1,\ldots,m+1, \\ &\qquad\qquad x_i \geqslant 0, \quad i \in N_2, \end{aligned}\right\} \tag{3.42}$$

take the form

$$\sum_{s=1}^{m} a_{si}\tilde{w}_s^{(\alpha)} + p_i\tilde{w}_{m+1}^{(\alpha)} = 0, \qquad i \in N_1,$$

$$\sum_{s=1}^{m} a_{si}\tilde{w}_s^{(\alpha)} + p_i\tilde{w}_{m+1}^{(\alpha)} \begin{cases} = 0, & \bar{x}_i^{(\alpha)} > 0, \\ \leqslant 0, & \bar{x}_i^{(\alpha)} = 0, \end{cases} \qquad i \in N_2$$

$$\tilde{w}_s^{(\alpha)} \begin{cases} = 0, & z_s(\bar{x}^{(\alpha)}) < 0, \\ \geqslant 0, & z_s(\bar{x}^{(\alpha)}) \geqslant 0, \end{cases} \qquad s \in M_1,$$

where

$$\bar{x}^{(\alpha)} = \bar{x}(b_{m+1}^{(\alpha)}), \quad \tilde{w}^{(\alpha)} = \bar{w}(b_{m+1}^{(\alpha)}),$$

$$z_s(\bar{x}^{(\alpha)}) = \sum_{i=1}^{n} a_{si}\bar{x}_i(b_{m+1}^{(\alpha)}) - b_s^{(\alpha)}).$$

Because of the inconsistency of system (3.42) it is clear that for each positive natural number α

$$\tilde{w}_{m+1}^{(\alpha)} < 0.$$

Dividing the equilibrium conditions by the positive number $|\tilde{w}_{m+1}^{(\alpha)}|$ and using the notation

$$w_s^{(\alpha)} = \frac{\tilde{w}_s^{(\alpha)}}{|\tilde{w}_{m+1}^{(\alpha)}|}, \tag{3.43}$$

we obtain the following equilibrium conditions analogous to (2.58)-(2.60)

$$\sum_{s=1}^{m} a_{si} w_i^{(\alpha)} - p_i = 0, \qquad i \in N_1,$$

$$\sum_{s=1}^{m} a_{si} w_s^{(\alpha)} - p_i \begin{cases} = 0, & \bar{x}_i^{(\alpha)} > 0, \\ \leqslant 0, & \bar{x}_i^{(\alpha)} = 0. \end{cases} \qquad i \in N_2,$$

$$w_s^{(\alpha)} \begin{cases} = 0, & z_s(\bar{x}^{(\alpha)}) < 0, \\ \geqslant 0, & z_s(\bar{x}^{(\alpha)}) \geqslant 0. \end{cases} \qquad s \in M_1,$$

In this manner the components of the feasible approximately optimal vector $\bar{w}^{(\alpha)}$ of the dual problem can be calculated according to (3.43) and (3.32) by the formulas

$$\begin{aligned} \bar{w}_s^{(\alpha)} &= \frac{\|a_{m+1}\|}{\|a_s\|} \frac{z_s(\bar{x}^{(\alpha)}) \pm [z_s(\bar{x}^{(\alpha)})]}{|z_{m+1}(\bar{x}^{(\alpha)})|}, \qquad s = 1,\dots,m \\ \bar{w}_s^{(\alpha)} &= \frac{\|a_{m+1}\|}{\|a_s\|} \frac{z_s(\bar{x}^{(\alpha)})}{|z_{m+1}(\bar{x}^{(\alpha)})|}, \qquad s = m_1+1,\dots,m, \end{aligned} \tag{3.44}$$

where

$$\|a_{m+1}\| = 1 + \sum_{i=1}^{n} |p_i|, \quad z_{m+1}(\bar{x}^{(\alpha)}) = \sum_{i=1}^{n} p_i x_i - b_{m+1}^{(\alpha)}.$$

And therefore according to Theorem 2.7 the following estimate holds

$$b_{m+1}^{(\alpha)} > \sum_{s=1}^{m} b_s \bar{w}_s^{(\alpha)} \geqslant \sum_{s=1}^{m} b_s w_s^* = \sum_{i=1}^{n} p_i x_i^*. \tag{3.45}$$

On the other hand if condition (2.95) holds then formulas (3.44) determine an exact solution w^* of the dual problem. If this happens for some value $\alpha = \alpha^*$, then, setting

$$b_{m+1} = \sum_{s=1}^{m} b_s w_s^*$$

we obtain by virtue of (3.45) a consistent system of the form (3.42) and all solutions of that system will be optimal solutions of the initial linear programming problem.

From (3.45) there results the possibility of speeding up the convergence of the algorithm described at the beginning of this section. This requires clearly that the sequence of numbers $b_{m+1}^{(0)}, b_{m+1}^{(1)}, \dots$ should be defined by the formula

$$b_{m+1}^{(\alpha+1)} = \sum_{s=1}^{m} b_s \bar{w}_s^{(\alpha)}. \tag{3.46}$$

Then there exists a finite neighbourhood of the optimal solution (or the set of optimal solutions) satisfying condition (2.95), the process becomes finite and the final result will be an optimal bivector (x^*, w^*) for the dual pair of linear programming problems.

Example. As an example we apply the algorithm of reduction of a linear programming problem to a series of equilibrium problems of physical models of inconsistent systems to the following problem

$$f(x) = 3x_1+x_2+2x_3+3x_4+x_5+2x_6+5x_7 \to \min, \tag{3.47}$$

$$\left.\begin{array}{l}
2x_1 \quad 3x_3 \quad +2x_6 \geqslant 10, \\
4x_2 \quad +3x_5+x_6 \geqslant 4, \\
x_2+5x_3 \quad +4x_5 \quad +x_7 \geqslant 2, \\
3x_1 \quad +2x_4 \quad +3x_6 \geqslant 5, \\
2x_3+4x_4 \quad +3x_7 \geqslant 3, \\
x_i \geqslant 0, \; i = 1,\dots,7.
\end{array}\right\} \tag{3.48}$$

The sequence of systems of linear inequalities corresponding to this system has the form

$$\left.\begin{array}{l}
2x_1 \quad +3x_3 \quad +2x_6 \geqslant 10, \\
4x_2 \quad +3x_5+x_6 \geqslant 4, \\
x_2+5x_3 \quad +4x_5 \quad +x_7 \geqslant 2, \\
3x_1 \quad +2x_4 \quad +3x_6 \geqslant 5, \\
2x_3+4x_4 \quad +3x_7 \geqslant 3, \\
-3x_1-x_2-2x_3-3x_4-x_5-2x_6-5x_7 \geqslant b^{(\alpha)}, \\
x_i \geqslant 0, \; i = 1,\dots,7.
\end{array}\right\} \tag{3.49}$$

Clearly $x = 0$ is not a feasible vector and $f(0) = 0 < f(x^*)$ the minimum of $f(x)$ over the domain (3.48). Therefore we can set $b^{(0)} = 0$, $b^{(1)} = f(x^{(0)}),\dots,b^{(k)} = f(x^{(k)})$, where $x^{(1)}$ is the equilibrium state vector of the physical model of system (3.49) with $b^{(\alpha)} = 0$ and $x^{(k)}$ is the equilibrium state vector of the physical model of (3.49) with $b^{(k)} = f(x^{(k-1)})$ and so on. The coordinates of the equilibrium state vectors of the physical models of each of the systems (3.49) were calculated by means of formula (3.38) with a precision given by the condition

$$\max_i |x_i^{(\alpha)\nu+1} - x_1^{(\alpha)\nu}| < 10^{-4}. \tag{3.50}$$

where α is the index of the iteration. As a criterion for terminating the whole process of solving the original problem (3.47)-(3.48) the following condition was used

$$|b^{(k+1)} - b^{(k)}| < 10^{-4}. \tag{3.51}$$

Denote with $\nu^*(\alpha)$ the number of iterations necessary to calculate the vector of the equilibrium state of problem (3.49) with the precision given by condition (3.50). The dependence $\nu^*(\alpha)$ for the problem under consideration is given in the following table

α	1	2	3	4	5	6	7	8	9
$\nu^*(\alpha)$	40	39	34	28	22	17	9	2	1

Thus to obtain a solution of problem (3.47)-(3.48) with a precision as determined by (3.51) only 192 iterations were needed and in this way the following solution was obtained

$$x^* = (0.3\times 10^{-4};\ 0.583325;\ 2.222167;\ 0;\ 0;\ 1.66691;\ 0).$$

$$f(x\ ) = 8.3611136.$$

Let us give the exact solution for purposes of comparison

$$x_{opt} = (0;\ 0.583(3);\ 2.2(2);\ 0;\ 0;\ 1.66(6);\ 0),$$

$$f(x_{opt}) = 8.361(1).$$

NOTES

1. Natura non operatur per faltum (translator's note).
2. For remarks concerning intensive and extensive variables see Section 2.2.
3. The brackets serve to distinguish between state vectors of subsystems and components of the state vector of the whole system x and w.
4. This means that in state $(x^{(0)}, w^{(0)})$ the state equations hold which connect the components of the vector $x^{(0)}$ and the vector $w^{(0)}$.
5. To determine $\beta^{(0)}$ in the manner indicated, it is advantageous to take a small value for the parameter $\tilde{q}_0$ because then it suffices to take only a small number of iterations, that means a coarse approximation of $\bar{x}(\tilde{q}_0)$.
6. In the following we shall consider a sequence of systems of the form (3.37) which differ only in the value of the parameter $b_{m+1}^{(\alpha)}$ and therefore the vector of the equilibrium state $\bar{x}$ can and will be considered as a function of that parameter.

Chapter IV

THE PRINCIPLE OF REMOVING CONSTRAINTS

> ... la physique ne nous donne pas seulement l'occasion de résoudre des problèmes; elle nous aide à en trouver les moyens, et cela de deux manières. Elle nous fait pressentir la solution; elle nous suggère des raisonnements.
>
> H. Poincaré [1)]

4.1. Introduction

Mathematically speaking an equilibrium problem for a mechanical system with perfect (that is without friction) bilateral and unilateral constraints in a conservative force field is equivalent to general mathematical programming problems. On the other hand, every maximalization problem with constraints can be interpreted as an equilibrium problem of a mechanical system in a conservative force field defined by the function to be maximized with the constraints expressed in terms of unilateral and bilateral constraints imposed on the system. If one uses this approach it is natural to base the theory and the solution methods on the fundamental principles of mechanics. As far as extremum problems are concerned these principles are the principle of virtual displacements (virtual work) and the principle of removing constraints (that is the principle of Lagrange multipliers); that is the fundamentals of Lagrange's analytical mechanics.

The present chapter examines various ways of putting to use the principle of removing constraints and the methods of solution which derive from them for mathematical programming problems. This principle says, grosso modo, that it is possible to consider parts of the system as isolated provided that one introduces extra exterior forces to match its interrelations with other parts of the system. In mechanical systems this amounts to the possibility of replacing certain constraints which force that part of the system to move only in certain ways with the forces which would be produced by those constraints. These forces are called constraint forces (reaction forces). Their magnitude, direction and point of action are defined by the condition that the free movement or equilibrium of the system subject to active forces and these constraint forces should be at all instants such that the constraints are satisfied. The constraints imposed on a system consisting of material points are usually expressible in terms of analytical relations between the coordinates of those points. These constraints are then said to be bilateral in the case of equality relations and unilateral in the case of inequality relations.

The author of the 'Mécanique analytique' bases himself on the principle of virtual displacements (Theorem 1.5) without stressing the exceptional role of the principle of removing constraints. But still he did mention this principle when he wrote: "... every constraint equation is equivalent to one or more forces acting on the system in given directions or, more generally, tending to vary the values of given functions; this in such a way that the equilibrium state of the system will remain the same if one takes into consideration these forces or if one takes into account the constraint equations.

"Conversely, these forces can take the place of the constraint equations which result from the nature of the given system; so that using these forces one could regard the bodies (making up the system) as entirely free without any constraints" [2] [38]. Bertrand was the first to appreciate these words of Lagrange to their true extent: "This important principle is as general as the principle of virtual velocities and it is often easier to apply".

Indeed the principle of virtual displacments makes it possible at equilibrium to eliminate from consideration the unknown reaction forces of the perfect constraints and the principle of removing constraints on the other hand is best suited to find those reaction forces. However, one wrongly understands Bertrand if one claims that the second principle can replace the first. The principle of removing constraints is a valuable complement to the principle of virtual displacements and together they form the foundations of Lagrange's edifice. The principle of removing constraints can be applied in various ways and certain of these applications will be studied in this chapter, Together with their impact on mechanics and optimization theory.

4.2. The method of generalized coordinates

Let us start with two procedures which are due to Lagrange himself: the method of generalized coordinates and the method of (Lagrange) multipliers. Let there be a mechanical system consisting of N material points $M_1, M_2, \ldots, M_N$ and let (ξ_i, η_i, ζ_i) be the coordinates of point M_i and $F_1, \ldots, F_n$ the forces acting on the corresponding points. Let us consider the case where the movements of the system are subject to constraints as expressed by the equations

$$\phi_s(\xi_1, \eta_1, \zeta_1, \ldots, \xi_N, \eta_N, \zeta_N) = 0, \quad s = 1, \ldots, m. \tag{4.1}$$

The general equilibrium condition which expresses the principle of virtual displacements can be written [38]:

$$\sum_{i=1}^{N} (F_{i\xi}\delta\xi_i + F_{i\nu}\delta\eta_i + F_{i\zeta}\delta\zeta_i) = 0, \tag{4.2}$$

where the components $\delta\xi_i$, $\delta\eta_i$, $\delta\zeta_i$, $i = 1, \ldots, N$ of the virtual displacement vectors δr_i are connected by the equalities

$$\sum_{i=1}^{N} \left(\frac{\partial\phi_s}{\partial\xi_i}\delta\xi_i + \frac{\partial\phi_s}{\partial\eta_i}\delta\eta_i + \frac{\partial\phi_s}{\partial\zeta_i}\delta\zeta_i\right) = 0, \quad s = 1, \ldots, m, \tag{4.3}$$

which follow from (4.1).

The idea behind the method of generalized coordinates is the following. Using the constraint Equations (4.1) one expresses m coordinates of the system as functions of the remaining 3N-m coordinates, or, and this is the essential point, one expresses all 3N coordinates as functions of $n = 3N-m$ arbitrary independent variables. The arbitrariness in the $x_1, \ldots, x_n$ opens up the possibility of adapting one's choice in each specific case in the best way possible to the nature of the system. The parameters $x_1, \ldots, x_n$ are given the name of the generalized coordinates and the method under consideration works if one can express each of the 3N coordinates of the points of the system by means of these parameters:

$$\begin{aligned} &\xi_i = \xi_i(x_1, \ldots, x_n), \quad \eta_i = \eta_i(x_1, \ldots, x_n), \\ &\zeta_i = \zeta_i(x_1, \ldots, x_n), \quad i = 1, \ldots, N. \end{aligned} \tag{4.4}$$

The variations in the $x_1,\ldots,x_n$ are not constrained in any way and once the $x_1,\ldots,x_n$ are given the state of the mechanical system is completely specified. In generalized coordinates the equilibrium condition (4.2) takes the form

$$\sum_{i=1}^{N}\sum_{\alpha=1}^{n}\left(F_{i\xi}\frac{\partial\xi_i}{\partial x_\alpha}+F_{i\eta}\frac{\partial\eta_i}{\partial x_\alpha}+F_{i\zeta}\frac{\partial\zeta_i}{\partial x_\alpha}\right)x_\alpha = 0 \qquad (4.5)$$

or

$$\sum_{\alpha=1}^{n} Q_\alpha\delta x_\alpha = 0, \qquad (4.6)$$

where the quantities

$$Q_\alpha = \sum_{i=1}^{N}\left(F_{i\zeta}\frac{\partial\xi_i}{\partial x_\alpha}+F_{i\eta}\frac{\partial\eta_i}{\partial x_\alpha}+F_{i\zeta}\frac{\partial\zeta_i}{\partial x_\alpha}\right),\ \alpha = 1,\ldots,n \qquad (4.7)$$

are said to be generalized forces. One regards $x_1,\ldots,x_n$ as rectangular coordinates in some n-dimensional space which is called configuration space. In the generalized coordinates approach one formally replaces a system of N material points with a vector in a configuration space of dimension n and N three dimensional force vectors $F_1,\ldots,F_n$ with an n-dimensional generalized force vector $Q = (Q_1,\ldots,Q_n)$. In the configuration space thus obtained the system is represented by a free point, which means that the possible variations $\delta x_1,\ldots,\delta x_n$ are independent so that (4.6) implies the following n equilibrium conditions

$$Q_\alpha = 0,\ \alpha = 1,\ldots,n. \qquad (4.8)$$

The method under examination is indeed an example of how the principle we are considering can work because it leads from the problem of determining movement or equilibrium of system of material points in three dimensional space with constraints to a problem of finding the movement or equilibrium of a free point in an n-dimensional configuration space.

Independently whether the system is free or constrained and independently of how many constraints there are, the idea of a configuration space will turn out to be useful in the following. The representation of a mechanical object by a system of material points using their orthogonal cartesian coordinates as state parameters is only useful in isolated cases. In general one chooses any parameters which best describe such an object, and this choice may or may not depend on the constraint equations. These parameters themselves also constitute a set of generalized coordinates $x_1,\ldots,x_n$ and they are 3N in number for a system of N material points (or less than 3N if, in passing to generalized coordinates one has made use of some of the constraint equations; usually those which can be eliminated without difficulty). The other constraints remain and describe constraints imposed on the coordinates $x_1,\ldots,x_n$ of the point in configuration space representing the system. In this case the equilibrium conditions are written

$$\sum_{s=1}^{n} Q_s \delta x_s = 0 \tag{4.9}$$

with the conditions

$$\sum_{s=1}^{n} \frac{\partial\phi_\alpha}{\partial x_s} \delta x_s = 0, \quad \alpha = 1,\ldots,m, \tag{4.10}$$

which result from $\phi_\alpha(x_1,\ldots,x_n) = 0$, $\alpha = 1,\ldots,m$.

The method of generalized coordinates enables one to eliminate constraints if (1) these constraints are bilateral, that is if they are given by equations of constraints; (2) the elimination of some coordinates by means of these constraint equations is not difficult.

The advantage of the method is that it reduces a mechanical problem for a system of material points to an analogous problem for one material point in a configuration space of dimension n and that it enables one to get rid of a part of the bilateral constraints during this reduction.

4.3. The method of multipliers

The method of multipliers of Lagrange is a dynamical manifestation of the principle of removing constraints. It is a procedure which is more general than the preceding one. This comes about not only because it avoids the often rather delicate problem of elimination of variables, but also because it can be applied with success to problems of mechanics of systems with unilateral constraints and non-holonomic constraints. The method of multipliers rests on two hypotheses which apply precisely in problems of statics and dynamics in mechanics. The first hypothesis is that the constraints are rigid and this means that they can be represented by means of equations or inequalities of constraint, that is the constraints are realized by perfect solids (no friction), inextensible threads, incompressible fluids, or other strongly idealized models of real objects. The second hypothesis is concerned with perfect restrictions (without friction): the constraint forces are always orthogonal to the surface described by a restriction equation or by the equation describing a certain fixed relation (between parts of the system) and the virtual work done by these feaction forces is thus zero. This hypothesis sees to it that the reaction force caused by $\phi(x_1,\ldots,x_n) = 0$ is parallel with the gradients of the function $\phi(x_1,\ldots,x_n)$ so that one has a formula $R = \lambda \operatorname{grad} \phi(x_1,\ldots,x_n)$ for the reaction force R where λ is a Lagrange multiplier. The equilibrium conditions for a system represented by a point in configuration space subject to forces and constrained to verify $\phi_s(x_1,\ldots,x_n) = 0$, $s = 1,\ldots,m$, can then be written

$$Q + \sum_{\alpha=1}^{m} R^{(\alpha)} = 0, \tag{4.11}$$

where

$$R^{(\alpha)} = \lambda^{(\alpha)} \operatorname{grad} \phi_\alpha(x_1,\ldots,x_n). \tag{4.12}$$

One chooses the multipliers $\lambda^{(1)},\ldots,\lambda^{(m)}$ such that $R^{(1)},\ldots,R^{(m)}$ guarantee that the constraint equations hold (i.e. such that (4.12) holds) and the following system of equilibrium equations results:

$$Q_s + \sum_{\alpha=1}^{n} \lambda^{(\alpha)} \frac{\partial \phi_\alpha}{\partial x_s} = 0, \quad s = 1,\ldots,n, \tag{4.13}$$

$$\phi_\alpha(x_1,\ldots,x_n) = 0, \quad \alpha = 1,\ldots,m, \tag{4.14}$$

which has n+m unknowns, viz. $x_1,\ldots,x_n$, $\lambda^{(1)},\ldots,\lambda^{(m)}$ and an equal number of equations.

A very important special case is the case where there are conservative driving forces for which there exists a function $U(x_1,\ldots,x_n)$ of the generalized coordinates such that

$$Q_s = \frac{\partial U}{\partial x_s}, \quad s = 1,\ldots,n. \tag{4.15}$$

$U(x_1,\ldots,x_n)$ is then a force function (negative potential) and the inequalities (4.13) say that the Lagrangian

$$L(x,\lambda) = U(x_1,\ldots,x_n) + \sum_{\alpha=1}^{m} \lambda^{(\alpha)}\phi_\alpha(x_1,\ldots,x_n). \quad (4.16)$$

must be stationary.

According to Lanczos the Lagrange multipliers measure the microscopic perturbations of the constraints equations [40]. In this interpretation the second term of the Lagrangian is the potential of a microscopic field generated by these perturbations. The function L thus represents the total force function for a field defined by active forces and reaction forces and the conditions

$$\frac{\partial L}{\partial x_s} = 0, \; s = 1,\ldots,m \quad (4.17)$$

express a principle that according to Lagrange was proposed by Maupertuis under the name 'Loi de repos' 3).
The author of the Mécanique Analytique has improved substantially on this by exhibiting the stable equilibria as those which realize the maximum of the force function or the minimum of the potential energy $\pi(x_1,\ldots,x_n) = -U(x_1,\ldots,x_n)$. Thus the stable equilibria are defined by

$$\max_x \{U(x_1,\ldots,x_n) + \sum_{\alpha=1}^{m} \lambda^{(\alpha)}\phi_\alpha(x_1,\ldots,x_n)\} \quad (4.18)$$

or by

$$\max_x U(x_1,\ldots,x_n)$$

over the set of vectors in configuration space satisfying $\phi_\alpha(x_1,\ldots,x_n) = 0$, $\alpha = 1,\ldots,m$.

The search for equilibrium states for systems with perfect unilateral constraints given by

$$\phi_\alpha(x_1,\ldots,x_n) \leq 0, \; \alpha = 1,\ldots,m, \quad (4.19)$$

turns out to be more complicated. The reaction forces are again colinear with the gradients of the associated forces $\phi_1,\ldots,\phi_m$, but they act in the opposite direction, that is the multipliers $\lambda^{(1)},\ldots,\lambda^{(m)}$ cannot be positive. One notes also that if the constraint $\phi_\alpha(x_1,\ldots,x_n) \leq 0$ becomes a strict one at an equilibrium state $(x_1,\ldots,x_n)$ then the corresponding reaction force is necessarily zero, that is $\lambda^{(\alpha)} = 0$, because then this particular constraint imposes no restrictions on the virtual displacements. Thus the conditions for equilibrium for a system subject to unilateral constraints are of the form

$$\left.\begin{aligned} &\max_x \{U(x_1,\ldots,x_n) + \sum_{\alpha=1}^{m} \lambda^{(\alpha)}\phi_\alpha(x_1,\ldots,x_n)\} \\ &\phi_\alpha(x_1,\ldots,x_n) \leqslant 0,\ \alpha = 1,\ldots,m, \\ &\lambda^{(\alpha)} \begin{cases} \leqslant 0 & \text{for } \phi_\alpha(x_1,\ldots,x_n) = 0, \\ = 0 & \text{for } \phi_\alpha(x_1,\ldots,x_n) < 0. \end{cases} \end{aligned}\right| \tag{4.20}$$

If one considers a more general case where the system is subject to both unilateral and bilaterial constraints, then the equilibrium problem can be stated as

$$\max_x U(x_1,\ldots,x_n) \tag{4.21}$$

over the set

$$\Omega = \left\{ x | \phi_\alpha(x_1,\ldots,x_n) \begin{cases} \leqslant 0, & \alpha = 1,\ldots,m_1 \\ = 0, & \alpha = m_1+1,\ldots,m. \end{cases} \right\} \tag{4.22}$$

With the method of multipliers one reduces this to a parametric maximization problem without constraints for the Lagrangian

$$\max_x \left[U(x_1,\ldots,x_n) + \sum_{\alpha=1}^{m} \lambda^{(\alpha)}\phi_\alpha(x_1,\ldots,x_n) \right], \tag{4.23}$$

with the parameters $\lambda^{(1)},\ldots,\lambda^{(m)}$ defined by the conditions

$$\phi_\alpha(x_1,\ldots,x_n) \left\{ \begin{array}{ll} \leqslant 0 & \text{for } \alpha = 1,\ldots,m_1, \\ = 0 & \text{for } \alpha = m_1+1,\ldots,m. \end{array} \right\} \tag{4.24}$$

$$\lambda^{(\alpha)} \left\{ \begin{array}{ll} \leqslant 0 & \text{for } \phi_\alpha(x_1,\ldots,x_n) = 0, \\ = 0 & \text{for } \phi_\alpha(x_1,\ldots,x_n) < 0, \end{array} \right\} \alpha = 1,\ldots,m_1. \tag{4.25}$$

The problem (4.21), (4.22) is the general problem of mathematical programming and it is the mathematical model of a vast class of optimum problems which turn up often both in theory and in practice. We have seen that it is equivalent to the search for the equilibrium of a mechanical system and that means that the methods for solving mathematical programming problems must be able to borrow substantially from analytical mechanics.

We conclude this section with a consideration of the linear programming problems

$$\max\{\sum_{i=1}^{n} p_i x_i \mid \sum_{i=1}^{n} a_s x_i \leqslant b_s, \quad s = 1,\ldots,m\}. \tag{4.26}$$

The mechanical interpretation will be easier if one supposes that the constraints have been normed in such a way that the gradient of each linear form $\Sigma a_{si}x_i$ is a unit vector. This can be assured by dividing each inequality by the corresponding $(\Sigma a_{si}^2)^{\frac{1}{2}}$. The Lagrange multiplier is then in absolute value equal to the reaction force which is oriented along the opposite gradient of the associated linear form.

$$\operatorname{grad}(\sum_{i=1}^{n} a_{si}x_i) = (a_{s1},\ldots,a_{sn}), \tag{4.27}$$

that is $a_{s1},\ldots,a_{sn}$ are the cosines of the angles formed by the gradient and the coordinate axes in configuration space. In this simple case of linear unilateral constraints, (4.23)-(4.25) imply the following maximum conditions

$$p_i + \sum_{s=1}^{m} \lambda^{(s)} a_{si} = 0, \quad i = 1,\ldots,n, \tag{4.28}$$

$$\lambda^{(s)} \begin{cases} \leqslant 0 & \text{for } (a_s,x) = b_s, \\ = 0 & \text{for } (a_s,x) < b_s. \end{cases} \tag{4.29}$$

The force $R^{(s)}$ produced by $(a_s,x) \leqslant b_s$ has the quantities $\lambda^{(s)}a_{s1},\lambda^{(s)}a_{s2},\ldots,\lambda^{(s)}a_{sn}$ as components and one has $|R^{(s)}| = |\lambda^{(s)}|$ in virtue of the conditions $(\sum_i a_{si}^2)^{\frac{1}{2}} = 1$.

Let $p = (p_1,\ldots,p_n)$ denote the vector of active forces. The system (4.28) can be written in terms of vectors as

$$p + \sum_{s=1}^{m} R^{(s)} = 0 \tag{4.30}$$

The conditions (4.29) do not signify anything other than that the reaction forces of the removed constraints are zero at equilibrium. Identifying (4.28)-(4.29) with the necessary and sufficient conditions for optimality (1.31)-(1.33) one sees that $\lambda^{(s)} = -w_s^*$. If one agrees to call positive the direction of the interior normal of $\phi_s(x_1,\ldots,x_n) = 0$ which has angle coefficients $-a_{s1},\ldots,-a_{sn}$ then the Lagrange multipliers become nonnegative and the reaction forces at equilibrium coincide with the components of optimal vector of the dual of (4.26); see Figure 4.1.

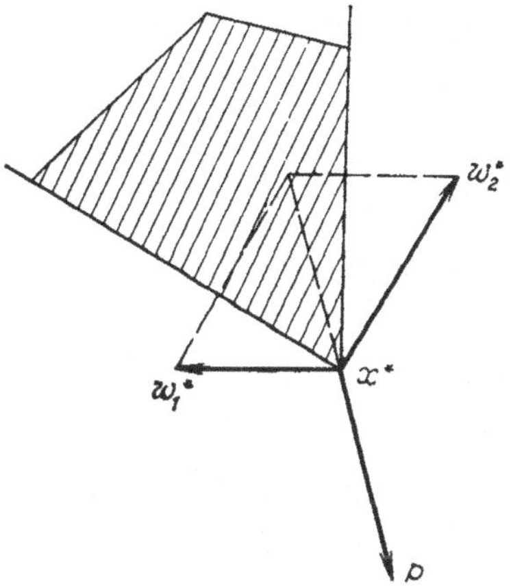

Fig. 4.1.

4.4. Elastic constraints. Penalty function methods

According to Lagrange constraint equations result from the nature of the mechanical system and one of the originators of the mechanics of deformable bodies would certainly be aware of the fact that in nature constraints are realized by deformable bodies and that the constraint equations only express the geometric properties of constraints as long as they are not active [4)]. The method of multipliers, one of Lagrange's great successes, has become a main pillar for statics, dynamics, the theory of maxima and minima and for the calculus of variations. It took a long time before one became aware of certain inconvenient aspects for practical applications. These technical (numerical) and analytical difficulties derive from one of the fundamental hypothesis (see Section 4.3). At the time when he wanted to justify the Raleigh-Ritz method, by trying to avoid to that end a fundamental difficulty deriving from the same hypothesis [23], Courant came up with an idea which **should have received the** name of a penalty function method. At that time the physical meaning of the method was not realized although the problem studied by Courant did afford a glimpse of it, and the method remained for a long time a procedure of approximation for constrained extremum problems. However, penalty functions are an appplication of the principle of removing constraints and the development of these methods is as important as that of the method of multipliers, though the first steps have not been as easy or as spectacular. This difference is clearly due to the genius and authority of Lagrange.

One thus relaxes the assumption that the constraints are rigid and now supposes them elastic, that means realized by elastic bodies which are deformable. During the evolution of the system or when the system is in equilibrium these constraints are deformed and the reactions are defined by active forces and the elasticity properties of the physical bodies which realize the constraints. Moreover for an equation which describes the form of a constraint which is under stress one must clearly know the law governing the dependence of the leastic force as a function of its deformations. Take for example the problem of Section 4.3 concerning the equilibrium of a system in a force field which admits a force function $U(x_1,\dots,x_n)$ and which is subject to elastic constraints

$$\phi_s(x_1,\dots,x_n) = 0, \quad s = 1,\dots,m. \tag{4.31}$$

The value of the function $\phi_s(x_1,\dots,x_n)$ at the point $(x_1,\dots,x_n)$ defines the magnitude of the deformation of this constraint in state $(x_1,\dots,x_n)$. Let us assume that the physical properties of the bodies realizing the elastic constraints determine a constant linear relation between the elastic force produced by the reaction and the deformation. The elastic constraint then generates a field with a force function of the form

$$-\frac{k_s}{2}\,\phi_s^2(x_1,\ldots,x_n), \tag{4.32}$$

where $k_s > 0$ is the elasticity coefficient.

All the elastic constraints together then generate a field of constraint forces given by the force function

$$\Phi(x_1,\ldots,x_n) = -\tfrac{1}{2}\sum_{s=1}^{m} k_s\phi_s^2(x_1,\ldots,x_n). \tag{4.33}$$

The quantity

$$E(x_1,\ldots,x_n) = -\Phi(x_1,\ldots,x_n) \tag{4.34}$$

is the potential energy of the deformations of the constraints at the state $(x_1,\ldots,x_n)$. It results from (4.33) that the formulas for the components of the force vector resulting from the constraints are:

$$R_i = \frac{\partial\Phi}{\partial x_i} = -\sum_{s=1}^{m} k_s\phi_s(x_1,\ldots,x_n)\frac{\partial\phi_s}{\partial x_i}\,. \tag{4.35}$$

An equilibrium state occurs in our example in the point $\bar{x}$ in configuration space which is defined by the condition

$$\max_x\,(U+\Phi) \tag{4.36}$$

or, because we are dealing with an unconstrained maximum problem, the system of equations

$$\frac{\partial}{\partial x_i}(U+\Phi) = 0, \quad i = 1,\ldots,m$$

which signify that at equilibrium the sum of the potential energy of the active forces and the deformation energy of the constraints is a free minimum. Our problem is thus significantly less hard than a search for equilibrium of a system with rigid constraints.

This new application of the principle of removing constraints is distinguished by the fact that the constraints are replaced not by forces which satisfy the constraints at their point of application but by a field of forces defined by the constraints.

Let us substitute $k_s = K_s\sigma$ in (4.32), (4.33), (4.35) and let us send σ to $+\infty$. The following limit formulas for the Lagrange multipliers result:

$$\lambda^{(s)} = -\lim_{\sigma\to\infty} \kappa_s \sigma \phi_s(\bar{x}(\sigma)), \tag{4.37}$$

with $\bar{x}(\sigma)$ an equilibrium vector. It follows that $\lim_{\sigma\to\infty} \phi_s(\bar{x}(\sigma)) = 0$.

It is interesting to remark that a very simple modification of the limit transition $\sigma \to \infty$ leads to a well known modification of the method of multipliers or the so called modified Lagrangian function. Introduce in the right hand side of expression (4.33) the parameter σ, not by substituting $k_s = K_s\sigma$ but by using the equally simple substitution $k_s = K_s(1+\sigma)$. Then problem (4.36) takes the form

$$\max_x\{U(x_1,\ldots,x_n) - \sum_{s=1}^m \kappa_s\phi_s^2(x_1,\ldots,x_n) - \sum_{s=1}^m \kappa_s\sigma\phi_s^2(x_1,\ldots,x_n)\}.$$

This time taking the limit and taking (4.37) into account we obtain a function to be maximized of the form

$$U(x_1,\ldots,x_n) + \sum_{s=1}^m \lambda^{(s)}\phi_s(x_1,\ldots,x_n) - \sum_{s=1}^m \kappa_s\phi_s^2(x_1,\ldots,x_n),$$

where, as before, the $k_1,\ldots,k_m$ are arbitrary positive numbers. In this manner we arrive in a very simple way at a function which is called the modified Lagrangian function.

Obviously there will be little difficulty in transferring the considerations of this section to the case of equilibrium problems in the presence of unilateral constraints of the form $\phi(x_1,\ldots,x_n) \leqslant 0$. Some ideas connected with the construction of numerical solutions methods in these cases are described in [63].

Penalty function methods (see Section 2.5) do result from an application of the principle of removing constraints provided one also relaxes the rigidity of the constraints. This statement of the problem of finding equilibria or maxima is no less general, and in fact it encompasses more. Indeed it includes, in addition to the constraint equations, the equations which generate the force field of reactions, the parameter vectors describing the elasticities of the constraints. The method of Lagrange becomes a limiting case of the model when the parameters grow indefinitely. One notes moreover that there exists a large class of important problems where the parameters are chosen arbitrarily within certain limits, that is they play the role of control parameters. This leads to optimization problems where the states and processes are controlled both by the exterior forces and by interior ones. In this form the principle of removing constraints corresponds to the real elasticities of economic constraints and one understands its most important role in the modelling of economic systems, a fact which we have already noted in the introduction.

4.5. Discussion

The reader is not supposed to infer from what he has read in the preceding section that the author denies the existence of real rigid constraints. In reality certain constraint equations express laws of nature 5) and they have thus the meaning of absolutely rigid constraints.

The author wants to say noting more than wat follows. The formulation of numerous problems contains a priori hypotheses which even if they do not cause analytical difficulties which are not really valid at the physical level, are still inadmissible a posteriori. Let us illustrate this by means of an example by considering the well known problem where one looks for the shortest time to go by sailing boat from port A to port B when facing the wind and with the current helping. First let us simplify the problem by supposing that there is no current. Every sailing enthousiast knows how to make use of the maximal width of the channel. If one neglects the time involved in every manoeuvre of changing direction, the optimal trajectory is any broken line of which the various segments form an optimal angle with the wind direction (see Figure 4.2). If there is a current one takes the drift of the ship into account. The velocity of the current is in general maximal in the middle of the river and diminishes to zero when approaching the borders (see Figure 4.3).
Assuming once more that the times involved in the tacking manoeuvres can be neglected, the optimal trajectory is evidently a broken line in the immediate neighbourhood of the line of maximal current velocity (see Figure 4.3). It would be more precise to say that the optimal path is a generalized curve consisting of an infinity of infinitesimal segments. We have taken this example from [52]. It shows that one may need to generalize the notion of a curve as certain variational problems do not have a solution in the class of curves which are piece-wise regular. One finds in Young [52; Section 6.1] another example of a generalized curve (Maxwell's problem).

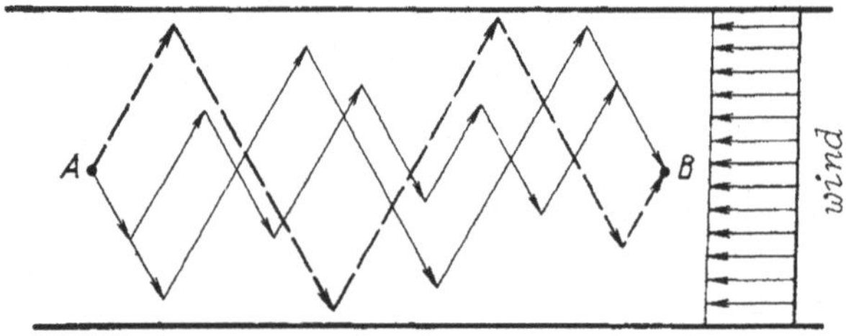

Fig. 4.2.

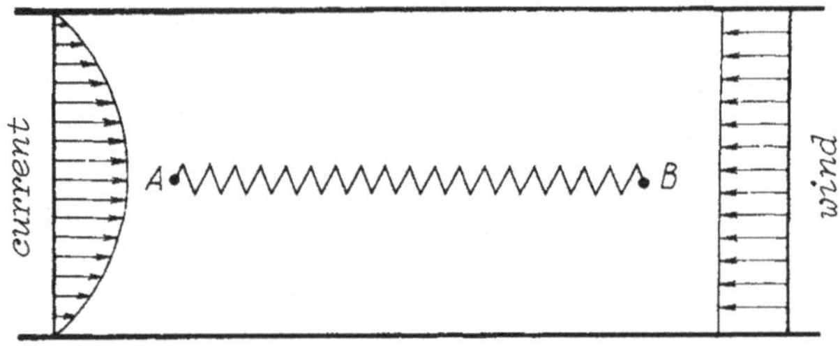

Fig. 4.3.

These models of two real problems make one conclude that it may be impossible to find the solution by means of the necessary conditions for a minimum in the form of the Euler-Lagrange equations. It seems that we have passed beyond the domain of application of the classical variational calculus. Is this indeed the case? Let us consider again our navigation problem, and let us examine its hypotheses to see whether they are admissible. It turns out that the model contains quite a number of arbitrary hypotheses which, although they seemed initially natural, become inadmissible once one has a solution. For example, was it not the case that the optimal trajectory was a broken line with an arbitrary number of segments when the current is absent and an infinity of segments in the presence of current and to reach this conclusion we neglected the times involved in the tacking manoeuvres, as well as the loss in velocity incurred when tacking. To do this one must suppose a priori that the optimal trajectory contains few such inflexion points. The solution of the problem thus has meaning if it confirms the a priori hypotheses and it is without significance if contradictions arise. One verifies without difficulty that if in the model one takes tacking time into account in the first case, then the optimal trajectory is given by the dashed line in Figure 4.2. In the second case the number of segments of the optimal trajectory is greater but finite (see Figure 4.4), because one has in the limit a simple drift of the sailing boat which describes a figure eight in a moving reference frame which has the velocity of the current (in the middle of the river).

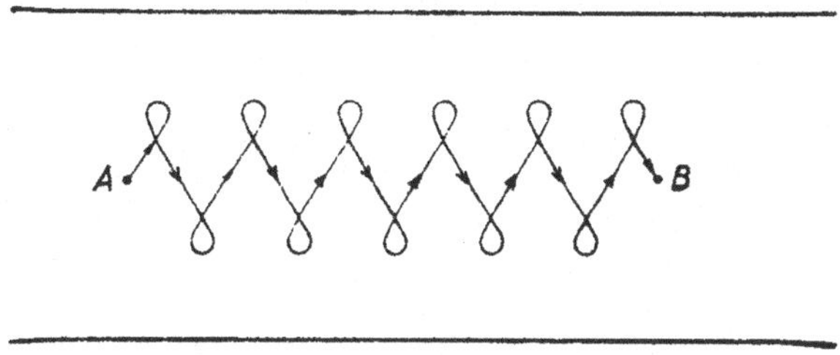

Fig. 4.4.

One encounters the same difficulties when tracing an optimal route between two given points. Here is the scheme for solving this: 1) find a geodesic AB; 2) as the slope may not exeed a certain optimal value, replace those parts where the slope is inadmissible by a zig-zag curve of optimal slope (see Figure 4.5).

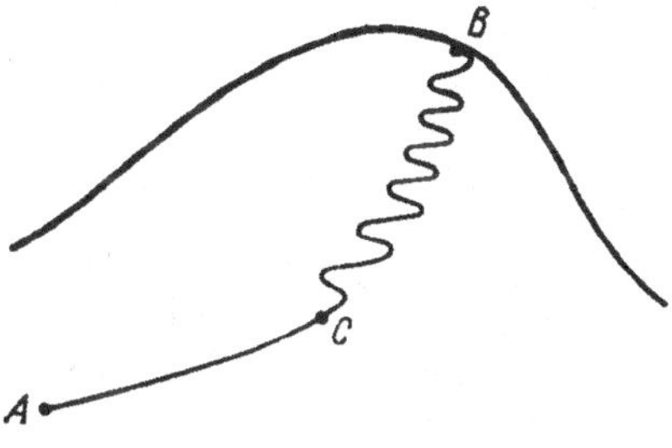

Fig. 4.5.

Provided one neglects the time or effort needed to make each turn the optimal trajectory between C and B may contain an infinity of zig-zags of finite total length. But if the model takes these parameters into account then such a solution is evidently no longer optimal either with respect to minimal time nor with respect to minimal loss of energy. Indeed, car drivers and mountaineers know very well how exhausting roads are which turn all the time.

Thus it does not suffice to solve a problem which comes from a simulation of real entity. Once the solution has been obtained one must return to the model to test the hypotheses which then must be perfected if they are no longer admissible. And it is important to note that these examples have not undermined the foundations of the classical variational calculus of Euler-Lagrange.

One can also approach the problem from a different point of view and take the standpoint that it is not important that the model is imperfect! The problem which results requires methods and theories which do not as yet exist, and thus there arises progress in mathematics, evolving in response to problems coming from the natural sciences and consequently from their intrinsic laws.

We want to stress several important results which come out of the mechanical interpretation of penalty methods. The penalty function measures at equilibrium the perturbations in the constraint conditions, but it also representes the deformation energy of the elastic constraints which is equal (by the law of conservation of energy, Theorem 1.6) to the amount of work done if one opposes the leastic forces and takes the system in a quasi static way from the state [6] which satisfies the constraint equations and inequalities to the equilibrium state $\bar{x}$. One can realize this transformation in thought if one replaces the function $U(x_1,\ldots,x_n)$ of problem (4.36) with the function $\gamma U(x_1,\ldots,x_n)$ and one follows the evolution of $\bar{x}(\gamma)$, the point where

$$\gamma U(x_1,\ldots,x_n) + \Phi(x_1,\ldots,x_n)$$

assumes its maximum, when γ grows from zero to one with infinitesimal speed. Because this transformation is formally equivalent with the transformation in the equilibrium state when the penalties grow indefinitely (see (4.37)) one clearly has $\bar{x}(0) = x^*$.

The amount of work mentioned above is done by the variable force γ grad $U(x_1,\ldots,x_n)$ when its point of application moves from x^* to $\bar{x}$. One sees without difficulty that it can be expressed by means of the integral [7]

$$A = \int_{U(\bar{x}^*)}^{U(\bar{x})} \gamma(U)dU.$$

The conservation of energy during a quasi static transformation (Theorem 1.6) now takes the form of the rule

$$\int_{U(x^*)}^{U(\bar{x})} \gamma(U)dU = \Phi(\bar{x}), \qquad (4.38)$$

where $\gamma(U)$ satisfies the limit conditions

$$\gamma(U(\bar{x})) = 1, \ \gamma(U(x^*)) = 0. \qquad (4.39)$$

In general one has considerably more information available about $\gamma(U)$. It is a convex function in the integration interval for a vast class of mathematical programming problems and it is piece-wise convex linear for linear programming problems. The convexity property enables to use (4.38) and to find a lower bound for the optimal value $U(x^*)$ of the objective function by replacing the real quasi static process by a given imaginary process $\gamma = \omega(U)$ which satisfies, as does the real one, condition (4.38).

$$\int_{U(x^*)}^{U(\bar{x})} \omega(U)dU = \Phi(\bar{x}) \qquad (4.40)$$

and the limit boundary conditions

$$\omega(U(\bar{x})) = 1, \ \omega(U(x^*)) = 0. \qquad (4.41)$$

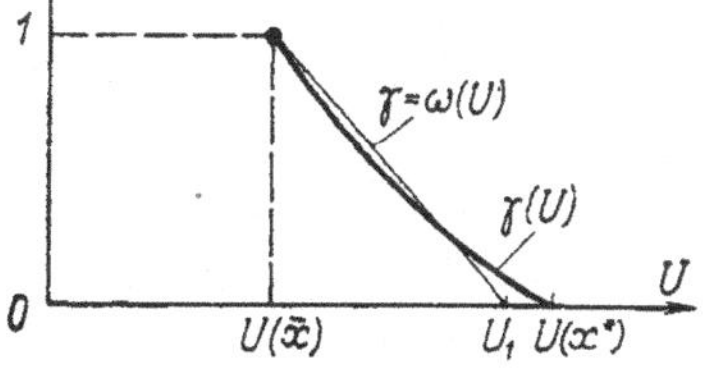

Fig. 4.6.

In Figure 4.6 there is represented a real quasi static process and an imaginary transformation $\omega(U)$ which satisfies in addition to (4.40) and (4.41) the two following conditions

1) $\omega(U)$ is linear in the interval $(U(\bar{x}),U_1)$

2) $\omega(U) \equiv 0$ for $U \in (U_1,U(x^*))$.

The equality (4.40) can now be rewritten

$$\tfrac{1}{2}(U_1-U(\bar{x})) = \Phi(\bar{x}), \tag{4.42}$$

with U_1, the lower bound for $U(x^*)$ which we are looking for, determined by the formula

$$U_1 = U(\bar{x}) + 2\Phi(\bar{x}). \tag{4.43}$$

There replacements of processes and the law of conservation of energy in the form (4.38) are at the basis of the numerical solution methods for optimization problems of Chapter IX.

NOTES

1. - physics not only gives us the occasion to solve problems; it aids us to find the means thereto, and that in two ways. It makes us foresee the solution; it suggests arguments to us. (From the authorized translation by G. Bruce Halsted, H. Poincaré. The value of science, on p. 82 of the Dover reprint, 1958). (Translator's note).
2. The constraint equations of Lagrange are equations of the form $\phi(\xi_1,\eta_1,\zeta_1,\ldots,\xi_N,\eta_N,\zeta_N) = 0$. The forces which tend to change the given functions $\phi(\xi,\eta,\zeta)$ are constraint forces.
3. In the opinion of the author Lanczos is too harsh with Maupertuis (see [40]; historical sketch), and Euler, of a generous, comprehensive and modest disposition, was right in attributing the principle of least action to Maupertuis.
4. i.e. not subject to pressures and such.
5. For example state equations, conservation laws, equations which govern transport phenomena. The differential equations of problems of dynamic movements under optimal control can be thought of as rigid non-holonomic constraints.
6. The optimal state vector for the primitive problem with rigid constraints.
7. (The scalar product of grad U with the path element ds of the quasi static process is equal to the differential dU.

Chapter V

THE HODOGRAPH METHOD

5.1. Introduction

Absolutely rigid ideal bilateral or unilateral constraints of which the mathematical models are respectively equations $\phi(x_1,\dots,x_n) = 0$ or inequalities $\phi(x_1,\dots,x_n) \leqslant 0$ are abstractions which have to do with idealized models of bodies subject to such constraints as are embodied by absolutely hard body, inextensible threads and incompressible fluids. In real equilibrium problems or problems of movement of physical systems the constraints are realized by deformable entities and the mathematical models of real constraints are force fields with a force function equal to the negative deformation energy of the bodies realizing the constraints. In this manner the absolutely rigid idealized constraints only present themselves as limits of force fields when the parameters characterizing the degrees of elasticity of the bodies realizing the constraints grow without bound. Such a physical and uniquely right treatment of the idea of ideal constraints yields much more understanding (see for example [55, 56]) and a great number of useful applications. Let us remark for instance that the case of non unique dual estimates of the restrictions in mathematical programming problems is included in the case of elastic constraints and turns out to have consequences for ideal constraints.

It is curious to remark that the situation indicated has an analogue for a class of equilibrium problems for mechanical systems which have unfortunately been called 'statically undetermined' and for which the equilibrium equations do not suffice for the unique determination of the reactions of the constraints. In fact, these problems have to do with a larger class of equilibrium problems for which models with absolutely rigid elements are not suitable. Of course it is possible for such problems to study the structure of the set of solutions of the equilibrium equations, but it would be more fruitful to take a look at nature and create a mechanics of deformable bodies.

In the present chapter we shall set out a new numerical method for solving linear programming problems which are effective and simple in that the trajectory of the quasi static process $\bar{x}(q)$ defined by increasing the parameter q from the value $q = q_0$ to $q = +\infty$ (see Figure 2.3 of Section 2.5)[1] can be represented by a broken line, consisting of a finite number of straight segments. The corner points of this line are completely determined by the hodograph of the vector $\bar{x}(q)$ and to find the direction vector of the following segments of the hodograph comes down to the solving of a closed system of linear algebraic equations.

An important property of the method is the fact that corresponding to the hodograph of $\bar{x}(q)$ there exists a hodograph for the vector $\bar{w}(q)$ consisting of feasible points for the dual problem and $\lim_{q\to\infty} w(q) = w^*$ as well as $\lim_{q\to\infty} x(q) = x^*$.

Thus the hodograph method turns out to be a finite method for solving simultaneously a primal and a dual linear programming problem and in applying it we shall not meet with the difficulties which characterize more primitive realizations of penalty function methods.

5.2. The hodograph method for linear programming problems

At the basis of the hodograph method lies the idea of modelling the constraints and restrictions by restraining force fields. The efficacy of the method lies in the fact that the optimal solution of the linear programming problem reduces to a finite number of unconstrained maximization problems of which all, except the first, turn out to be simple problems of minimizing a quadratic function of class $C^{(\infty)}$. In the following section it will be shown that the hodograph method leads at the same time to a solution of the primal and the dual problem.

Below we shall consider a linear programming problem of the form

$$\max\{(p,x) \mid (a,x) \leq b_s,\ s \in M\}, \tag{5.1}$$

where

$$x = (x_1,\dots,x_n),\ p = (p_1,\dots,p_n),$$
$$a_s = (a_{s1},\dots,a_{sn}),\ s \in M = \{1,\dots,m\}.$$

Together with problem (5.1) we shall consider the unconstrained maximization problem

$$\max\{(p,x) - \tfrac{1}{2}q \sum_{s\in M} y_s^2(x)\}, \tag{5.2}$$

where $q > 0$ is a penalty parameter and

$$y_s(x) = \max\{0,(a_s,x)-b_s\} \tag{5.3}$$

Let the vector $x^{(0)}$ be a solution of problem (5.2) for some value $q^{(0)} > 0$ of the parameter q, and let $\bar{x}(t)$ be the solution of the same problem (5.2) for $q = q^{(0)} + t$. It is well known [28] that

$$\lim_{t\to\infty} \bar{x}(t) = x^*, \tag{5.4}$$

where x^* is the optimal vector for problem (5.1). Moreover each monotonically growing sequence of positive numbers $t_0,t_1,\ldots,t_{\alpha+1} > t_\alpha$, results in a monotonically decreasing sequence $(p,\bar{x}(t_0))$, $(p,\bar{x}(t_1))$, ... of values for the obejctive function $(p,x(t))$ of problem (5.1). As the parameter t grows continuously from $t = 0$ to $t = +\infty$ the vector function $\bar{x}(t)$ defines a continuous curve joining the point $x^{(0)}$ to x^*. This line is called the hodograph of the vector function $\bar{x}(t)$. An important pecularity of linear programming problems is that the hodograph of the vector function $\bar{x}(t)$ presents itself as a broken line consisting of a finite number of straight segments (see Figure 5.1).

Problem (5.1) admits a mechanical interpretation as an equilibrium problem for a mechanical system consisting of a point x of a configuration space in a force field defined by the force function (p,x) in the presence of ideal linear unilateral constraints $(a_s,x) \leqslant b_s$, $s \in M$. These constraints should be considered as limits of the potential force fields

$$U_s(x) = -\tfrac{1}{2}qy_s^2(x)$$

and when the parameter q grows without bound the stresses of the force field $U_s(x)$ model the constraints $(a_s,x) \leqslant b_s$. From this presentation it follows that the solution of the equilibrium problem (5.1) is a limit of a sequence of

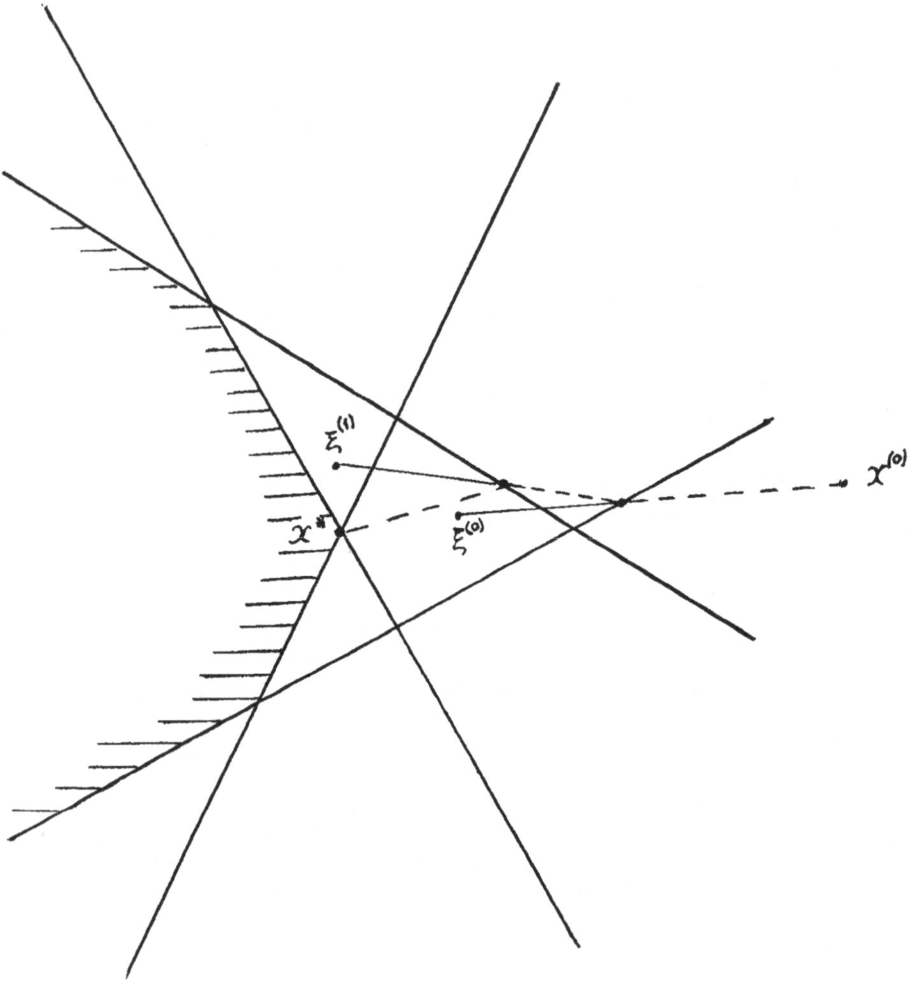

Fig. 5.1.

maxima of force functions

$$(p,x) - \tfrac{1}{2}q \sum_{s \in M} y_s^2(x)$$

(or a sequence of equilibrium state of the system) in the field of the given external forces and the elastic reaction forces. Therefore the hodograph of the vector function x(q) is the trajectory of quasi static process defined by the continuous but unbounded growing of the resilience parameter of the elastic constraints. The simplicity and efficacy of the hodograph method result from the fact that it leads to a finite number of unconstrained minimizations of quadratic functions.

The corner points of the hodograph are its points of intersection with the hyperplanes $(a_s,x) = b_s$. Of course it is possible to present the hodograph itself as the trajectory defined by the solution $\bar{x}(q)$ of problem (5.2) considered as a parametrized problem by the scalar parameter q with values in the set $[q_0,\infty]$. In that manner one meets the well known numerical difficulties of all primitive realizations of the fundamental idea of penalty functions. The value and essence of the method expounded below is that the construction of the hodograph reduces to a finite number of unconstrained minimizations of convex quadratic functions belonging to the class $C^{(\infty)}$. It is known [57] that the method of conjugate gradients leads to a solution of all such problems in n steps and estimates may be obtained for the number of elementary operations of gradient descent leading to the solution x^* of problem (5.1).

Let us proceed to the method under study. Let $\bar{M}^{(0)} \subset M$ be the set of indices of constraints determined by the condition

$$\bar{M}^{(0)} = \{s \mid y_s(x^{(0)}) > 0\} \qquad (5.5$$

and let $\bar{x}^{(0)}(t) = x^{(0)} + \xi^{(0)}(t)$ be the solution of problem (5.2) for $q = q^{(0)}+t$. There exists a sufficiently small number $\varepsilon^{(0)} > 0$ such that for all $t \in [0,\varepsilon^{(0)}]$ the condition holds that $y_s(\bar{x}^{(0)}(t)) = 0$ for $s \in M \setminus \bar{M}^{(0)}$ and $y_s(\bar{x}^{(0)}(t)) \geqslant 0$ for $s \in \bar{M}^{(0)}$. The vector $\xi^{(0)}(t)$ is the solution of a system of linear equations satisfying the conditions for a maximum for the problem (5.2) for

$q = q^{(0)}+t,\ t \in [0,\varepsilon^{(0)}]$ and

$$p_i = (q^{(0)}+t) \sum_{s\in \bar{M}^{(0)}} a_{si}[y_s(x^{(0)})+(a_s,\xi^{(0)}(t))],\qquad i = 1,\dots,n \qquad (5.6)$$

For t = 0 the system (5.6) goes over into the conditions defining the vector $x^{(0)}$, viz.

$$p_i = q^{(0)} \sum_{s\in \bar{M}^{(0)}} a_{si}y_s(x^{(0)}),\qquad i = 1,\dots,n. \qquad (5.7)$$

Subtracting equation (5.7) from equation (5.6) we obtain the system

$$(q^{(0)}+t) \sum_{s\in \bar{M}^{(0)}} a_{si}(a_s,\xi^{(0)}(t))+t \sum_{s\in \bar{M}^{(0)}} a_{si}y_s(x^{(0)}) = 0,\qquad i = 1,\dots,n, \qquad (5.8)$$

which can be conveneintly written in the form

$$\sum_{s\in \bar{M}^{(0)}} a_{si}(a_s,\xi^{(0)}(t)) = -\tau^{(0)}(t) \sum_{s\in \bar{M}^{(0)}} a_{si}y_s(x^{(0)}), \qquad (5.9)$$

where

$$\tau^{(0)}(t) = \frac{t}{q^{(0)}+t}. \qquad (5.10)$$

It is easy to convince oneself that the solution of system (5.9) has the form

$$\xi^{(0)}(t) = \tau^{(0)}(t)C^{(0)}, \qquad (5.11)$$

where the vector $C^{(0)} = (C_1^{(0)},\dots,C_n^{(0)})$ which is independent of the parameter t is the solution of the system of linear equations

$$\sum_{s\in \bar{M}^{(0)}} a_{si}(a_s,C^{(0)}) = - \sum_{s\in \bar{M}^{(0)}} a_{si}y_s(x^{(0)}),\qquad i = 1,\dots,n \qquad (5.12)$$

and it is the direction vector of the initial straight line segement of the hodograph of $\bar{x}^{(0)}$.

From formula (5.11) it follows that for sufficiently small values of the parameter t or $\tau^0(t)$, that is in a neighbourhood of the point $x^{(0)}$ the hodograph of $\bar{x}^{(0)}(t)$ is a straight segment. More precisely speaking this property of the hodograph holds for $0 \leqslant \tau^{(0)}(t) \leqslant \tau_1$, where

$$\tau_1 = \max\left\{\tau \,\middle|\, (a_s,x^{(0)})+\tau(a_s,C^{(0)})\begin{cases} > b_s & \text{for } s \in \bar{M}^{(0)} \\ \leqslant b_s & \text{for } s \in M\backslash\bar{M}^{(0)}\end{cases}\right\}. \tag{5.13}$$

From problem (5.13) it follows that

$$\tau_1 = \min\left\{\frac{b_s-(a_s,x^{(0)})}{(a_s,C^{(0)})} \,\middle|\, \begin{matrix} s \in \bar{M}^{(0)} \text{ and } (a_s,C^{(0)}) < 0 \\ s\in M\backslash\bar{M}^{(0)} \text{ and } (a_s,C^{(0)}) \geqslant 0\end{matrix}\right\} \tag{5.14}$$

Let $s_1 \in M$ be the value of an index defined by the quantity τ_1. This means that according to (5.14)

$$\tau_1 = \frac{b_{s_1}-(a_{s_1},x^{(0)})}{(a_{s_1},C^{(0)})} \tag{5.15}$$

and there are two cases possible: $s_1 \in \bar{M}^{(0)}$ and $s_1 \in M\backslash\bar{M}^{(0)}$. For $s_1 \in \bar{M}^{(0)}$ the hodograph of $\bar{x}^{(0)}(t)$ intersects one of the hyperplanes $(a_{s_1},x) = b_{s_1}$ of the set $\bar{M}^{(0)}$ and in this case

$$y_{s_1}(x^{(0)}+\tau C^{(0)})\begin{cases} > 0 & \text{for } \tau < \tau_1 \\ = 0 & \text{for } \tau \geqslant \tau_1.\end{cases} \tag{5.16}$$

For $s_1 \in M\backslash\bar{M}^{(0)}$ the hodograph intersects one of the hyperplanes $(a_{s_1},x) = b_{s_1}$ of the set $M\backslash\bar{M}^{(0)}$ and in this case

$$y_{s_1}(x^{(0)}+\tau C^{(0)})\begin{cases} = 0 & \text{for } \tau \leqslant \tau_1 \\ > 0 & \text{for } \tau > \tau_1.\end{cases} \tag{5.17}$$

In both cases $x^{(1)} = x^{(0)} + \tau_1 C^{(0)}$ is a corner point of the hodograph of $\bar{x}(t)$ and we proceed to the problem of finding the following straight line segment of the hodograph. From (5.10) it follows that the increase t_1 of the penalty parameter leading to the point $x^{(1)} = \bar{x}(q^{(0)}+t_1)$ is equal to

$$t_1 = q^{(0)} \frac{\tau_1}{1-\tau_1}, \tag{5.18}$$

and the vector $x^{(1)} = \bar{x}(q^{(1)})$, where $\bar{x}(q^{(1)})$ is the solution of problem (5.2) for

$$q = q^{(1)} = q^{(0)}+t_1 = q^{(0)} \frac{1}{1-\tau_1}. \tag{5.19}$$

We now proceed by defining the set $\bar{M}_1$ by the conditions

$$\bar{M}^{(1)} = \begin{cases} \bar{M}^{(0)}\backslash s_1 & \text{if } s_1 \in \bar{M}^{(0)} \\ \bar{M}^{(0)} \cup s_1 & \text{if } s_1 \in M\backslash\bar{M}^{(0)}. \end{cases} \tag{5.20}$$

Thus finding the next straight line segment of the hodograph leads to the determination of the direction vector $C^{(1)}$ satisfying the system of linear equations

$$\sum_{s\in\bar{M}^{(1)}} a_{si}(a_s,C^{(1)}) = -\sum_{s\in\bar{M}^{(1)}} a_{si}y_s(x^{(1)}), \tag{5.21}$$

$$i = 1,\ldots,n.$$

The next straight line segment of the hodograph is determined by the formula

$$\bar{x}^{(1)} = x^{(1)} + \tau^{(1)}(t)C^{(1)}, \tag{5.22}$$

where, analogously as in (5.10)

$$\tau^{(1)}(t) = \frac{t}{q^{(1)}+t} \tag{5.23}$$

The quantity τ_2 is defined analogously to (5.13) by the condition

$$\tau_2 = \max\left\{(a_s,x^{(1)})+\tau(a_s,C^{(1)})\begin{cases} \geqslant b_s & \text{for } s \in \bar{M}^{(1)} \\ < b_s & \text{for } s \in M\backslash\bar{M}^{(1)} \end{cases}\right. \tag{5.24}$$

or the analogue of (5.14)

$$\tau_2 = \min\left\{\frac{b_s-(a_s,x^{(1)})}{(a_s,C^{(1)})} \;\middle|\; \begin{array}{ll} s \in \bar{M}^{(1)} & \text{and } (a_s,C^{(1)}) \leqslant 0 \\ s \in M\backslash\bar{M}^{(1)} & \text{and } (a_s,C^{(1)}) > 0 \end{array}\right\}.$$

Therefore $\bar{x}(t_2) = \bar{x}(q^{(1)}+t_2) = \bar{x}(q^{(2)})$, hwere

$$t_2 = q^{(1)} \frac{\tau_2}{1-\tau_2},\; q^{(2)} = q^{(1)} \frac{1}{1-\tau_2} = q^{(0)} \frac{1}{(1-\tau_1)(1-\tau_2)}\cdot \tag{5.25}$$

In this manner the next corner point of the hodograph is the point given by the vector

$$x^{(2)} = x^{(1)} + \tau_2 C^{(1)}. \tag{5.26}$$

It is clear how one should continue, so that the hodograph method consists of the construction of a finite sequence of vectors $x^{(0)},x^{(1)},\ldots$ defined by recursive relations

$$x^{(\alpha+1)} = x^{(\alpha)} + \tau_{\alpha+1}C^{(\alpha)}, \quad \alpha = 0,1,\ldots, \tag{5.27}$$

where the direction vectors $C^{(\alpha)}$ are solutions of the system of linear equations

$$\sum_{s\in\bar{M}^{(\alpha)}} a_{si}(a_s,C^{(\alpha)}) = -\sum_{s\in\bar{M}^{(\alpha)}} a_{si}y_s(x^{(\alpha)}),\; i = 1,\ldots,n, \tag{5.28}$$

and the scalar $\tau_{\alpha+1}$, the solution of the simple problem

$$\tau_{\alpha+1} = \max\left\{\tau \;\middle|\; (a_s,x^{(\alpha)})+\tau(a_s,C^{(\alpha)})\left\{\begin{array}{ll} \geqslant b_s, & s \in \bar{M}^{(\alpha)} \\ < b_s, & s \in M\backslash\bar{M}^{(\alpha)} \end{array}\right.\right\}, \tag{5.29}$$

can be calculated by the formula

$$\tau_{\alpha+1} = \min\left\{\frac{b_s-(a_s,x^{(\alpha)})}{(a_s,C^{(\alpha)})} \;\middle|\; \begin{array}{ll} s \in \bar{M}^{(\alpha)} & \text{and } (a_s,C^{(\alpha)}) \leqslant 0 \\ s \in M\backslash\bar{M}^{(\alpha)} & \text{and } (a_s,C^{(\alpha)}) > 0 \end{array}\right\},$$

and the set $\bar{M}^{(\alpha)}$ is defined by the conditions

$$\bar{M}^{(\alpha)} = \begin{cases} \bar{M}^{(\alpha-1)} \setminus s_\alpha & \text{for } s_\alpha \in \bar{M}^{(\alpha-1)} \\ \bar{M}^{(\alpha-1)} \cup s_\alpha & \text{for } s_\alpha \in M \setminus \bar{M}^{(\alpha-1)}, \end{cases} \tag{5.30}$$

where s_α is the index from the set of inequalities of problem (5.29) which for $\tau = \tau_{\alpha+1}$ turns into an equality and

$$\tau_{\alpha+1} = \frac{b_{s_\alpha} - (a_{s_\alpha}, x^{(\alpha)})}{(a_{s_\alpha}, C^{(\alpha)})} . \tag{5.31}$$

The solution of problem (5.1) terminates at a finite $\alpha = \alpha^*$ namely when $\bar{M}^{(\alpha^*-1)} = M^*$ is the set of constraints $(a_s, x) \leqslant b_s$ which turn into equalities at the optimal point x^* of problem (5.1). The vector x^* therefore is determined by the equality

$$x^* = x^{(\alpha^*-1)} + \tau_{\alpha^*} C^{(\alpha^*-1)} \tag{5.32}$$

Therefore the set M^* of restrictions which are active in the optimal solution is defined by the condition

$$M^* = \left\{ s \,\middle|\, \frac{b_s - (a_s, x^{(\alpha^*-1)})}{(a_s, C^{(\alpha^*-1)})} = \tau_{\alpha^*} \right\} .$$

It is useful to remark that the system of linear equations (5.28) which determines the direction vector $C^{(\alpha)}$ expresses the minimum conditions for the quadratic function and vector $C^{(\alpha)}$ connected with the solution $\xi^{(\alpha)}$ of the problem

$$\min \sum_{s \in \bar{M}^{(\alpha)}} [(a_s, \xi) - b_s]^2 \tag{5.33}$$

by the equation

$$C^{(\alpha)} = \xi^{(\alpha)} - x^{(\alpha)} . \tag{5.34}$$

Thus to find the direction vector $C^{(\alpha)}$ it may turn out convenient to solve the unconstrained minimization problem (5.33), using, for example, the conjugate gradient method which is finite for such problems. In virtue of equality (5.34) the iterative relations (5.27) take the form

$$x^{(\alpha+1)} = (1-\tau_{\alpha+1})x^{(\alpha)} + \tau_{\alpha+1}\xi^{(\alpha)}, \tag{5.35}$$

where

$$\tau_{\alpha+1} = \min\left\{\frac{b_s-(a_s,x^{(\alpha)})}{(a_s,x^{(\alpha)})+(a_s,\xi^{(\alpha)})}\;\middle|\; \begin{array}{ll} s \in \bar{M}^{(\alpha)} & \text{and } (a_s,c^{(\alpha)}) < 0 \\ s \in M\backslash\bar{M}^{(\alpha)} & \text{and } (a_s,c^{(\alpha)}) > 0 \end{array}\right\}$$

Now let us turn to problem (5.29) and satisfy ourselves that its solution $\tau_{\alpha+1}$ exists and fulfuls the condition $0 < \tau_{\alpha+1} \leqslant 1$. The correctness of this assertion will have been proved if among the hyperplanes $(a_s,x) = b_s$, $s \in \bar{M}^{(\alpha)}$ there exists at least one separating the points $x^{(\alpha)}$ and $\xi^{(\alpha)}$ (see Figure 5.1). In fact by the definition of the set $\bar{M}^{(\alpha)}$

$$y(x^{(\alpha)}) = (a_s,x^{(\alpha)}) - b_s > 0,\ s \in \bar{M}^{(\alpha)} \tag{5.36}$$

and consequently the hyperplane $(a_s,x) = b_s$ from the set $\bar{M}^{(\alpha)}$ which separates $x^{(\alpha)}$ and $\xi^{(\alpha)}$ will be the one for which the following condition holds

$$(a_s,\xi^{(\alpha)}) - b_s \leqslant 0. \tag{5.37}$$

The following assertion holds:

THEOREM 5.1. <u>If the linear programming problem</u> (5.1) <u>is solvable, then for each</u> $\alpha = 0,1,\ldots$ <u>the solution</u> $\xi^{(\alpha)}$ <u>of problem</u> (5.33) <u>satisfies the following condition: either</u> $(a_s,\xi^{(\alpha)}) - b_s = 0$ <u>for all</u> $s \in \bar{M}^{(\alpha)}$ <u>or among the quantities</u> $(a_s,\xi^{(\alpha)}) - b_s$ <u>there exists at least one negative one. Moreover if the first case applies, that is</u> $(a_s,\xi^{(\alpha)}) - b_s = 0$ <u>for all</u> $s \in \bar{M}^{(\alpha)}$ <u>and none of the hypersurfaces</u> $(a_s,x) = b_s$, $s \in M\backslash\bar{M}^{(\alpha)}$ <u>separates the points</u> $x^{(\alpha)}$ <u>and</u> $\xi^{(\alpha)}$ <u>then</u> $\xi^{(\alpha)} = x^*$ <u>the optimal vector of problem</u> (5.1).

Proof. Let $\Omega = \{x \mid (a_s,x) < b_s,\ s \in M\}$ and $\Omega^{(\alpha)} = \{x \mid (a_s,x) \leqslant b_s,\ s \in \bar{M}^{(\alpha)}\}$. From $M \supset \bar{M}^{(\alpha)}$ it follows that $\Omega^{(\alpha)} \supset \Omega$ and if problem (5.1) is solvable, then $\Omega \neq \emptyset$

and $\Omega^{(\alpha)} \neq \emptyset$ and consequently the system of linear inequalities

$$(a_s,x) \leqslant b_s, \quad s \in \bar{M}^{(\alpha)}. \tag{5.38}$$

is solvable. In virtue of the alternative theorems (see Section 2.3) the following system is then not solvable

$$\begin{aligned} &\sum_{s\in\bar{M}^{(\alpha)}} a_{si}w_s = 0, \quad i = 1,\ldots,n, \\ &w_s \geqslant 0, \quad s \in \bar{M}^{(\alpha)}, \\ &\sum_{s\in\bar{M}^{(\alpha)}} b_s w_s < 0. \end{aligned} \tag{5.39}$$

Now consider the first case when the system

$$(a_s,\xi) = b_s, \quad s \in \bar{M}^{(\alpha)} \tag{5.40}$$

is inconsistent to that consequently

$$\sum_{s\in\bar{M}^{(\alpha)}} [(a_s,\xi^{(\alpha)})-b_s]^2 > 0. \tag{5.41}$$

In this case there exists among the quantities

$$V_s^{(\alpha)} = (a_s,\xi^{(\alpha)}) - b_s, \quad s \in \bar{M}^{(\alpha)} \tag{5.42}$$

at least one which is different from zero. The minimum conditions for problem (5.33) have the form

$$\sum_{s\in\bar{M}^{(\alpha)}} V_s^{(\alpha)} a_{si} = 0, \quad i = 1,\ldots,n. \tag{5.43}$$

Now suppose that the assertion of the theorem is not true and that vector $V^{(\alpha)}$ is semi positive. Then multiplying equations (5.43) respectively with $\xi_1^{(\alpha)},\ldots,\xi_n^{(\alpha)}$ and summing we obtain

$$\sum_{s\in\bar{M}^{(\alpha)}} V_s^{(\alpha)}(a_s,\xi^{(\alpha)}) = 0. \tag{5.44}$$

On the other hand from (5.42) and the assumption $V^{(\alpha)} \geqslant 0$ it follows that

$$(a_s,\xi^{(\alpha)}) \begin{cases} > b_s & \text{for } V_s^{(\alpha)} > 0 \\ = b_s & \text{for } V_s^{(\alpha)} = 0. \end{cases} \tag{5.45}$$

From (5.44) and (5.45) we obtain the inequality

$$\sum_{s\in\bar{M}^{(\alpha)}} b_s V_s^{(\alpha)} < 0 \tag{5.46}$$

Setting $W_s = V_s^{(\alpha)}$ in system (5.39) we can see that from the hypothesis $V^{(\alpha)} \geqslant 0$, conditions (5.43) and (5.46) there follows the solvability of system (5.39). The contradiction thus obtained means that the hypothesis $V^{(\alpha)} \geqslant 0$ is not true and that consequently there is among the quantities $V_s^{(\alpha)}$, $s \in \bar{M}^{(\alpha)}$ at least one negative one. If nevertheless for some α the vector $V^{(\alpha)}$ is semipositive then, in virtue of the theorems on alternatives (see Section 2.3), it follows that the system (5.38) is unsolvable and also problem (5.1).

It remains to consider the case where $V^{(\alpha)}$ is the zero vector and there does not exist among the hyperplanes $(a_s,x) = b_s$, $s \in M\backslash\bar{M}^{(\alpha)}$ one which separates the point $x^{(\alpha)}$ and $\xi^{(\alpha)}$. In that case $\tau_{\alpha+1} = 1$ and $\xi^{(\alpha)} = x^*$ the optimal vector for problem (5.1). Indeed in the case under considera consideration

$$\tau_{\alpha+1} = \tau^{(\alpha)}(\infty) = 1 \tag{5.47}$$

and, consequently, the vector $\xi^{(\alpha)} = \lim_{\tau\to\infty} \bar{x}(q^{(\alpha)}+t)$ where $\bar{x}(q^{(\alpha)}+t)$ is the solution of the problem

$$\max\{(p,x)-\tfrac{1}{2}(q^{(\alpha)}+t)\sum_{s\in M} y_s^2(x)\} \tag{5.48}$$

It is known that $\lim_{t\to\infty} \bar{x}(q^{(\alpha)}+t) = x^*$. This proves the theorem.

Using equality (5.34) it is easy to obtain formulas for the calculation of the quantities $\tau_1,\tau_2,\ldots$

$$\tau_{\alpha+1} = \min\left\{\frac{z_s(x^{(\alpha)})}{z_s(x^{(\alpha)})-z_s(\xi^{(\alpha)})} \;\middle|\; \begin{array}{l} s \in \bar{M}^{(\alpha)}, \quad z_s(\xi^{(\alpha)}) < 0 \\ s \in M\backslash\bar{M}^{(\alpha)}, z_s(\xi^{(\alpha)}) > 0 \end{array}\right\},$$

(5.49)

where

$$z_s(x) = (a_s,x) - b_s. \tag{5.50}$$

5.3. Solution of the dual problem

The hodograph method turns out to be a method which simultaneously solves the primal and the dual linear programming problem. This comes about because the sequence of vectors $x^{(0)},x^{(1)},\ldots$ corresponds to an equally finite sequence of m-dimensional vectors $w^{(0)},w^{(1)},\ldots$ which are feasible vectors for the problem

$$\min\{(b,w) \mid \sum_{s\in M} a_{si}w_s = p_i,\ i = 1,\ldots,n,\ w_s \geqslant 0,\ s \in M\},$$

(5.51)

the dual problem of (5.1), and the vector $w^{(\alpha^*-1)} = w^*$ is an optimal vector for problem (5.51). Indeed the vector $x^{(\alpha)}$ is the solution of the system of equations

$$q^{(\alpha)} \sum_{s\in M} a_{si}y_s(x^{(\alpha)}) = p_i,\ i = 1,\ldots,n, \tag{5.52}$$

where according to (5.3), $y_s(x^{(\alpha)}) \geqslant 0$, $s \in M$. Consequently the vector $w^{(\alpha)}$ consisting of the components

$$w_s^{(\alpha)} = q_s^{(\alpha)} y_s(x^{(\alpha)}),\ s \in M,\ \alpha = 1,2,\ldots \tag{5.53}$$

is a feasible vector for the dual problem (5.51). Further because $M^{(\alpha^*-1)} = M^*$ we have

$$y_s(x^{(\alpha^*-1)})\begin{cases} > 0 & \text{for} \quad s \in M^* \\ = 0 & \text{for} \quad s \in M\backslash M^*. \end{cases} \tag{5.54}$$

It follows that the quantities

$$w_s^* = q^{(\alpha^*-1)} y_s(x^{(\alpha^*-1)}), \quad s \in M^*, \tag{5.55}$$

are the components of an optimal vector for problem (5.51).

It is easy to check that the value of the parameter $q^{(\alpha)}$ is determined by the formula

$$q^{(\alpha)} = \frac{q^{(0)}}{(1-\tau_1)(1-\tau_2)\ldots(1-\tau_\alpha)}. \tag{5.56}$$

The algorithm for the hodograph method consists of an initial stage and then a finite sequence of identical steps. It is necessary to remark on the property of the algorithm that the objective function (p,x) and the penalty parameter q only intervene in the first initial problem step and play no role in the later steps. The calculation of the quantities $q^{(\alpha)}$, $\alpha = 1,2,\ldots\alpha^*-1$ may be done after one has obtained the optimal vector x* and these quantities are only necessary to find the finite sequence (5.53) of feasible vectors and the optimal vector (5.55) for the problem (5.51) dual to the problem (5.1). We also remark that an extension of the hodograph method to linear programming problems with mixed constraints of the form

$$\max\left\{(px)\,\middle|\,(a_s x)\begin{cases}\leqslant b_s & \text{for} \quad s \in M_1\\ = b_s & \text{for} \quad s \in M_2\end{cases}\right\} \tag{5.57}$$

presents no difficulties and that in that case for each α, $M_2 \subset \bar{M}^{(\alpha)}$.

5.4. Results of numerical experiments

Numerical experiments with the hodograph method were performed on the calculating machine EC-1040. The test programs were written in PL/1. The program to solve a linear programming problem with constraints in the form of inequalities contained 256 operations and the program for solving a problem in canonical form 242 operations.

To solve the unconstrained minimization problems the method of conjugate directions was used. To solve the problem (5.2) a strategy of searching the constraints was followed, according to which the process of solving problem (5.2) divided itself into cycles of maximizations with the complexity of solving a problem of Chebysev type (5.33).

The test problems were generated by means of a list of random numbers. The general form of these problems was the following

$$\max\{(p,x)\,|\,(a_s,x) \leqslant b_s,\ s \in M,\ x_i \geqslant 0,\ i \in N\} \tag{5.58}$$

The experiments were performed with the following three examples of dimensions

1) $m = n = 15$,

2) $m = n = 20$,

3) $m = n = 25$.

For the solution by means of the first program all problems were transformed into a form with constraints of inequality type only and for solution by means of the second program into canonical form.

In Table (5.1) are listed the results of solving the examples by means of the first and second programs. The symbols have the following meanings:

k_1 = number of cycles of maximalizations to solve problem (5.2);

k_2 = number of problems (5.33) to solve;

r_Σ = total number of iterations for the method of conjugate gradients to solve the problems by the first program;

l_1 = number of cycles of maximalizations to solve problem (5.2);

l_2 = number of problems (5.33) to solve;

S_Σ = total number of iterations of the method of conjugate gradients to solve the problems by the second program.

TABLE 5.1

Problem number	k_1	k_2	r_Σ	l_1	l_2	S_Σ
1	3	3	130	3	3	68
2	6	5	227	5	5	139
3	6	7	287	6	7	152

A comparison of the results shows that the parameters k_1 and l_1 and k_2 and l_2 practically coincide but the total number of iterations of the method of conjugate gradients is significantly less in the second case. It is also interesting to consider the dependence of the computational complexity of the problem in terms of the values of the initial penalty parameters $q^{(0)}$. For this reason the same problem (5.58) was solved five times for different values of $q^{(0)}$. The problem had m = 15 and n = 10. The results of solving this problem by means of the first program are listed in Table 5.2. The quantities k_1, k_2, r_Σ have the previous meaning.

r_1 = the number of iterations of the method of conjugate gradients required for all maximalization cycles for solving problem (5.2).

r_2 = number of iterations required for solving all problems of Chebysev type (5.33).

TABLE 5.2.

$q^{(0)}$	k_1	r_1	k_2	r_2	r_Σ
10^{-3}	4	50	2	23	73
3×10^{-4}	3	44	4	48	92
6×10^{-5}	2	33	6	65	98
3×10^{-5}	2	24	7	75	99
10^{-5}	2	21	7	70	91

The results say that with diminishing penalty parameter the computational complexity of solving (5.2) decreases while the number of problems (5.33) to solve increases.

NOTE

1. This trajectory bears the name of the hodograph of the vector $\bar{x}(q)$, hence the title of this chapter.

Chapter VI

THE METHOD OF DISPLACEMENT OF ELASTIC CONSTRAINTS

6.1. Introduction

The present chapter contains an exposition of numerical methods of calculating solutions to linear programming problems which are based on the idea of displacing elastic constraints. A general property of these methods is the optimality conditions for finite values of the penalty para-optimality conditions or finite values of the penalty parameter which defines the degree of resilience (elasticity) of the constraints. The essentials of the methods expounded below consist in the following.

A linear programming problem of the form

$$\max\{(p,x)\,|\,(a_s,x) \leqslant b_s,\ s = 1,\dots,m\} \tag{6.1}$$

is put in correspondence with an unconstrained maximization problem

$$\max\{(p,x) - \tfrac{1}{2}\sum_{s=1}^{m} q_s z_s^2(x,\beta)\}, \tag{6.2}$$

where $q_1,\dots,q_m$ are nonnegative and $\beta = (\beta_1,\dots,\beta_m)$ are positive parameters with $\beta_s < b_s$, $s = 1,\dots,m$ and

$$z_s(x,\beta) = \max[0,(a_s,x)-\beta_s]. \tag{6.3}$$

If $\bar{x}(q)$ is a solution of problem (6.2), which as we know will be an approximate solution of the problem

$$\max\{(p,x)\,|\,(a_s,x) \leqslant \beta_s,\ s = 1,\dots,m\}, \tag{6.4}$$

then, as we know (see 2.4) the vector $\bar{x}(q)$ will be an optimal vector for the following linear programming problem

$$\max\{(p,x)\,|\,(a_s,x) \leqslant \beta_s + z_s(\bar{x}(q),\beta)\},\ s = 1,\dots,m \tag{6.5}$$

From (6.5) there result several different approaches to

formulate methods for the numerical solution of (6.1). Indeed if there exist quantities $\beta_1^*,\ldots,\beta_m^*$, $q_1^*,\ldots,q_m^*$ which satisfy the conditions

$$\beta_s^* + z_s(\bar{x}(q^*),\beta^*) = b_s, \quad s = 1,\ldots,m \tag{6.6}$$

then a solution of (6.1) becomes a solution of the unconstrained maximization problem (6.2) for $q = q^*$, $\beta = \beta^*$ and therefore $\bar{x}(q^*) = x^*$ the optimal solution of (6.1). It is easy to see that we have available a redundant set of 2m parameters to satisfy the m Equations (6.6) and that opens up great possibilities for numerical realizations of the idea of displacing elastic constraints. For example we can assume that the quantities $q_1,\ldots,q_m$ are given and the quantities $\beta_1^*,\ldots,\beta_m^*$ to be determined, or, inversely we can consider the problem of choosing the parameters $q_1^*,\ldots,q_m^*$ of elasticity of the constraints $(a_s,x) - \beta_s^* \leqslant 0$ in such a way that the conditions (6.6) are satisfied. Obviously combined algorithms are also possible. It is important to note that the idea of displacing elastic constraints turns out to be a basic method for solving dynamical optimal control problems and the last chapter of this book will be devoted to that.

6.2. The first algorithm

In problem (6.2) let $q = q^*$ be an arbitrary given positive vector and let us search for the quantities $\beta_1^*,\ldots,\beta_m^*$ or $c_1^*,\ldots,c_m^*$ which are connected to the latter by the conditions

$$c_s^* = b_s - \beta_s^*, \quad s = 1,\ldots,m. \tag{6.7}$$

The maximum conditions for problem (6.2) have the form

$$p_i = \sum_{s=1}^{m} q_s^* a_{si} z_s(\bar{x}(q^*), \beta^*), \quad i = 1,\ldots,n. \tag{6.8}$$

Recall that β^* is that vector for which $\bar{x}(q^*) = x^*$ and compare (6.8) to the conditions for an optimal solution for problem (6.1), which are

$$p_i = \sum_{s=1}^{m} w_s^* a_{si}, \quad i = 1,\ldots,n, \tag{6.9}$$

where w^* is an optimal vector for the dual problem. This leads to the equation

$$w_s^* = q_s^* z_s(\bar{x}(q^*),\beta^*), \quad s = 1,\dots,m \tag{6.10}$$

From the conditions

$$w_s^* \begin{cases} > 0, & s \in M^* \\ = 0, & s \in M\backslash M^* \end{cases} \qquad (a_s,x^*) \begin{cases} = b_s, & s \in M^* \\ < b_s, & s \in M\backslash M^* \end{cases}$$

and Equation (6.7) there follow the equations

$$z_s(\bar{x}(q^*),\beta^*) = z_s(x^*,\beta^*) = \begin{cases} c_s^*, & s \in M \\ 0, & s \in M\backslash M^* \end{cases} \tag{6.11}$$

$$w_s^* = q_s^* c_s^*, \quad s = 1,\dots,m. \tag{6.12}$$

We now consider the sequence of m-dimensional vectors $c^{(0)},c^{(1)},\dots$ defined by the recursive formulas

$$c^{(0)} = 0, \quad c^{(\nu+1)} = z(x^{(\nu)}(q^*),\ \beta^{(\nu)}), \quad \nu = 1,2,\dots \tag{6.13}$$

where $x^{(\nu)}(q^*)$ is a solution of the problem

$$\max\{(p,x) - \tfrac{1}{2}\sum_{s=1}^{m} q_s z_s^2(x,\beta_s^{(\nu)})\}. \tag{6.14}$$

The sequence (6.13) is finite and the following assertion holds.

THEOREM 6.1. <u>The sequence</u> (6.13) <u>converges and there exists a real number</u> ν^* <u>such that</u> $c^{(\nu)} = c^*$ <u>for</u> $\nu \geqslant \nu^*$ <u>and</u> $x^{(\nu)} = x^*$ <u>for</u> $\nu > \nu^*$.

<u>Proof</u>. In the point $x^{(\nu)}(q^*)$ the following condition holds

$$p_i = \sum_{s=1}^{m} a_{si} q_s^* z_s(x^{(\nu)}(q^*),\beta_s^{(\nu)}), \quad i = 1,\dots,n, \quad \nu = 0,1,\dots \tag{6.15}$$

and in virtue of (6.13) this takes the form

$$p_i = \sum_{s=1}^{m} a_{si} q_s^* c_s^{(\nu)}, \quad i = 1,\dots,n; \ \nu = 0,1,\dots\ . \tag{6.16}$$

From condition (6.9) there follow the equalities

$$\sum_{s=1}^{m} a_{si}(q_s^* c_s^{(\nu)} - w_s^*) = 0, \; i = 1,\dots,n, \; \nu = 0,1,\dots \quad (6.17)$$

Now multiply the equations of (6.17) respectively with the components of $x^{(\nu-1)}$ and those of x^* and sum the result. This yields

$$\sum_{i=1}^{n} x_i^{(\nu-1)} \sum_{s=1}^{m} a_{si}(q_s^* c_s^{(\nu)} - w_s^*) = 0 \quad (6.18)$$

$$\sum_{i=1}^{n} x_i^* \sum_{s=1}^{m} a_{si}(q_s^* c_s^{(\nu)} - w_s^*) = 0, \; \nu = 0,1,\dots . \quad (6.19)$$

Subtracting (6.19) from (6.18) we arrive at the condition

$$\sum_{s=1}^{m} (q_s^* c_s^{(\nu)} - w_s^*)[(a_s, x^{(\nu-1)}) - (a_s, x^*)] = 0. \quad (6.20)$$

The Equation (6.20) is evidently equivalent to the equation

$$\sum_{s=1}^{m} (q_s^* c_s^{(\nu)} - w_s^*)[(a_s, x^{(\nu-1)}) - b_s + c_s^{(\nu-1)} - (a_s, x^*) + b_s - c_s^{(\nu-1)}] \quad (6.21)$$

The following relations will be necessary in what follows:

(a) One of the conditions of optimality

$$\sum_{s=1}^{m} w_s^*[(a_s, x^*) - b_s] = 0. \quad (6.22)$$

(b) A consequence of definitions (6.13) and (6.4)

$$\sum_{s=1}^{m} c_s^{(\nu)}[(a_s, x^{(\nu-1)}) - b_s + c_s^{(\nu-1)}] = \sum_{s=1}^{m} (c_s^{(\nu)})^2. \quad (6.23)$$

(c) A consequence of the properties $c_s^{(\nu)} \geqslant 0$, $(a_s, x^*) - b_s \leqslant 0$

$$\sum_{s=1}^{m} c_s^{(\nu)}[(a_s, x^*) - b_s] \leqslant 0. \quad (6.24)$$

d) A consequence of the property $w_s^* \geqslant 0$ and definition (6.13)

$$\sum_{s=1}^{m} w_s^*[(a_s, x^{(\nu-1)}) - b_s + c_s^{(\nu-1)}] \leqslant \sum_{s=1}^{m} w_s^* c_s^{(\nu)}. \quad (6.25)$$

Using conditions (6.22)-(6.25) one easily obtains from (6.21) the inequality

$$\sum_{s=1}^{m} (q_s^* c_s^{(\nu)} - q_s^* c_s^{(\nu-1)})(q_s c_s^{(\nu)} - w_s^*) \leqslant 0, \qquad (6.26)$$

which, as is easily verified, is equivalent to the inequality

$$\| q^* c^{(\nu)} - w^* \|^2 + \| q^* c^{(\nu)} - q^* c^{(\nu-1)} \|^2 \leqslant \| q^* c^{(\nu-1)} - w^* \|^2 \qquad (6.27)$$

where

$$\| u \| = \left(\sum_{s=1}^{m} u_s^2 \right)^{\frac{1}{2}}$$

Now sum the last inequalities with respect to ν from $\nu = 1$ to $\nu = k$ where k is an arbitrary natural number. One obtains

$$\sum_{\nu=1}^{k} \| q^* c^{(\nu)} - q^* c^{(\nu-1)} \|^2 \leqslant \| q^* c^{(0)} - w^* \|^2 - \| q^* c^{(k)} - w^* \|^2 \leqslant \| q^* c^{(0)} - w^* \|^2. \qquad (6.28)$$

From (6.28) it follows that the series of positive terms

$$\sum_{\nu=1}^{\infty} \| q^* c^{(\nu)} - q^* c^{(\nu-1)} \|^2$$

converges and there follows the equality

$$\lim_{\nu \to \infty} \| c^{(\nu)} - c^{(\nu-1)} \| = 0.$$

From definitions (6.13) and (6.4) there results the equation

$$c_s^* = \max\{0, (a_s, x^*) - b_s + c_s^*\},$$

from which we obtain

$$c_s^* \begin{cases} > 0, & s \in M^* \\ = 0, & s \in M \backslash M^*, \end{cases}$$

where

$$M^* = \{ s \mid (a_s, x^*) = b_s \}$$

Moreover because the sequence (6.13) converges, which in turn implies the convergence of the sequences $x^{(0)}, x^{(1)}, \ldots,$

and because of the equilibrium (maximum) condition (6.16) there exists a number $\bar{\nu}$ big enough for the following condition to hold

$$(a_s,x^{(\bar{\nu})})-b_s \begin{cases} \geqslant 0 & s \in M \\ < 0 & s \in M \setminus M^*. \end{cases}$$

If ε is an arbitrarily small positive number such that $b_s \geqslant \varepsilon + (a_s,x^*)$ for $s \in M\setminus M^*$ then there exists an integer $\bar{\nu}$ such that also $b_s \geqslant \frac{\varepsilon}{2} + (a_s,x^{(\nu)})$ for all $s \in M\setminus M^*$ and $\nu \geqslant \bar{\nu}$. This last fact means that $c_s^{(\nu+1)} \leqslant c_s^{(\nu)} - \frac{\varepsilon}{2}$, $s \in M\setminus M^*$ and $\nu \geqslant \bar{\nu}$ and from this by the condition that $c_s^{(\nu)} \geqslant 0$ there follows the existence of an integral number ν^* such that $c_s^{(\nu)} = 0$ for all $\nu \geqslant \nu^*$ and $s \in M\setminus M^*$. Therefore one has the equality $c^{(\nu)} = c^*$ for all $\nu \geqslant \nu^*$ because any of the vectors $c^{(\nu^*)}, c^{(\nu^*+1)},\ldots$ is a solution of the equilibrium equations (6.16) and these solutions are unique under the conditions $c_s^{(\nu)} = 0$ for $\nu \geqslant \nu^*$, $s \in M\setminus M^*$. This proves the theorem.

Thus the first algorithm based on displacements of elastic constraints consists in the reducing of problem (6.1) to a finite number of unconstrained maximizations of type (6.14) where

$$\beta^{(\nu)} = b-c^{(\nu)},\ c^{(0)} = 0,\ c^{(\nu)} = z(x^{(\nu-1)}(q^*),\beta^{(\nu-1)}),$$
$$\nu = 0,1,\ldots,\nu^*.$$

From this it follows that problem (6.14) leads to a problem of maximizing a simple quadratic function.

6.3. The second algorithm

In this section there will be outlined another numerical algorithm for solving linear programming problems of type (6.1). It differs from the algorithm of Section 6.2 in that now the amounts of displacements of the constraints $c_1^*,\ldots,c_m^*$ are supposed to be given positive numbers and the basis of the algorithm is a procedure for choosing the parameters $q_1^*,\ldots,q_m^*$ of elasticity of the constraints which have been displaced in the amounts $c_1^*,\ldots,c_m^*$. The algorithm described below also follows from the analogy between linear programming problems and equilibrium problem of physical systems which was described in Chapter I (see Figures 1.8, 1.10).

Consider the unconstrained maximization problem (6.2) in which $c_1,\ldots,c_m$ are given positive numbers. The nonnegative vector $q^* = (q_1^*,\ldots,q_m^*)$ which we are looking for will be called optimal if for $q = q^*$ the solutions to (6.1) and (6.2) coincide. Let x^* be a solution of problem (6.1) and $\bar{x}(q)$ a solution of (6.2) for some $q > 0$. We define a partition of the set $M = (1,\ldots,m)$ in two subsets

$$M_1(q) = \{s|\bar{z}_s(\bar{x}(q)) > 0\},\ M_2(q) = \{s|\bar{z}_s(\bar{x}(q)) = 0\} \tag{6.29}$$

where

$$\bar{z}_s(x) = \max[0,(a_s,x)-b_s+c_s] \tag{6.30}$$

For $q = q^*$ the corresponding partition is denoted by M_1^* and M_2^* by setting

$$\begin{aligned} M_1^* &= \{s|\bar{z}_s(\bar{x}(q^*)) = \bar{z}_s(x^*) > 0\} \\ M_2^* &= \{s|\bar{z}_s(\bar{x}(q^*)) = \bar{z}_s(x^*) = 0\} \end{aligned} \tag{6.31}$$

If $M^* = \{s|(a_s,x^*) = b_s\}$ then clearly, $M_1^* \supseteq M^*$ because $\bar{z}_s(x^*) = c_s > 0$, $s \in M^*$. The vector q^* for which $\bar{x}(q^*) = x^*$ is generally speaking not unique because the maximum conditions for problem (6.2) are the system of equations

$$p_i = \sum_{s=1}^{m} a_{si}q_s^*\bar{z}_s(x^*) = \sum_{s\in M_1^*} a_{si}q_s^*\bar{z}_s(x^*),\ i = 1,\ldots,n \tag{6.32}$$

If, assuming that x^* is a known vector, we find the vector q^* as a solution of (6.32) then for $M_1^* \supset M^*$ the solution of this system is non unique. This difficulty is easily avoided by supplementing the system (6.32) with the conditions

$$q_s^* = 0 \text{ for } s \in M_1^* \backslash M^* \qquad (6.33)$$

Then the equilibrium conditions (6.32) take the form

$$p_i = \sum_{s \in M^*} a_{si} q_s^* \bar{z}_s(x^*), \quad i = 1,\dots,n$$

or, because of the condition $\bar{z}_s(x^*) = c_s^*$ for $s \in M^*$

$$p_i = \sum_{s \in M^*} a_{si} q_s^* c_s, \quad i = 1,\dots,n. \qquad (6.34)$$

It follows that the quantities

$$w_s = \begin{cases} q_s^* c_s, & s \in M^* \\ 0, & s \in M \backslash M^* \end{cases} \qquad (6.35)$$

are the components of an optimal vector for the linear programming problem dual to problem (6.1). Below we shall make sure that a procedure for constructing the sequence $q^{(0)}, q^{(1)}, \dots$ has a limit vector q^* such that condition (6.33) holds and such that the sequence $q^{(0)}, q^{(1)}, \dots$ belongs to a sequence $\bar{x}^{(0)}, \bar{x}^{(1)}, \dots$ of solution of problems of type (6.2) and a sequence of vectors $w^{(0)}, w^{(1)}, \dots,$ $w^{(\nu)} = (q_1^{(\nu)} c_1, \dots, q_m^{(\nu)} c_m)$ satisfying the conditions

$$\lim_{\nu \to \infty} \bar{x}^{(\nu)} = x^*, \quad \lim_{\nu \to \infty} w^{(\nu)} = w^* \qquad (6.36)$$

A fundamental role in the construction of the algorithm will be played by a physical model of problem 6.1 with which the reader is already acquainted and which is analogous to the model sketched in Figure 1.8 (see Section 1.4). It is only necessary to keep in mind the sole difference that the coordinates of the fixed pistons of the model must be equal to the quantity $b_s - c_s$ and not b_s as is indicated in Figure 1.8.

Let $q^{(0)} > 0$ be an arbitrary positive vector, and $\bar{x}^{(0)}$ a solution of problem (6.2) for $q = q^{(0)}$. The quantities $\bar{z}_s(x^{(0)})$ define a partition of the set M into two subsets

$$M_1^{(0)} = \{s \mid \bar{z}_s(x^{(0)}) > 0\}, \quad M_2^{(0)} = \{s \mid \bar{z}_s(x^{(0)}) = 0\}.$$

Define the vector $q^{(1)}$ by means of the following simple formulas

$$q_s^{(1)} = \begin{cases} q_s^{(0)} \dfrac{\bar{z}_s(x^{(0)})}{c_s}, & s \in M_1^{(0)} \\ q_s^{(0)}, & s \in M_2^{(0)} \end{cases}$$

In the following the sequence of vectors $q^{(0)}, q^{(1)}, \ldots$ is defined by the recursive formulas

$$q_s^{(\nu+1)} = \begin{cases} q_s^{(\nu)} \dfrac{\bar{z}_s(x^{(0)})}{c_s}, & s \in M_1^{(\nu)} \\ q_s^{(\nu)}, & s \in M_2^{(\nu)}. \end{cases} \tag{6.37}$$

where $x^{(\nu)}$ is the solution of problem (6.2) for $q = q^{(\nu)}$ and

$$M_1^{(\nu)} = \{s \,|\, \bar{z}_s(x^{(\nu)}) > 0\}, \quad M_2^{(\nu)} = \{s \,|\, \bar{z}_s(x^{(\nu)}) = 0\} \tag{6.38}$$

<u>Remark</u>. The defining choice $q_s^{(\nu+1)} = q_s^{(\nu)}$ for $s \in M_2^{(\nu)}$ is not necessarily required. The only thing that is necessary is that $q_s^{(\nu+1)} > 0$ for $s \in M_2^{(\nu)}$, and consequently, as the components $q_s^{(\nu+1)}$, $s \in M_2^{(\nu)}$ of the vector $q^{(\nu+1)}$ one can choose arbitrary positive numbers. Only the choice $q_s^{(\nu+1)} = 0$ is not admissible, because then in the following step the corresponding index s may enter into the set $M_1^{(\nu+1)}$.

Let us review the physical meaning of the algorithmic procedure defined by formulas (6.37). For that we turn to the physical model of problem (6.2) and consider an equilibrium state $x^{(\nu)}$ for $q = q^{(\nu)}$ of that model. In that state let us select from the model a block of communicating volumes as sketched in figure 6.1. In order that this block should remain in the same equilibrium state $x^{(\nu)}$ it is necessary by virtue of the principle of removing constraints (see Chapter IV) to replace the links between the blocks, represented by the bars, by equivalent forces $R_{s1}, \ldots, R_{sn}$. These forces are determined by the equilibrium conditions

$$a_{si}(q_s^{(-)} - q_s^{(+)}) = R_{si}, \quad i = 1, \ldots, n,$$

where

$$q_s^{(-)} - q_s^{(+)} = q_s^{(\nu)}\bar{z}_s(x^{(\nu)}).$$

Consequently

$$R_{si} = q_s^{(\nu)} a_{si}\bar{z}_s(x^{(\nu)}). \tag{6.39}$$

We remark that among the m blocks of the model there may be some such that in the state $x^{(\nu)}$ the quantities $\bar{z}_s(x^{(\nu)})$ are zero, and correspondingly $R_{si} = 0$. After choosing the block we consider the quasi static process corresponding to the change of the variable q_s from the value $q_s^{(\nu)}$ to some other value $q_s^{(\nu+1)}$ <u>without changing the exterior forces</u> $R_{s1},\ldots,R_{sn}$ defined by equations (6.39). Of course the coordinates of the pistons of the block selected can change independently of the coordinates of the pistons in the other blocks and they are necessarily denoted by quantities $x_{s1},\ldots,x_{sn}$.

Choose the quantities $q_s^{(\nu+1)}$ in such a way that in the equilibrium state $x_s^{(\nu+1)} = (x_{s1}^{(\nu+1)},\ldots,x_{sn}^{(\nu+1)})$ the conditions $\bar{z}_s(x_s^{(\nu+1)}) = c_s$ are satisfied. Because $x_s^{(\nu+1)}$ is an equilibrium state vector of a block for the same forces $R_{s1},\ldots,R_{sn}$ there must hold the equilibrium conditions

$$R_{si} = q_s^{(\nu+1)} a_{si}\bar{z}_s(x_s^{(\nu+1)})$$

or, by virtue of the condition $\bar{z}_s(x_s^{(\nu+1)}) = c_s$

$$\begin{aligned} R_{si} &= q_s^{(\nu+1)} a_{si} c_s = q_s^{(\nu)} a_{si}\bar{z}_s(x^{(\nu)}) \quad \text{for } s \in M_1^{(\nu)}, \\ R_{si} &= 0 \quad \text{for } s \in M_2^{(\nu)}. \end{aligned} \tag{6.40}$$

From (6.40) there follows formula (6.37) for $s \in M_1^{(\nu)}$ or for $\bar{z}_s(x^{(\nu)}) > 0$. In the case where $\bar{z}_s(x^{(\nu)}) = 0$ obviously no change in the quantity q_s will change the set of equilibrium states of the block and for all these cases the condition

$\bar{z}_s(x_s) = 0$ will remain valid for every value of the parameter $q_s \geq 0$. This evident physical property was expressed above by the remark that for $s \in M_2^{(\nu)}$ the quantity $q_s^{(\nu+1)} > 0$ can be arbitrarily chosen.

The division of the model of problem (6.2) into m isolated blocks and the realization of the quasi static processes by means of the change of the parameter q_s from the value $q_s^{(\nu)}$ to $q_s^{(\nu+1)}$ show themselves as a consequence of replacing the constraints $x_{si} = x_s$, $s = 1,\ldots,m$ by their reaction forces in state $x^{(\nu)}$ of the model. Therefore the equilibrium state of any of the blocks is not unique, and this applies to any state x_s which only satisfies the one condition $\bar{z}_s(x_s^{(\nu+1)}) = c_s$. This fact follows from the equilibrium conditions

$$R_{si} = q_s^{(\nu+1)} a_{si} \bar{z}_s(x^{(\nu+1)}) = q_s^{(\nu+1)} a_{si} c_s. \tag{6.41}$$

It is also clear that to restore the constraints $x_{si} = x_i$, $s = 1,\ldots,m$ requires a strictly positive amount of work of the exterior forces on the system in the case where the system of linear equations

$$\begin{aligned} &z_s(x_s^{(\nu+1)}) = c_s, \quad s = 1,\ldots,m, \\ &x_{si} = x_i, \qquad\qquad s = 1,\ldots,m;\ i = 1,\ldots,n \end{aligned} \tag{6.42}$$

is not solvable and the amount of work involved will be zero if the system (6.42) is solvable. It can easily be checked that in the latter case by virtue of the obvious condition

$$\sum_{s=1}^{m} R_{si} = p_i \quad i = 1,\ldots,n \tag{6.43}$$

each solution of the system (6.42) is also an optimal solution of problem (6.1). In the general case that system (6.42) is inconsistent the minimal amount of work by the exterior forces necessary to restore the constraints $x_{si} = x_i$ leading to a state $x^{(\nu+1)}$ of the physical model turns out to be the solution of problem (6.2) for $q = q^{(\nu+1)}$ and satisfies the maximum conditions

$$p_i = \sum_{s=1}^{m} q_s^{(\nu+1)} a_{si} \bar{z}_s(x^{(\nu+1)}), \ i = 1,\dots,n. \tag{6.44}$$

On the other hand, summing the Equations (6.40) we obtain by virtue of (6.43) that

$$p_i = \sum_{s\in M_1^{(\nu)}} q_s^{(\nu+1)} a_{si} c_s, \ i = 1,\dots,n. \tag{6.45}$$

Now subtract (6.45) from (6.44); this leads to the equation

$$\sum_{s\in M_1^{(\nu)}} q_s^{(\nu+1)} a_{si} [z_s(x^{(\nu+1)}) - c_s] +$$

$$\sum_{s\in M_2^{(\nu)}} q_s^{(\nu+1)} a_{si} z_s(x^{(\nu+1)}) = 0. \tag{6.46}$$

The last system is a system of minimum conditions for the problem

$$\min_x \tfrac{1}{2}\{ \sum_{s\in M_1^{(\nu)}} q_s^{(\nu+1)} [\bar{z}_s(x) - c_s]^2 + \sum_{s\in M_2^{(\nu)}} q_s^{(\nu+1)} z_s^2(x)\}. \tag{6.47}$$

The problem (6.47) can be considered as a problem to find the minimal sum of squares of the errors with weights $q_1^{(\nu+1)},\dots,q_m^{(\nu+1)}$ of the system of equations

$$\begin{aligned} \bar{z}_s(x) &= c_s, \ s \in M_1^{(\nu)} \\ \bar{z}_s(x) &= 0 \ , \ s \in M_2^{(\nu)} . \end{aligned} \tag{6.48}$$

Thus the sequence of problems (6.2) for $\nu = 1,2,\dots$ corresponding to the sequence of vectors $q^{(1)}, q^{(2)}, \dots$ defined by formula (6.37) is equivalent to the sequence of problems (6.47) for $\nu = 0,1,2,\dots$.

For the initial problem, determining the vector $x^{(0)}$ and the partition $M_1^{(0)}, M_2^{(0)}$ there remains problem (6.2) for $q = q^{(0)}$. It is easy to check that the system (6.48) is solvable under the conditions

$$\bar{z}_s(x(q^*)) - c_s \begin{cases} = 0, & q_s^* > 0, \ s \in M^* \\ < 0, & q_s^* = 0, \ s \in M_1^* \backslash M^* \end{cases} \tag{6.49}$$

$$\bar{z}_s(x(q^*)) = 0, \qquad q_s^* > 0, \ s \in M_2^*$$

and that its solution is an optimal vector for problem (6.1).

6.4. Combining the algorithms

A numerical difficulty with the algorithms expounded in Sections 6.1 and 6.3 is connected with the necessity of solving a sequence of problems (6.2) with discontinuous elements of the Hessian of the function to be maximized. In this section there will be described an algorithm which does not suffer from this difficulty and problem (6.1) will lead to a sequence of unconstrained maximization problems of ordinary quadratic functions with continuous partial derivatives of all orders. The algorithm is called a combined one because on the way to the solution both the quantities $c_1,\ldots,c_m$ of displacements of the constraints and the parameter $q_1,\ldots,q_m$ of elasticity of the displaced constraints will be changed.

Again let $q^{(0)}$ and $r^{(0)}$ be arbitrary given strictly positive m-dimensional vectors. Together with problem (6.2) we consider the problem

$$\max\{(p,\xi) - \tfrac{1}{2}\sum_{s=1}^{m} q_s^{(0)} y_s^2(\xi,\gamma_s^{(0)})\}, \tag{6.50}$$

where

$$y_s(\xi,\gamma_s^{(0)}) = (a_s,\xi) - \gamma_s^{(0)}, \tag{6.51}$$

$$\gamma_s^{(0)} = b_s - r_s^{(0)}, \quad r_s^{(0)} = c_s^{(0)}, \quad s = 1,\ldots,m. \tag{6.52}$$

Let $\xi^{(0)}$ be a solution of (6.50), which, as is well known [57,58], can be obtained as a result of a finite number of operations. Among the quantities $y_s(\xi^{(0)},\gamma_s^{(0)})$, $s = 1,\ldots,m$ there may be both positive and negative ones, and the initial step for this algorithm (just as for all the last ones) consists in finding quantities $\gamma_1,\dots,\gamma_m$ for which the errors in the conditions $(a_s,\xi) - \gamma_s = 0$, $s = 1,\dots,m$ will be non-negative. Let $M_1^{(0)}$ and $M_2^{(0)}$ be the subsets of indices $s \in M$ defined by the conditions

$$M_1^{(0)} = \{s|y_s(\xi^{(0)},\gamma_s^{(0)}) > 0\},$$
$$M_2^{(0)} = \{s|y_s(\xi^{(0)},\gamma_s^{(0)}) \leqslant 0\}. \tag{6.53}$$

We determine the next approximation $\gamma^{(1)}$ of the vector γ we are looking for by the equalities

$$\gamma_s^{(1)} = \begin{cases} \gamma_s^{(0)}, & s \in M_1^{(0)} \\ (a_s,\xi^{(0)}), & s \in M_2^{(0)} \end{cases} \tag{6.54}$$

and find the solution $\xi^{(1)}$ of the following problem

$$\max\{(p,\xi) - \tfrac{1}{2}\sum_{s=1}^{m} q_s^{(0)} y_s^2(\xi,\gamma_s^{(1)})\}. \tag{6.55}$$

Here, by virtue of (6.54), it is obvious that $y_s(\xi^{(0)},\gamma_s^{(1)}) \geqslant 0$, $s = 1,\dots,m$, but in the point $\xi^{(1)}$ some of the quantities $y_s(\xi^{(1)},\gamma_s^{(1)})$ may be negative. As there are negative components in the vector $y(\xi^{(1)},\gamma^{(1)})$ it is necessary to calculate the next iteration for the vector γ^* we are looking for

$$\gamma_s^{(2)} = \begin{cases} \gamma_s^{(1)}, & s \in M_1^{(1)} \\ (a_s,\xi^{(1)}), & s \in M_2^{(1)}, \end{cases}$$

where

$$M_1^{(1)} = \{s|y_s(\xi^{(1)},\gamma_s^{(1)}) > 0\},$$
$$M_2^{(1)} = \{s|y_s(\xi^{(1)},\gamma_s^{(1)}) \leqslant 0\}.$$

It is clear how to proceed and the initial stage of the algorithm leads to the solving of a sequence of problems

$$\max\{(p,\xi) - \tfrac{1}{2}\sum_{s=1}^{m} q_s^{(0)} y_s^2(\xi,\gamma_s^{(\alpha)})\}$$

$$\gamma_s^{(\alpha)} = \begin{cases} \gamma_s^{(\alpha-1)}, & s \in M_1^{(\alpha-1)} \\ (a_s,\xi^{(\alpha-1)}), & s \in M_2^{(\alpha-1)} \end{cases}$$

$$M_1^{(\alpha-1)} = \{s \mid y_s(\xi^{(\alpha-1)},\gamma_s^{(\alpha-1)}) > 0\},$$
$$M_2^{(\alpha-1)} = \{s \mid y_s(\xi^{(\alpha-1)},\gamma_s^{(\alpha-1)}) \leqslant 0\}.$$

The convergence of the sequences $\xi^{(0)},\xi^{(1)},\ldots$, $\gamma^{(0)},\gamma^{(1)},\ldots$ follows from the obvious property of the monotonic increase of the function being maximized considered as a function of m + n arguments $\xi_1,\ldots,\xi_n$, $\gamma_1,\ldots,\gamma_m$. It is also important to note that the process is finite and it could be a useful exercise for the reader to convince himself of this fact.

Denoting with

$$x^{(0)} = \lim_{\alpha\to\infty} \xi^{(\alpha)}, \quad \beta^{(0)} = \lim_{\alpha\to\infty} \gamma^{(\alpha)}, \quad c^{(0)} = b - \beta^{(0)}$$

we conclude the initial stage by calculating the starting point for the next first step, the vector $q^{(1)}$ of parameters of elasticity of constraints $(a_s,x) = \beta_s^{(0)}$, $s = 1,\ldots,m$ by the formulas (6.37)

$$q_s^{(1)} = \begin{cases} q_s^{(0)} \dfrac{y_s(x^{(0)},c_s^{(0)})}{c_s^{(0)}} & \text{for } y_s(x^{(0)},c_s^{(0)}) > 0, \\ q_s^{(0)} & \text{for } y_s(x^{(0)},c_s^{(0)}) = 0. \end{cases}$$

The next stage is essentially analogous to the previous one.

In the previous section of the present chapter we restricted our considerations of the method of displacing constraints to applications to linear programming problems. Obviously this approach must also lead to effective algorithms for the numerical solution of nonlinear programming problems. The difficulties which one could thereby encounter are caused by the fact that for nonlinear problems the hodograph is a continuous nonsmooth curve with corner points coinciding (as in the case of linear problems) with points of intersection of the hodograph with boundaries of the set of feasible vectors.

Therefore precautions are necessary, because displacements of the boundaries in the amount of the errors in an equilibrium state may lead to unstable numerical processes. Such instabilities may occur also for changes of the elasticity parameters in accordance with formulas (6.37). Various realizations of the idea of displacing the constraints can be found by the reader in papers [58-61]. It is important to remark that in a sufficiently small neighbourhood of the solution the difficulties just mentioned disappear.

Chapter VII

DECOMPOSITION METHODS FOR LINEAR PROGRAMMING PROBLEMS

7.1. Introduction

The problem of how to decompose linear programming problems and economic planning rpoblems has excited investigators and attracted their attention not only because of the necessity of solving high dimensional problems on calculating machines with limited storage space. Another no less important fact is that decomposition techniques are in essence mathematical models for the processes of control and planning, implementing the division of these functions between a central planning organisation and the remaining sections of the economy.

In this respect the problem of decomposition is connected to the problem of finding a reasonable demarcationline between centralized planning and control and the semi-autonomous remaining parts of the economic system. That is the problem of finding a reasonable planning and control structure.

The decomposition problem admits an obvious physical interpretation and the method of redundant constraints not only leads to various approaches for the construction of decomposition models but also to the construction of numerical solution algorithms of very high dimensional problems.

There is not space in this book to give a sufficiently complete survey of the large amount of work devoted to decomposition problems. The author finds himself compelled to make only a few remarks and to refer the reader to a number of treatises on the problem which also contain substantial bibliographies [2,4,10,12,14,15,24,31,35]. The first decomposition method for linear programming problems was developed by Dantzig and Wolfe [25]. Since then there has been a whole series of investigations devoted to formulation of blockwise programming as a separate branch of linear programming, see for example [24,31]. In 1962 Kornai and Liptak [35] proposed an iterative decomposition method based on a idea of divinding the constraint matrix into blocks consisting of column vectors of that matrix. In this way the original problem splits into a control centre problem consisting of allocating resources to the various blocks and a set of problems, one for each block, consisting of dual problems of determining the costs of the resources allocated. The next control centre problem reallocates resources using the information on the prices for the blocks thus obtained.

This approach can also be viewed as a sort of game between the centre and the blocks in which the strategies of the centre are matrices of admissible allocations of resources and the strategies of the blocks are admissible dual prices. Various aspects of this approach to decomposition problems are considered in many treatises, see for example [2,10,12].

An essential difficulty with the method just described is that it is necessary for the constraints for the separate blocks, defined by an admissible allocation of resources matrix, to be consistent.[1)] The construction of physical models and the study of the analogy between linear programming problems and equilibrium problems for physical models [14, 15] radically did away with these difficulties. The decomposition method developed on 1967 in [14] and elaborated in [15] has a clear physical interpretation. It turns out to be the description of a spontaneous transition process of the physical system to an equilibrium state controlled by a series of redundant constraints which replace each other cyclically. Thus the equilibrium problem at the first stage turns into a number of independent equilibrium problems of isolated physical systems obtained by imposing a number of redundant constraints of the impermeable membrane type. In the second stage the impermeable membrane type restrictions are replaced by geometric ones compatible with the state just found and one solves the very simple problem of equalizing the pressures, that is one maximizes the entropy of the system. The problems of the first stage type are problems of optimal use for the blocks of the resources allocated to them and the second stage problems are problems of allocating recources to the blocks, that is centre (control) problems. The role of the dual prices of the resources assigned to each block is played by the quantities $w_s^{(\alpha)}$, $s \in M$, the pressure differences between the volumes $V_s^{(\alpha)-}$ and $V_S^{(\alpha)+}$ in the model representing the constraints on the resources of the block in question. It is important that the $w_s^{(\alpha)}$ are in general not only the dual prices of the resources. They also contain information on the degree of incompatibility between the block resources and consequently the solution of the problem of equalizing the pressure differences leads to a new allocation of resources matrix, that is an allocation which at the same time results in equalization of the dual block prices and diminishes the incompatibility of the constraints of the blocks. The economic interpretation of this decomposition method is very fruitful and the spontaneous nature of the physical processes involved combined with the second principle of thermodynamics guarantees monotonic convergence, independently of the choice of decomposition into blocks.

Also for decomposition problems we can in this manner use spontaneous physical processes where we restrict our interference in the evolution of these processes to suitable changes in the exterior conditions. Recall (see Chapter III) that changes in the exterior conditions by means of imposing extra constraints do not affect the spontaneous nature of the processes, provided these extra constraints are stationary (that is time indepedent), and compatible with the state of the system before the constraints are imposed. This restriction does not really diminish the choices one has for redundant constraints and leaves lots of possibilities for finding inventive ways of solving decomposition problems.

7.2. Decomposition algorithms

The first algorithm. The subject of the investigations below will be a general linear programming problem.

$$\sum_{i=1}^{n} p_i x_i \to \max, \tag{7.1}$$

$$\sum_{i=1}^{n} a_{si} x_i \begin{cases} \leqslant b_s, & s \in M_1, \\ = b_s, & s \in M_2, \end{cases} \tag{7.2}$$

$$x_i \geqslant 0, \ i \in N_2,$$

where

$$M_1 = \{1,\dots,m_1\}, \ M_2 = \{m_1+1,\dots,m\},$$
$$N_1 = \{1,\dots,n_1\}, \ N_2 = \{n_1+1,\dots,n\}.$$

Let there be given a partition into k non intersecting parts of the set $n = N_1 \cup N_2$ of columns of the matrix (a_{si}); $N = N^{(1)} \cup \ldots \cup N^{(k)}$. Because there may be among the elements of $N^{(\alpha)}$ both elements of N_1 and N_2, one has a natural presentation

$$N^{(\alpha)} = N_1^{(\alpha)} \cup N_2^{(\alpha)}$$

where

$$N_1^{(\alpha)} = \{i \in N_1 \cap N^{(\alpha)}\},$$
$$N_2^{(\alpha)} = \{i \in N_2 \cap N^{(\alpha)}\}.$$

so that one has the formulas

$$N = N^{(1)} \cup N^{(2)} \cup \ldots \cup N^{(k)},$$
$$N_1 = N_1^{(1)} \cup N_1^{(2)} \cup \ldots \cup N_1^{(k)},$$
$$N_2 = N_2^{(1)} \cup N_2^{(2)} \cup \ldots \cup N_2^{(k)}.$$

We shall use economic ideas and terminology and call the vector b with components $b_1,\ldots,b_m$ a vector of resources.

Below we shall see that the choice of the way of partitioning N is not subject to any restriction. The only important thing for the choice of one or another partition is the structure of matrix (a_{si}) of the initial problem. We

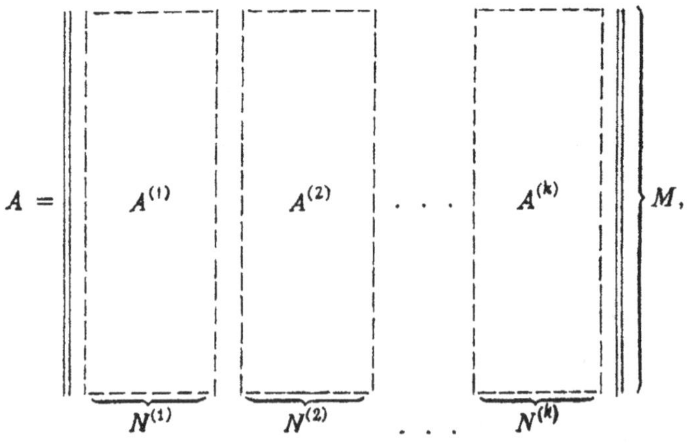

also remark that the partition indicated above is a partition of the matrix $A = (a_{si})$ into blocks with one and the same number of rows (viz. m rows).

where $A^{(\alpha)}$ is the matrix containing the elements

$$a_{si}, \ s = 1,2,\dots,m, \ i \in N^{(\alpha)}.$$

Under this partition of the set N the problem (7.1):(7.2) can be rewritten in the following form:

$$\sum_{\alpha=1}^{k} \sum_{i \in N^{(\alpha)}} p_i x_i \to \max, \tag{7.3}$$

$$\sum_{\alpha=1}^{k} \sum_{i \in N^{(\alpha)}} a_{si} x_i \begin{cases} \leqslant b_s, & s \in M_1, \\ = b_s, & s \in M_2. \end{cases} \tag{7.4}$$

$$x_i \geqslant 0, \ i \in N_2^{(\alpha)}, \ \alpha = 1,\dots,k.$$

Let us introduce the following definitions which will be useful in the sequel.

DEFINITION 7.1. Each matrix $(b_{s\alpha})$ of size m×k whose elements satisfy the conditions

$$\sum_{\alpha=1}^{k} b_{s\alpha} \begin{cases} \leqslant b_s, & s \in M_1, \\ = b_s, & s \in M_2, \end{cases} \tag{7.5}$$

will be called an admissible allocation of resources matrix.

DEFINITION 7.2. An admissible allocation of resources matrix B* whose elements $b_{s\alpha}$ satisfy

$$b^*_{s\alpha} = \sum_{i \in N^\alpha} a_{si} x^*_i, \ s \in M_1 \cup M_2 = M; \ i = 1,\dots,k, \tag{7.6}$$

where x* is an optimal vector for problem (7.1)-(7.2) is called an optimal resource allocation matrix.

Regard the quantities $b_{s\alpha}$ as unknowns which are only required to satisfy condition (7.5). Then we can write problem (7.1)-(7.2) in the following equivalent form

$$\sum_{\alpha=1}^{k} \sum_{i \in N^{(\alpha)}} p_i x_i \to \max, \tag{7.7}$$

$$\sum_{i\in N^{(\alpha)}} a_{si}x_i - b_{s\alpha} = 0,\ s \in M, \tag{7.8}$$

$$x_i \geqslant 0,\ i \in N_2^{(\alpha)},\ \alpha = 1,2,\dots,k.$$

$$\sum_{\alpha=1}^{k} b_{s\alpha} \begin{cases} \leqslant b_s, & s \in M_1, \\ = b_s, & s \in M_2. \end{cases} \tag{7.9}$$

It is easy to present the physical model of the linear programming problem (7.7)-(7.9) in such a way that there are visibly k sub-models of the problem

$$\sum_{i\in N^{(\alpha)}} p_i x_i \to \max, \tag{7.10}$$

$$\left.\begin{aligned} &\sum_{i\in N^{(\alpha)}} a_{si}x_i - b_{s\alpha} = 0,\ s \in M, \\ &x_i \geqslant 0, \qquad i \in N_2^{(\alpha)}, \end{aligned}\right\} \tag{7.11}$$

connected with one another by a system of communicating volumes modelling condition (7.9). This model is depicted in Figure 7.1. And in Figures 7.2 and 7.3 there are shown a model of the systems of equations and inequalities (7.9) and a model of one of the blocks (7.10)-(7.11). The structure of the physical model of the linear programming problem (7.1)-(7.2), written in the form (7.7)-(7.9) naturally leads to the idea of the decomposition method expounded below. The essentials consist in the following.

Let $(b_{s\alpha}^{(0)})$ be an admissible resource allocation matrix. Such matrices can be found in various ways. One natural way to obtain one, for example, is to choose the quantities $b_{s\alpha}^{(0)}$ in the following manner:

$$b_{s\alpha}^{(0)} = \frac{\sum_{i\in N^{(\alpha)}} |a_{si}|}{\sum_{\alpha=1}^{k} \sum_{i\in N^{(\alpha)}} |a_{si}|}\, b_s,\ s = 1,\dots,m \tag{7.12}$$

$$s = 1,\dots,m;\ \alpha = 1,\dots,k.$$

With this choice for the matrix $(b_s^{(0)})$ one has that $b_s^{(0)} = 0$

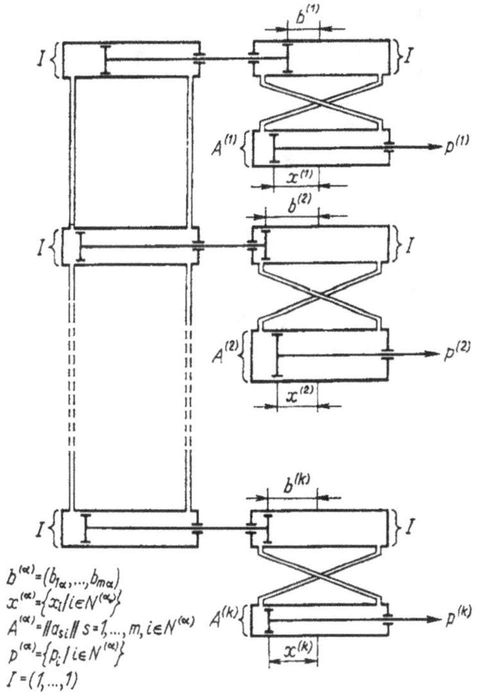

Fig. 7.1.

for those blocks which do not use the resources of type s. For such blocks, clearly, it follows that the vector $a_s^{(\alpha)} = \{a_{si}^{(\alpha)} | i \in N^{(\alpha)}\}$ is the zero vector. We shall see in the following that for all s and α for which $b_{s\alpha}^{(0)} = 0$ also

$$0 = b_{s\alpha}^{(0)} = b_{s\alpha}^{(1)} = b_{s\alpha}^{(2)} = \dots .$$

For fixed values of $b_s = b_s^{(0)}$, that is after the corresponding rods have been fixed in position, the physical model of problem (7.7)-(7.9) splits into k isolated blocks, the physical models for the problems (7.10)-(7.11) with parameter $\tilde{q}_0 > 0$. The models for these problems are by now well known to the reader from the preceding chapters and to find equilibrium states for these models one can use, for example, the algorithms of Chapter III.

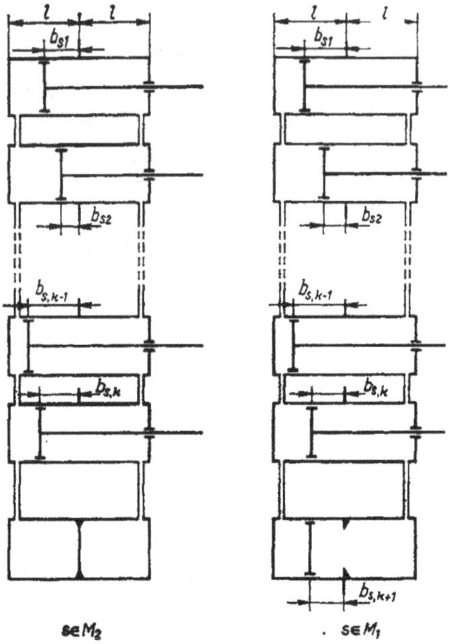

Fig. 7.2.

It is worthwhile to remark that for an arbitrary choice of the matrix $b_{s\alpha}^{(0)}$ it may well be that the conditions (7.11) are consistent but as we already know (see Section 2.3 cf. Chapter II) for a fixed value of $\tilde{q}_0$ the consistency of the system is not a necessary condition for the existence of an equilibrium state for the physical model with its containers filled with an ideal gas. This circumstance is even advantageous for the use of decomposition methods.

Let $x^0(q_0)$ be the equilibrium state vector for the physical model (7.7)-(7.9) subject to redundant constraints of the form $b_{s\alpha} = b_{s\alpha}^{(0)}$, $s \in M$, $\alpha = 1,\dots,k$. Thus the components of the vector $x^0(\tilde{q}_0)$ are the components of the vectors $x^{(\alpha)0}(\tilde{q}_0)$, $\alpha = 1,\dots,k$ of the equilibrium states of the k physical models of the k problems (7.10)-(7.11). The

$$x^{(\alpha)} = (x_1^{(\alpha)}, \ldots, x_{n_\alpha}^{(\alpha)}) = \{x_i / i \in N^{(\alpha)}\}$$

$$p^{(\alpha)} = (p_1^{(\alpha)}, \ldots, p_{n_\alpha}^{(\alpha)}) = \{p_i / i \in N^{(\alpha)}\}$$

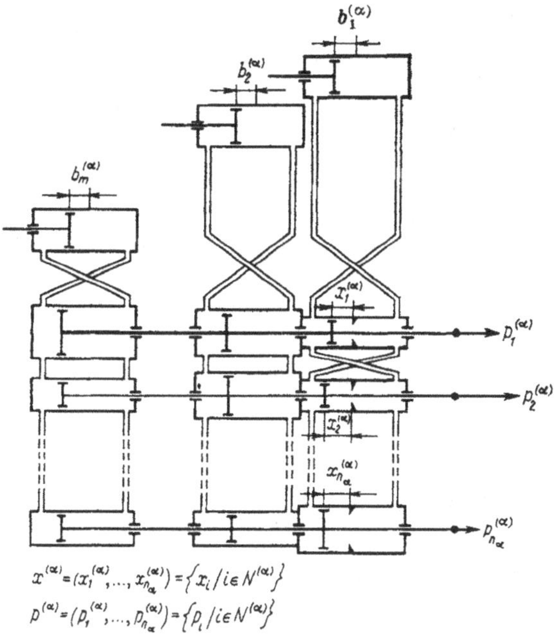

Fig. 7.3.

next step in the algorithm turns out to be a problem of
reallocating the resources between the blocks, that is a
problem of finding the next approximation $(b_{s\alpha}^{(1)})$ to an optimal
resource allocation matrix $(b_{s\alpha}^*)$. To solve this problem it is
necessary to replace the redundant constraints $b_{s\alpha} = b_{s\alpha}^{(0)}$,
$s \in M$, $\alpha = 1, \ldots, k$ by the redundant constraints $x_i = x_i^0(\tilde{q}_0)$,
$i = 1, \ldots, n$, that is to fix the positions of the corresponding
rods in the equilibrium positions found at the end of the
first stage of the algorithm. Obviously such a replacement
of the redundant constraints $b_{s\alpha} = b_{s\alpha}^{(0)}$ with the redundant
constraints $x_i = x_i^0$ which are compatible with the equilibrium
state just found, does not involve work done by exterior
forces on the model and consequently the basic condition is
fulfilled for convergence of the algorithm, namely that the

processes taking place in the system should be spontaneous.

As is clear from (7.7)-(7.9) under the extra constraints $x = x^0(\tilde{q}_0)$ the problem of finding the matrix $(b_{s\alpha}^{(1)})$, that is the problem of finding the next equilibrium, splits itself into m simpler problems each of which turns out to be either an equilibrium problem for a model of a system

$$\sum_{\alpha=1}^{k} b_{s\alpha} \leqslant b_s. \tag{7.13}$$

$$b_{s\alpha} = \phi_{s\alpha}(x^{(\alpha)0}), \ \alpha = 1,\ldots,k. \tag{7.14}$$

If $s \in M_1$, or an equilibrium problem for a model of a system

$$\sum_{\alpha=1}^{k} b_{s\alpha} = b_s, \tag{7.15}$$

$$b_{s\alpha} = \phi_{s\alpha}(x^{(\alpha)0}), \ \alpha = 1,\ldots,k \tag{7.16}$$

if $s \in M_2$. In these problems the

$$\phi_{s\alpha}(x^{(\alpha)0}) = \sum_{i \in N^{(\alpha)}} a_{si}x_i^0 \tag{7.17}$$

are known numbers

The problems (7.13)-(7.14) or (7.15)-(7.16) are so simple that the elements of the matrix $(b_{s\alpha}^{(1)})$ can be calculated by finite explicit formulas. It is important for this that the conditions (7.13) or (7.15) can be modelled by systems consisting of containers filled with an incompressible liquid, which guarantees that these conditions are satisfied at an equilibrium state so that the new matrix $(b_{s\alpha}^{(1)})$ is in fact an admissible resource allocation matrix.

A model of one of the problems (7.13)-(7.14) is shown in Figure 7.2. From this model we obtain a model of a problem of type (7.15)-(7.16) by fixing the piston of the lowest cylinder of the left column of cylinders in the position b_s, indicated by the stops. The volumes of the left column, the model for conditions (7.13) or (7.15) are filled with an incompressible liquid and the volumes of the right column of cylinders contain ideal gases in such amounts that if conditions (7.14) hold then the pressure in these volumes is equal to the given pressures $\tilde{q}_0$.

Let us denote with w_s the force acting on each of the pistons from the left column of cylinders filled with an incompressible liquid. Obviously the force w_s is numerically equal to the pressure difference in two systems of containers modelling condition (7.13) or condition (7.15) [2)]. Let $w_{s\alpha}$ be the force acting on the piston whose position indicates the quantity $b_{s\alpha}$ and which is located inside the cylinder block α with containers filled with ideal gases. The conditions for equilibrium of the model under the extra constraints $x = x^0$ have the following truly simple form

$$w_{s\alpha}^{(1)} = w_s, \ \alpha = 1,\dots,k. \tag{7.18}$$

where

$$w_s \begin{cases} = 0 \text{ for } \sum_{\alpha=1}^{k} b_{s\alpha}^{(1)} < b_s, \\ > 0 \text{ for } \sum_{\alpha=1}^{k} b_{s\alpha}^{(1)} = b_s, \end{cases} \quad s \in M_1. \tag{7.19}$$

For $s \in M_2$ the quantity w_s can be positive as well as negative. [3)] To obtain the dependence of the quantities w_s on the $w_{s\alpha}^{(1)}$ quantities $b_{s\alpha}^{(1)}$ to be determined we make use of the following by now well known consequence of the state equation for ideal gases.

$$w_{s\alpha}^{(1)} = \frac{\tilde{q}_0}{\| a_s^{(\alpha)} \|} y_{s\alpha}(b_{s\alpha}^{(1)}, x^{(\alpha)0}) \tag{7.20}$$

where

$$\| a_s^{(\alpha)} \| = 1 + \sum_{i \in N^{(\alpha)}} a_{si}$$

and in analogy with (2.24) because the constraints (7.11) are equalities the quantities y_s $(b_s^{(1)}, x^{(\alpha)0})$ are defined by the formulas

$$y_{s\alpha}(b_{s\alpha}^{(1)}, x^{(\alpha)0}) = z_{s\alpha}(b_{s\alpha}^{(1)}, x^{(\alpha)0}) = \sum_{i \in N^{(\alpha)}} a_s x_i^{(\alpha)0} - b_{s\alpha}^{(1)}. \tag{7.21}$$

Using (7.17) we obtain

$$y_{s\alpha}(b_{s\alpha}^{(1)0}) = \phi_{s\alpha}(x^{(\alpha)0}) - b_{s\alpha}^{(1)}. \tag{7.22}$$

From the equilibrium conditions (7.18) and formulas (7.20) and (7.22) there follows the following lemma:

LEMMA 7.1. If in the equilibrium state $b_{s1}^{(1)}, b_{s2}^{(1)}, \ldots, b_{sk}^{(1)}$ for each $\alpha = \alpha_1 \in \{1,\ldots,k\}$ and each $s \in M_1$ the following inequality holds

$$y_{s_1\alpha_1}(b_s^{(1)}, x^{(\alpha)0}) = \phi_{s_1\alpha_1}(x^{(\alpha_1)0}) - b_{s_1\alpha_1}^{(1)} \leqslant 0, \qquad (7.23)$$

then the condition $w_{s_1\alpha}^{(1)} = 0$, $\alpha = 1,\ldots,k$ holds.

The lemma has a simple economic meaning. If in the equilibrium state one or another kind of the resources is present in a superfluous amount for one of the subsystems then that kind of resource cannot be deficient for any of the other subsystems.

Let us consider the various possible cases:

(1) $s \in M_1$ and there exists a number $\alpha_1 \in \{1,\ldots,k\}$ such that in the equilibrium state $(b_{s1}^{(1)},\ldots,b_{sk}^{(1)})$, $y_{s_1\alpha_1}(b_s^{(1)}, x^{(\alpha)}) \leqslant 0$. Then by virtue of the lemma

$$b_{s\alpha}^{(1)} \geqslant \phi_{s\alpha}(x^{(\alpha)0}), \quad \alpha = 1,2,\ldots,k,$$

and according to (7.18)-(7.19)

$$w_{s\alpha}^{(1)} = w_s = 0, \quad \alpha = 1,\ldots,k.$$

Therefore because of (7.19)

$$\sum_{\alpha=1}^{k} b_{s\alpha}^{(1)} < b_s,$$

and in the case under consideration one can take as elements for the s-th row of the matrix $b_{s\alpha}^{(1)}$ any one from the nonempty set of solutions of the system of inequalities

$$\sum_{\alpha=1}^{k} b_{s\alpha}^{(1)} < b_s, \quad b_{s\alpha}^{(1)} \geqslant \phi_{s\alpha}(x^{(\alpha)0}), \quad \alpha = 1,\ldots,k. \qquad (7.24)$$

It is simplest to choose the obvious solution of system (7.24), namely

$$b_{s\alpha}^{(1)} = \phi_{s\alpha}(x^{(\alpha)0}).$$

(2) $s \in M_1$ and for all $\alpha \in \{1,\ldots,k\}$ one has in the equilibrium state $(b_{s1}^{(1)},\ldots,b_{sk}^{(1)})$ the inequality

$$y_{s\alpha}(b_{s\alpha}^{(1)}, x^{(\alpha)0}) > 0, \quad \alpha = 1,\ldots,k$$

In this case it follows from (7.20) that

$$w_{s\alpha}^{(1)} = \frac{\tilde{q}_0}{\| a_s^{(\alpha)} \|} (\phi_{s\alpha}(x^{(\alpha)0}) - b_{s\alpha}^{(1)}) = w_s > 0, \quad \alpha = 1,\ldots,k. \tag{7.25}$$

On the other hand from (7.19) for $w_s > 0$ it follows that

$$\sum_{\alpha=1}^{k} b_{s\alpha}^{(1)} = b_s.$$

This condition means that the quantity w_s can be eliminated from the system (7.25). Indeed mulitplying equations (7.25) respectively with the $\| a_s^{(\alpha)} \|$ and summing the results one obtains

$$\tilde{q}_0 \{ \sum_{\alpha=1}^{k} \phi_{s\alpha}(x^{(\alpha)0}) - b_s \} = w_s \sum_{\alpha=1}^{k} \| a_s^{(\alpha)} \| ,$$

and from this

$$w_s^{(0)}(\tilde{q}_0) = \frac{\tilde{q}_0}{\sum_{\alpha=1}^{k} \| a_s^{(\alpha)} \|} [\sum_{\alpha=1}^{k} \phi_{s\alpha}(x^{(\alpha)0}) - b_s] =$$

$$= \frac{\tilde{q}_0}{\sum_{\alpha=1}^{k} \| a_s^{(\alpha)} \|} z_s(x^0). \tag{7.26}$$

Substituting this in (7.25) one obtains by simple manipulations finite (closed) formulas for calculating the elements of the matrix $b_{s\alpha}^{(1)}$ in the case under consideration

$$b_{s\alpha}^{(1)} = \phi_{s\alpha}(x^{(\alpha)0}) - \frac{\| a_s^{(\alpha)} \|}{\sum_{\alpha=1}^{k} \| a_s^{(\alpha)} \|} z_s(x^0). \qquad)7.27)$$

(3) $s \in M_2$. In this case the quantities $y_{s\alpha}(b_{s\alpha}^{(1)}, x^{(\alpha)0})$ can evidently be both positive and negative and it follows from the equilibrium conditions that the quantities

$w_{s\alpha}^{(1)}$ are all equal and consequently of the same sign. Repeating the arguments of the second case one obtains again the formulas (7.27) for the quantities $b_{s\alpha}^{(1)}$.

It remains to analyse the results obtained. The starting point for all considerations was the known vector $x^0 = (x^{(1)0},\dots,x^{(k)0})$ which is an ordered set of equilibrium state vectors for the models of problems (7.10)-(7.11) for given $b_{s\alpha} = b_{s\alpha}^{(0)}$. In this state x^0, not knowing the quantities $b_{s\alpha}^{(1)}$ one can calculate the quantities $z_s(x^0)$ using the formula

$$\sum_{\alpha=1}^{k} \phi_{s\alpha}(x^{(\alpha)0}) - b_s = z_s(x^0). \tag{7.28}$$

Here if $s \in M_1$ two alternatives are possible. Either $z_s(x^0) \leqslant 0$ or $z_s(x^0) > 0$. If $z_s(x^0) \leqslant 0$ for $s \in M_1$ then it follows from (7.27) that $b_{s\alpha}^{(1)} \geqslant \phi_{s\alpha}(x^{(\alpha)})$, that is the first case applies. If $z_s(x^0) > 0$ then (7.27) says that $b_{s\alpha}^{(1)} < \phi_{s\alpha}(x^{(\alpha)0})$ that is we are concerned with the second of the cases considered. If $s \in M_2$ the sign of the errors $y_{s\alpha}(b_{s\alpha}^{(1)},x^{(\alpha)0})$ is not relevant.

In this way one obtains a general formula for determining the elements of the matrix $(b_{s\alpha}^{(1)})$ which according to various possibilities takes the form

$$b_{s\alpha}^{(1)} = \begin{cases} \phi_{s\alpha}(x^{(\alpha)0}) - \dfrac{\| a_s^{(\alpha)} \|}{\sum\limits_{\alpha=1}^{k} \| a_s^{(\alpha)} \|} z_s(x^{(0)}) \underline{\underline{1}}[z_s(x^{(0)})], & s \in M_1, \\[2ex] \phi_{s\alpha}(x^{(\alpha)0}) - \dfrac{\| a_s^{(\alpha)} \|}{\sum\limits_{\alpha=1}^{k} \| a_s^{(\alpha)} \|} z_s(x^{(0)}), & s \in M_2. \end{cases} \tag{7.29}$$

One now proceeds again to the consideration of the k equilibrium problems of the physical models for problems (5.10)-(5.11) but now for

$$b_{s\alpha} = b_{s\alpha}^{(1)}, \quad s = 1,\dots,m; \quad \alpha = 1,\dots,k,$$

that is one proceeds to find, using any of the algorithms of Chapter III vectors $x^{(\alpha)1}$, $\alpha = 1,\dots,k$, representing equilibrium states for these models with one and the same parameter $\tilde{q}_0$. A second approximation of an optimal resource allocation matrix is now found by analogy with (7.29) by means of the formulas

$$b_{s\alpha}^{(2)} = \begin{cases} \phi_{s\alpha}(x^{(\alpha)1}) - \dfrac{\|a_s^{(\alpha)}\|}{\sum_{\alpha=1}^{k} \|a_s^{(\alpha)}\|} z_s(x^1)\underline{\underline{1}}[z_s(x^1)], & s \in M_1, \\ \phi_{s\alpha}(x^{(\alpha)1}) - \dfrac{\|a_s^{(\alpha)}\|}{\sum_{\alpha=1}^{k} \|a_s^{(\alpha)}\|} z_s(x^1), & s \in M_2. \end{cases} \tag{7.30}$$

The sequel is obvious and this iterative decomposition algorithm thus has the form of a sequence of cycles each of which consists of two steps. Here follows the description of the r-th cycle of the algorithm.

<u>First stage of the r-th cycle</u>. One calculates the coordinates of the equilibrium state vectors $\bar{x}^{(\alpha)r}(\tilde{q}_0)$, $\alpha = 1,\dots,k$ of the physical models of the k problems of type (7.10)-(7.11) for a given value of the parameter $\tilde{q}_0$. To solve these problems one can use the first or the second of the algorithms of Chapter III, see (3.14)-(3.15) of Section 3.3 or (3.31) in Section 3.4. In these problems of the first step

$$b_{s\alpha} = b_{s\alpha}^{(r)}, \quad s = 1,\dots,m; \quad \alpha = 1,\dots,k$$

(are known). For the problems of the first step of the zero-th cycle the $b_s^{(0)}$ are determined by formulas (7.12).

<u>Second step of the r-th cycle</u>. This step consists of the calculation of the elements of the matrix $(b_{s\alpha}^{(r+1)})$, that is

the next approximation to an optimal resource allocation matrix and this is done by the finite closed form formulas analogous to (7.29) and (7.30)

$$b_s^{(r+1)}(\tilde{q}_0) = \begin{cases} \phi_{s\alpha}(\bar{x}^{(\alpha)r}(\tilde{q}_0)) - \dfrac{\| a_s^{(\alpha)} \|}{\sum\limits_{\alpha=1}^{k} \| a_s^{(\alpha)} \|} \, z_s(\bar{x}^r(\tilde{q}_0)) \underline{\underline{1}} \, z_s(x^r(\tilde{q}_0))], & \\ & s \in M_1, \\ \phi_{s\alpha}(\bar{x}^{(\alpha)r}(\tilde{q}_0)) - \dfrac{\| a_s^{(\alpha)} \|}{\sum\limits_{\alpha=1}^{k} \| a_s^{(\alpha)} \|} \, z_s(\bar{x}^r(\tilde{q}_0)), & s \in M_2, \end{cases}$$

$$s = 1,\ldots,m; \quad \alpha = 1,\ldots,k \quad (7.31)$$

The quantities

$$\phi_{s\alpha}(\bar{x}^{(\alpha)r}(\tilde{q}_0)), \quad z_s(\bar{x}^r(\tilde{q}_0)), \quad \| a_s^{(\alpha)} \|$$

are determined by the formulas

$$\left.\begin{aligned} \phi_{s\alpha}(\bar{x}^{(\alpha)r}(\tilde{q}_0)) &= \sum_{i \in N^{(\alpha)}} a_{si} \bar{x}_i^{(\alpha)r}(\tilde{q}_0), \\ z_s(\bar{x}^r(\tilde{q}_0)) &= \sum_{\alpha=1}^{k} \phi_{s\alpha}(\bar{x}^{(\alpha)r}(\tilde{q}_0)) - b_s, \\ \| a_s^{(\alpha)} \| &= 1 + \sum_{i \in N^{(\alpha)}} |a_{si}|, \\ s = 1,\ldots,m;& \quad \alpha = 1,\ldots,k. \end{aligned}\right\} \quad (7.32)$$

Evidently the algorithm is a description of spontaneous transition processes in the model of problem (7.3)-(7.4) to an equilibrium state which are controlled by means of a sequence of alternating stationary redundant constraints. Such processes give rise to a monotonically decreasing thermodynamic potential in view of the second principle of thermodynamics. It is also clear that in each cycle all virtual displacements of the bodies making up the model have been given their chance. Therefore the general condition of convergence, as formulated in Section 3.2 of Chapter III is fulfilled and the following equalities hold

$$\left.\begin{aligned} &\lim_{r\to\infty} \bar{x}^r(\tilde{q}_0) = \bar{x}(\tilde{q}_0), \quad \lim_{\tilde{q}_0\to\infty} \bar{x}(\tilde{q}_0) = x^*, \\ &\lim_{r\to\infty} b_s^{(r)}(\tilde{q}_0) = \bar{b}_{s\alpha}(\tilde{q}_0), \quad \lim_{\tilde{q}_0\to\infty} \bar{b}_{s\alpha}(\tilde{q}_0) = b^*_{s\alpha}, \\ &s = 1,\ldots,m;\ \alpha = 1,\ldots,k. \end{aligned}\right\} \tag{7.33}$$

Remarks:

1) From formula (7.31) it follows that the following assertion is true: if for some pair of indices $s_1 \in \{1,\ldots,m\}$, $\alpha_1 \in \{1,\ldots,k\}$ the quantity $b^{(0)}_{s_1,\alpha_1}$ as calculated by formula (7.12) turns out to be zero then $b^{(r)}_{s_1,\alpha_1} = 0$ for all values of the index $r = 0,1,2,\ldots$.

2) Applying the algorithm leads to a simultaneous solution of the primal and the dual problem. A sequence of approximating optimal vectors of the dual problem of problem (7.1)-(7.2) is defined by the formulas analogous to formulas (7.27):

$$\bar{w}_s^{(r)}(\tilde{q}_0) = \frac{\tilde{q}_0}{\sum_{\alpha=1}^{k} \| a_s^{(\alpha)} \|} z_s(\bar{x}^r(\tilde{q}_0)), \quad s = 1,\ldots,m, \tag{7.34}$$

and the following condition holds

$$\lim_{r\to\infty} \bar{w}^{(r)}(\tilde{q}_0) = w(\tilde{q}_0), \quad \lim_{\tilde{q}_0\to\infty} \bar{w}(\tilde{q}_0) = w\ . \tag{7.35}$$

3) It is clearly possible to construct a decomposition algorithm by using ideas from the method of displacing constraints, see Chapter VI. In that way one obtains a finite number of equilibrium problems for physical models which are models of problems of the form

$$\begin{aligned} &\sum_{i=1}^{n} p_i x_i \to \max, \\ &\sum_{i=1}^{n} a_{si} x_i \begin{cases} \leqslant b_s - \beta_s^{(\omega)}, & s \in M_1, \\ = b_s - \beta_s^{(\omega)}, & s \in M_2, \end{cases} \\ &\omega = 0,1,\ldots,\omega^* \end{aligned} \tag{7.36}$$

where ω^* is the number of active constraints. Just as in Chapter VI the parameter $\tilde{q}_0$ for all of these models is positive and given in advance and thus one avoids the necessity of choosing a growing sequence of values of that parameter. The algorithm described above does not change itself; it is only necessary for problem (7.36) to replace the quantity b_s by the quantities $b_s - \beta_s^{(\omega)}$. Thus, because $\tilde{q}_0$ keeps one and the same value all quantities are functions only of the index number ω of the problem and the following conditions hold

$$\left.\begin{array}{l} \lim\limits_{r\to\infty} \bar{x}^{(r)}(\omega) = \bar{x}(\omega), \quad \bar{x}(\omega^*) = x^*, \\ \lim\limits_{r\to\infty} b_{s\alpha}^{(r)}(\omega) = b_{s\alpha}(\omega), \; b_{s\alpha}(\omega^*) = b_{s\alpha}^*, \\ \lim\limits_{r\to\infty} w_s^{(r)}(\omega) = w_s(\omega), \quad w_s(\omega^*) = w_s^*, \\ s = 1,\ldots,m; \; \alpha = 1,\ldots,k. \end{array}\right\} \qquad (7.37)$$

4) It is always possible to divide the matrix $A = (a_{si})$ in so many blocks $A^{(\alpha)}$ that the solution of the problems of the first stage (of each cycle) can be obtained by finite (closed form) formulas. This, for example, is easy for $k = n$, that is in the case where each block $A^{(\alpha)}$ of the matrix is the column vector with the elements $a_{s\alpha}$, $s = 1,\ldots,m$. Then one easily obtains the finite formulas looked for by a passage to the limit for $\nu \to \infty$ in the formulas of the first or second algorithm of Chapter III, that is by setting in them

$$x^{\nu+1} = x^{\nu} = x^{(\infty)} = \bar{x}(\tilde{q}_0).$$

The derivation of these formulas will be a useful exercise for the reader. In this simple decomposition method the dimensions of the matrices (a_{si}) and $(b_{s\alpha})$ coincide and according to (7.6) zero elements of the optimal resource allocation matrix $(b_{s\alpha}^*)$ correspond to zeros of the matrix (a_{si}). The algorithm must take account of this especially because in economic planning problems the matrices (a_{si}) contain lots of zeros.

Finite expressions for the coordinates $\bar{x}^{(\alpha)r}(\tilde{q}_0)$ of the equilibrium state vector of the physical model of problem

(7.10)-(7.11) are also sufficiently simple to obtain in the case where the blocks $A^{(\alpha)}$ each consist of two or three columns. The obtaining of such formulas will also be a nice exercise for the reader.

Economic interpretation. The algorithm just described admits a suggestive economic interpretation. If $b_1,\dots,b_m$ are the amounts of resources used in a linear economy and problem (7.1)-(7.2) is interpreted as a problem about the optimal use of these resources then the decomposition proposed consists of a partition of the economy into a finite number k of subsystems between which there is a central planning organisation which allocates the resources. A given resource allocation matrix $(b_{s\alpha})$ leads to a partition of the economy into k isolated subsystems for which the problem of the optimal use of the resources allocated to them can be solved independently of one another. That is the first step (of each cycle) of the algorithm. It is interesting to note that the centre need not know the objective functions of the various subsystems and only receives information on their plans. This information is sufficient for the reallocation of resources and the economic meaning of the second step (of each cycle) consists in an adjustment of the (dual) proces (for each subsystem) of all those kinds of resources used by those subsystems. Below we shall return to the matter in chapter X when we shall be considering certain models of economic equilibrium.

The second algorithm. Another way to approach the decomposition problem consists in partitioning the matrix (a_{si}) of a general linear programming problem (7.1)-(7.2) into blocks consisting of row vectors of this matrix. [4)]

Let there be given a partition of the set M of rows of the matrix $A = (a_{si})$ into r on intersecting subsets $M^{(1)},\dots,M^{(r)}$ where

$$M^{(\beta)} = M_1^{(\beta)} \cup M_2^{(\beta)} \subset M, \; M_1^{(\beta)} \subset M_1, \; M_2^{(\beta)} \subset M_2,$$
$$\beta = 1,\dots,r.$$

Thus

$$M = M^{(1)} \cup M^{(2)} \cup \dots \cup M^{(r)},$$

$$M_1 = M_1^{(1)} \cup M_1^{(2)} \cup \dots \cup M_1^{(r)},$$

$$M_2 = M_2^{(1)} \cup M_2^{(2)} \cup \dots \cup M_2^{(r)}.$$

If $m^{(\beta)}$ is the number of elements of the subset $M^{(\beta)}$, then the block $A^{(\beta)}$ contains $m^{(\beta)}$ rows and n columns.

$$A^{(\beta)} = \left\| \begin{array}{c} \\ \\ \end{array} \right\| \Big\} m^{(\beta)}.$$

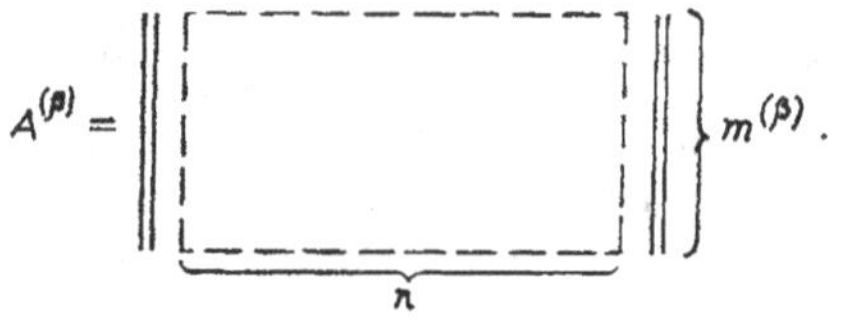

The linear programming problem can now be written in the following equivalent form

$$\sum_{i=1}^{n} p_i x_i \rightarrow \max. \tag{7.38}$$

$$\sum_{i=1}^{n} a_{si} x_{i\beta} \begin{cases} \leqslant b_s, & s \in M_1^{(\beta)}, \\ = b_s, & s \in M_2^{(\beta)}, \end{cases} \tag{7.39}$$

$$\left. \begin{array}{l} x_i - x_{i\beta} = 0, \ i \in N, \ \beta = 1,2,\ldots,r, \\ x_i \geqslant 0, \ i \in N_2. \end{array} \right\} \tag{7.40}$$

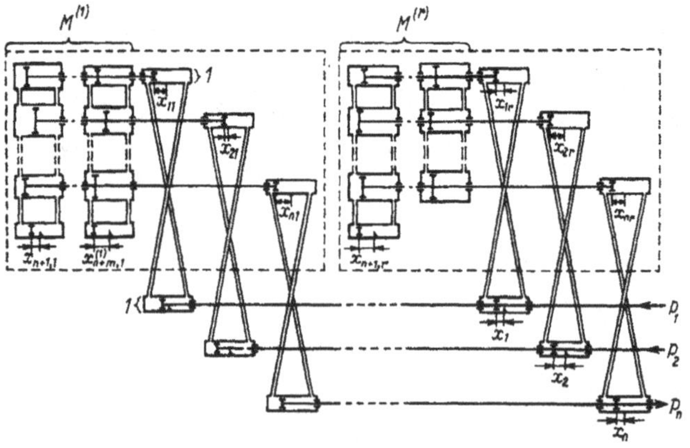

Fig. 7.4.

A physical model of problem (7.38)-(7.40) is shown in Figure 7.4. If the containers of the model are filled with ideal gases then all states $\{x_i, x_{i\beta} \mid i \in N,\ \beta = 1,\ldots,r;$ $x_i \geqslant 0,\ i \in N_2\}$ are physically possible and the values $\bar{x}_1(\tilde{q}_0),\ldots,\bar{x}_n(\tilde{q}_0)$ of the coordinates $x_1,\ldots,x_n$ of the model in an equilibrium state are as close as wished to an optimal solution of problem (7.1)-(7.2) for sufficiently high values of the model parameter $\tilde{q}_0$.

Let $x_1^{(0)},\ldots,x_n^{(0)}$ be an arbitrary choice of n number which are only required to satisfy the conditions $x_i^{(0)} \geqslant 0$, $i \in N_2$. Under the extra (redundant) constraints $x_i = x_i^{(0)}$, $i \in N$ the equilibrium problem splits into r equilibrium problems for the r blocks

$$\sum_{i=1}^{n} a_{si} x_{i\beta} \begin{cases} \leqslant b_s, & s \in M_1^{(\beta)}, \\ = b_2, & s \in M_2^{(\beta)}, \end{cases} \tag{7.41}$$

$$x_{i\beta} = x_i^{(0)},\ i \in N. \tag{7.42}$$

To solve these problems one can use the algorithms of Chapter III. Let $x_{i\beta}^{(0)}$, $i \in N$, $\beta = 1,\ldots,r$ be the coordinates of the equilibrium state under the redundant constraints $x_i = x_i^{(0)}$, $i \in N$. [5] The next step of the algorithm is to solve the equilibrium problems for the model subject to redundant constraints of the form

$$x_{i\beta} = x_{i\beta}^{(0)},\ i \in N,\ \beta = 1,2,\ldots,r. \tag{7.43}$$

It is easily seen that we thus encounter an equilibrium problem for a physical model of a very simple kind of linear programming problem

$$\sum_{i=1}^{n} p_i x_i \to \max,$$

$$x_i = x_{i\beta}^{(0)},\ i \in N,\ \beta = 1,2,\ldots,r.$$

$$x_i \geqslant 0,\ i \in N_2.$$

The last problem, evidently, splits into n still simpler equilibrium problems

$$
\begin{aligned}
&p_i x_i \to \max, \\
&x_i = x_{i\beta}^{(0)}, \quad \beta = 1,2,\ldots,r, \\
&x_i \geqslant 0 \ , \ i \in N_2.
\end{aligned}
\tag{7.44}
$$

which can be solved in terms of finite (closed) formulas. In fact in the case under consideration a passage to the limit for $\nu \to \infty$ in the formulas for the second algorithm of Chapter III leads to the explicit formulas

$$
x_i^{(1)} = \begin{cases} \theta_i(x^{(0)},\tilde{q}_0), & i \in N_1, \\ \theta_i(x^{(0)},\tilde{q}_0)\underline{1}[\theta_i(x^{(0)},\tilde{q}_0)], & i \in N_2, \end{cases}
\tag{7.45}
$$

where

$$
\theta_i(x^{(0)},\tilde{q}_0) = \frac{1}{r}\Big[\sum_{\beta=1}^{r} x_{i\beta}^{(0)} + \frac{2p_i}{\tilde{q}_0}\Big], \quad i = 1,\ldots,n.
\tag{7.46}
$$

Once the coordinates $x_1^{(1)},\ldots,x_n^{(1)}$ have been calculated one again meets r equilibrium problems for physical models of the blocks in which the physical model of the original problem splits; this time under the redundant constraints $x_i = x_i^{(1)}$.

In this manner we again arrive at a problem analogous to the problem (7.41)-(7.42) about equilibria of the physical model of a system of equations and inequalities

$$
\sum_{i=1}^{n} a_{si}x_{i\beta} \begin{cases} \leqslant b_s, & s \in M_1^{(\beta)}, \\ = b_s, & s \in M_2^{(\beta)}. \end{cases}
\tag{7.47}
$$

$$
x_{i\beta} = x_i^{(1)}, \qquad i \in N.
\tag{7.48}
$$

Thus, the structure of the algorithm is clear. It turns out to be a cyclical affair of which each cycle consists of the following two steps.

<u>First step of the ν-th cycle</u>. The solution of r independent equilibrium problems models of systems of linear equations and inequalities

$$\sum_{i=1}^{n} a_{si}x_{i\beta} \begin{cases} \leqslant b_s, & s \in M_1^{(\beta)}, \\ = b_s, & s \in M_2^{(\beta)} \end{cases} \tag{7.49}$$

$$x_{i\beta} = x_i^{(\nu)}, \; i \in N, \; \beta = 1,2,\ldots,r. \tag{7.50}$$

<u>Second step of the ν-th cycle</u>. Let $(x_{1\beta}^{(\nu)},\ldots,x_{n\beta}^{(\nu)})$, $\beta = 1,\ldots,r$ be the vectors of the equilibrium states of the r physical models of the r systems (7.49)-(7.50). The problem of the second step now consists of the calculation of the coordinates of the vector $x^{(\nu+1)}$ by means of the analogous formulas of (7.45)-(7.46), that is the formulas

$$x_i^{(\nu+1)} = \begin{cases} \theta_i(x^{(\nu)},\tilde{q}_0), & i \in N_1, \\ \theta_i(x^{(\nu)},\tilde{q}_0)\underline{\underline{1}}[\theta_i(x^{(\nu)},\tilde{q}_0)], & i \in N_2, \end{cases} \tag{7.51}$$

where

$$\theta_i\,(x^{(\nu)},\tilde{q}_0) = \frac{1}{r}\Big[\sum_{\beta=1}^{r} x_{i\beta}^{(\nu)} + \frac{2p_i}{\tilde{q}_0}\Big], \; i = 1,\ldots,n. \tag{7.52}$$

Also in this case the convergence of the algorithm follows from the second principle of thermodynamics. Both the first and the second decomposition algorithm can be used by taking as the initial problem the dual of problem (7.1)-(7.2).

7.3. Allocation of resources problems

Let us consider a problem of allocating resources for the example of an economy consisting of two firms A_1 and A_2 and a centre C which has the task of dividing the total assortment of resources $B^{(0)} = (B_1^{(0)},\ldots,B_{m_0}^{(0)})$ between the two firms A_1 and A_2 with the aim of maximizing some efficiency criterion. Let $N^{(1)} = (N_1^{(1)},\ldots,N_{n_1}^{(1)})$ and $N^{(2)} = (N_1^{(2)},\ldots,N_{n_2}^{(2)})$ be the collection of technological processes available respectively to the firm A_1 and the firm A_2, which determine the production capacities of these two firms.

We shall use the following notations

1) $b^{(0)} = (b_1^{(0)},\ldots,b_{m_0}^{(0)})$ is the m_0-vector whose components

designate the amounts of resources which the centre C has available for allocation.

2) $b^{(1)} = (b_1^{(1)},\dots,b_{m_1}^{(1)})$ is the m_1-vector of resources of firm A_1.

3) $b^{(2)} = (b_1^{(2)},\dots,b_{m_2}^{(2)})$ is the m_2-vector of resources of firm A_2.

4) $(a_{si}^{(0,1)})$, $s = 1,\dots,m_0$; $i = 1,\dots,n_1$ is a matrix of technological coefficients where $a_{si}^{(0,1)}$ is the amount of resource $B_1^{(0)}$ used by technological process $N_i^{(1)}$ at intensity of production 1.

5) $(a_{si}^{(0,2)})$, $s = 1,\dots,m_0$; $i = 1,\dots,n_2$ is a matrix of technological coefficients where $a_{si}^{(0,2)}$ is the amount needed by technological process $N_i^{(2)}$ at intensity of production 1.

6) $(a_{si}^{(1)})$, $s = 1,\dots,m_1$; $i = 1,\dots,n_1$ is a matrix of technological coefficients where $a_{si}^{(1)}$ denotes the amount of resource of type $B_s^{(1)}$ required by firm A_1 for operating technological process $N_i^{(1)}$ at intensity 1.

7) $(a_{si}^{(2)})$, $s = 1,\dots,m_2$; $i = 1,\dots,n_2$ is a matrix of technological coefficients where $a_{si}^{(2)}$ denotes the amount of resource of type $B_1^{(2)}$ required by firm A_2 to run production process $N_i^{(2)}$ at intensity one.

8) $x^{(1)} = (x_1^{(1)},\dots,x_{n_1}^{(1)})$ and $x^{(2)} = (x_1^{(2)},\dots,x_{n_2}^{(2)})$ denote vectors of intensities with which the corresponding technological processes are used.

9) $p^{(1)} = (p_1^{(1)},\dots,p_{n_1}^{(1)})$ and $p^{(2)} = (p_1^{(2)},\dots,p_{n_2}^{(2)})$ are vectors whose components are equal to the yields per year resulting from using the corresponding technological processes at unit intensity.

The problem consists in maximizing the total yearly yield, that is in finding two vectors $x^{(1)}$ and $x^{(2)}$ which satisfy the conditions

$$\sum_{i=1}^{n_1} p_i^{(1)} x_i^{(1)} + \sum_{i=1}^{n_2} p_i^{(2)} x_i^{(2)} \to \max, \tag{7.53}$$

$$\left.\begin{aligned}
&\sum_{i=1}^{n_1} a_{si}^{(0,1)} x_i^{(1)} + \sum_{i=1}^{n_2} a_{si}^{(0,2)} x_i^{(2)} \leqslant b_s^{(0)}, \quad s=1,\ldots,m_0,\\
&\sum_{i=1}^{n_1} a_{si}^{(1)} x_i^{(1)} \leqslant b_s^{(1)}, \qquad s=1,\ldots,m_1,\\
&\sum_{i=1}^{n_2} a_{si}^{(2)} x_i^{(2)} \leqslant b_s^{(2)}, \qquad s=1,\ldots,m_2,\\
&x_i^{(1)} \geqslant 0,\ i = 1,\ldots,n_1,\ x_i^{(2)} \geqslant 0,\ i = 1,\ldots,n_2.
\end{aligned}\right\} \tag{7.54}$$

In problem (7.53)-(7.54) it is natural to divide things into two blocks and one can rewrite the problem in the following equivalent form

$$\sum_{i=1}^{n_1} p_i^{(1)} x_i^{(1)} + \sum_{i=1}^{n_2} p_i^{(2)} x_i^{(2)} \to \max, \tag{7.55}$$

$$\sum_{i=1}^{n_1} a_{si}^{(0,1)} x_i^{(1)} = b_s^{(0,1)}, \quad \sum_{i=1}^{n_2} a_{si}^{(0,2)} x_i^{(2)} = b_s^{(0,2)} \tag{7.56}$$

$$b_s^{(0,1)} + b_s^{(0,2)} \leqslant b_s^{(0)}, \quad s = 1,\ldots,m_0, \tag{6.57}$$

$$\left.\begin{aligned}
&\sum_{i=1}^{n_1} a_{si}^{(1)} x_i^{(1)} \leqslant b_s^{(1)}, \quad s = 1,\ldots,m_1,\\
&x_i^{(1)} \geqslant 0, \qquad i = 1,\ldots,n_1,
\end{aligned}\right\} \tag{7.58}$$

$$\left.\begin{aligned}
&\sum_{i=1}^{n_2} a_{si}^{(2)} x_i^{(2)} \leqslant b_2^{(2)}, \quad s = 1,\ldots,m_2,\\
&x_i^{(2)} \geqslant 0, \qquad i = 1,\ldots,n_2.
\end{aligned}\right\} \tag{7.59}$$

From the last description of problem (7.53)-(7.54) it is

evident that it takes the form of two problems, connected by means of the condition (7.57). If one chooses as an initial allocation of resources the partition $b_1^{(0,1)(1)}$, $b_s^{(0,2)(2)}$, $s = 1,\ldots,m_0$ of the central resources B^0, which is only required to satisfy (7.57), then one obviously finds two independent linear programming problems of dimensions $(m_0+m_1)\times n_1$ and $(m_0+m_2)\times n_2$, of which the solution can be obtained by means of the method of physical models. A physical model of problem (7.53)-(7.59) is depicted in Figure 7.5. It seems unnecessary to write down explicitly the iterative decomposition algorithm for this particular case of problem (7.7)-(7.8). It is, however, of interest to note that for fixed values of the coordinates $x_i^{(1)}$, $i = 1,\ldots,n_1$; $x_i^{(2)}$, $i = 1,\ldots,n_2$ the equilibrium conditions involve only the pressures $q_s^{(0,1)(+)}$, $q_s^{(0,1)(-)}$, $q_s^{(0,2)(+)}$, $q_s^{(0,2)(-)}$ in the volumes $V_s^{(0,1)(+)}$, $V_s^{(0,1)(-)}$, $V_s^{(0,2)(+)}$, $V_s^{(0,2)(-)}$ modelling conditions (7.56). These equilibrium conditions which are analogous to (7.18)-(7.19) have the form

$$w_q^{(0)} = q_s^{(1,0)(-)} - q_s^{(1,0)(+)} = q_s^{(2,0)(-)} - q_s^{(2,0)(+)}, \tag{7.60}$$

$$w_s^{(0)} \begin{cases} = 0, & b_s^{(0,1)} + b_s^{(0,2)} < b_s^{(0)}, \\ \geqslant 0, & b_s^{(0,1)} + b_s^{(0,2)} = b_s^{(0)}, \end{cases} \tag{7.61}$$

$$s = 1,\ldots,m_0,$$

where, as before, $w_s^{(0)}$ is the pressure difference (between right and left of the piston) in the fluid filling the containers in the model of condition (7.57). The equilibrium conditions just mentioned have a simple economic interpretation; the optimal plans $x^{(1)}$ and $x^{(2)}$ of the separate blocks and the optimal allocation of resources by the centre are such that the (dual) prices of these resources as seen by the separate blocks are identical.

Let us now show that important results in the theory of linear economic models concerned with allocation of resources are simple consequences of well known principles of

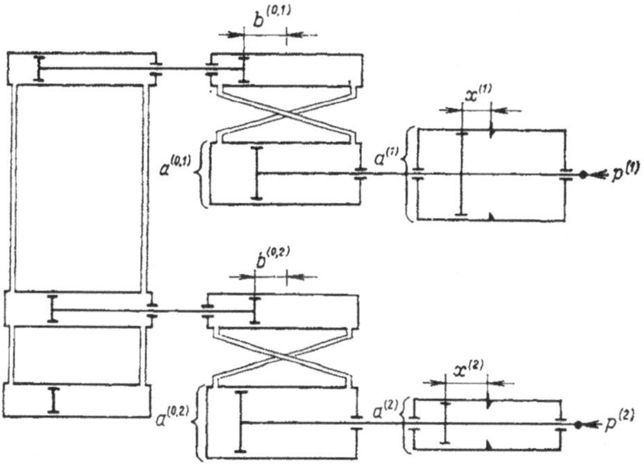

Fig. 7.5.

analytical mechanics. For this we turn our attention to models (filled with an incompressible liquid) of problems of allocation of resources as depicted by Figure 7.5 and we consider an equilibrium state of such a model given by vectors

$$x^{(1)*} = (x_1^{(1)*}, \dots, x_{n_1}^{(1)*}), b^{(0,1)*} = \begin{pmatrix} b_1^{(0,1)*} \\ \cdot \\ \cdot \\ \cdot \\ b_{m_0}^{(0,1)*} \end{pmatrix}$$

$$x^{(2)*} = (x_1^{(2)*}, \dots, x_{n_2}^{(2)*}), b^{(0,2)*} = \begin{pmatrix} b_1^{(0,2)*} \\ \cdot \\ \cdot \\ \cdot \\ b_{m_0}^{(0,2)*} \end{pmatrix}$$

Here $x^{(1)*}$ and $x^{(2)*}$ are optimal plans of the first and the second firm, $b^{(0,1)*}$ and $b^{(0,2)*}$ are the column vectors of a (corresponding) optimal resource allocation matrix. In the following, we shall make use of two well known prinicples of analytical mechanics, as we did earlier.

1. An equilibrium state of a mechanical of physical system remains as it was if one imposes on the system additional (redundant) stationary constraints compatible with this equilibrium state [39].

Recall that a constraint

$$f(x^{(1)},x^{(2)},b^{(0,1)},b^{(0,2)} = 0$$

is called compatible with the state $x^{(1)*}$, $x^{(2)*}$, $b^{(0,1)*}$, $b^{(0,2)*}$ if

$$f(x^{(1)*},x^{(2)*},b^{(0,1)*},b^{(0,2)*}) = 0. \qquad (7.62)$$

As potential constraints for the model of problem (7.55)-(7.59) one could take the following:

$$b^{(0,1)} = b^{(0,1)*}, \; b^{(0,2)} = b^{(0,2)*}.$$

This constraint is evidently compatible with the equilibrium state and if one imposes it it splits the model into two isolated ones:

$$\sum_{i=1}^{n_1} p_i^{(1)} x_i^{(1)} \to \max, \qquad (7.63)$$

$$\sum_{i=1}^{n_1} a_{si}^{(0,1)} x_i^{(1)} = b_s^{(0,1)*}, \; s = 1,\dots,m_0, \qquad (7.64)$$

$$\left.\begin{array}{l} \sum_{i=1}^{n_2} a_{si}^{(1)} x_i^{(1)} \leqslant b_s^{(1)}, \; s = 1,\dots,m_1, \\ x_i \geqslant 0, \; i = 1,\dots,n \;\; i = 1,\dots,n_1, \end{array}\right\} \qquad (7.65)$$

$$\sum_{i=1}^{n_2} p_i^{(2)} x_i^{(0)} \to \max, \qquad (7.66)$$

$$\sum_{i=1}^{n_2} a_{si}^{(0,2)} x_i^{(2)} = b_s^{(0,2)*}, \; s = 1,\dots,m_0, \qquad (7.67)$$

$$\left.\begin{aligned} &\sum_{i=1}^{n_2} a_{si}^{(2)} x_i^{(2)} \leqslant b_s^{(2)}, \quad s = 1,\ldots,m_2, \\ &x_i^{(2)} \geqslant 0, \qquad\qquad\quad i = 1,\ldots,n_2. \end{aligned}\right\} \tag{7.68}$$

Thus the principle of conservation of equilibrium states under additional compaticle constraints corresponds precisely with an optimal central allocation of resources between firms.

2. A mechanical or physical system remains in equilibrium if one replaces certain constraints imposed on the system by the reaction forces produced by those constraints at equilibrium.

This principle bears the name of the principle of removing constraints (and it has been mentioned and used before, Chapter IV). Let us apply this principle of removing constraints to our model replacing the constraints (7.57) by the reaction forces at equilibrium which are obviously equal to $w_1^{(0)},\ldots,w_m^{(0)}$. In this way again the model of problem (7.53)-(7.59) splits into two isolated models and correspondingly the equilibrium problem splits into two independent problems.

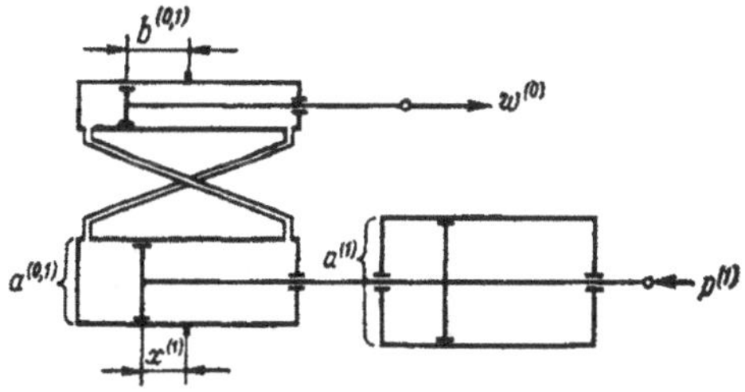

Fig. 7.6.

Comparing the model depicted in Figure 7.6 with the model of a general linear programming problem (Figure 1.8 in Chapter I) we easily convince ourselves that it is a model of the following problem

$$\sum_{i=1}^{n_1} p_i^{(1)} x_i^{(1)} - \sum_{s=1}^{m_0} w_s^{(0)} b_s^{(0,1)} \to \max, \tag{7.69}$$

under the conditions

$$\sum_{i=1}^{n_1} a_{si}^{(0,1)} x_i^{(1)} - b_s^{(0,1)} = 0, \quad s = 1,\ldots,m_0, \tag{7.70}$$

$$\left.\begin{aligned} &\sum_{i=1}^{n_1} a_{si}^{(1)} x_i^{(1)} \leqslant b_s^{(1)}, \quad s = 1,\ldots,m_1, \\ &x_i^{(1)} \geqslant 0, \qquad\qquad i = 1,\ldots,n_1, \end{aligned}\right\} \tag{7.71}$$

In this problem, clearly, both the quantities $b_s^{(0,1)}$, $s = 1,\ldots,m_0$, and the quantities $x_i^{(1)}$, $i = 1,\ldots,n_1$ are to be determined. Eliminating the quantities $b_s^{(0,1)}$ from the objective function, with the help of the constraint equalities (7.70), one obtains

$$\sum_{i=1}^{n_1} (p_i^{(1)} - \sum_{s=1}^{m_0} w_s^{(0)} a_{si}^{(0,1)}) x_i^{(1)} \to \max, \tag{7.72}$$

$$\left.\begin{aligned} &\sum_{i=1}^{n_1} a_{si}^{(1)} x_i^{(1)} \leqslant b_s^{(1)}, \quad s = 1,\ldots,m_1, \\ &x_i^{(1)} \geqslant 0, \qquad\qquad i = 1,\ldots,n_1. \end{aligned}\right\} \tag{7.73}$$

Analogously the equilibrium state of the other problem is the solution of the following problem

$$\sum_{i=1}^{n_2} (p_i^{(2)} - \sum_{s=1}^{m_0} w_s^{(0)} a_{si}^{(0,2)}) x_i^{(2)} \to \max, \tag{7.74}$$

$$\left.\begin{aligned} &\sum_{i=1}^{n_2} a_{si}^{(0)} x_i^{(2)} \leqslant b_s^{(2)}, \quad s = 1,\ldots,m_2, \\ &x_i^{(2)} \geqslant 0, \qquad\qquad i = 1,\ldots,n_2. \end{aligned}\right\} \tag{7.75}$$

The objective functions (7.72) and (7.74) have an evident economical meaning: they are the net profits of the first and second firm, that is they are the total income from the production processes minus the total cost of the resources bought from the centre at the equilibrium proces $w_s^{(0)}$, $s = 1,\ldots,m_0$. In this manner the principle of removing constraints leads to another well known approach to the solution of allocation of resources (Gale [29]). In this approach the resources are not simply allocated but they are sold against prices fixed by the central organisation.

Thus we have seen that the vectors $x^{(1)}$ and $x^{(2)}$ at an equilibrium of the model of problem (7.55)-(7.59) and the vector $w^{(0)}$, with components equal to the pressure differences in the volumes modelling condition (7.57), constitute a competitive equilibrium in the sense of Gale [29]. In the case of the problem under consideration this definition is as follows: the vectors $x^{(1)},x^{(2)},w^{(0)}$ constitute a competitive equilibrium if they satisfy the conditions (7.57), (7.72)-(7.75) and if the condition

$$\sum_{i=1}^{n_2} a_{si}^{(0,1)}x_i^{(1)} + \sum_{i=1}^{n_2} a_{si}^{(0,2)}x_i^{(2)} < b_s^{(0)} \qquad (7.76)$$

implies that $w_s^{(0)} = 0$.

We have seen above that conditions (7.57), (7.72)-(7.75) hold at equilibrium. Conditions (7.76) follow from conditions (7.61) and the equivalence of (7.76) and (7.57) holds by virtue of condition (7.56). Thus it has been established that the quantities $w_1^{(0)},\ldots,w_{m_0}^{(0)}$ at an equilibrium state of the model are the components of an equilibrium price system. It is clear that all the results thus obtained easily extend to the case of economics having an arbitrary number of firms.

NOTES

1. In the case of inconsistency the objective function of the dual problem is unbounded.
2. The surface area of each of the pistons is such that the pressure difference is equal numerically to the force acting on the piston.
3. In Figures (7.1)-(7.3) the positive direction goes from left to right.

4. The variant to be described below was proposed by A.I. Golikov.

5. Note that the quantities $x_{i\beta}^{(0)}$ do not depend on the model parameter $\tilde{q}_0$.

Chapter VIII

NONLINEAR PROGRAMMING

8.1. Introduction

In this chapter nonlinear programming problems will be considered. These also can be interpreted as equilibrium problems for mechanical or physical systems subject to unilateral and bilateral constraints. In this way each solution becomes an equilibrium state vector of its mechanical or physical model and necessary and sufficient conditions for optimality should follow from the principle of virtual displacements of Lagrange (see Theorem 1.5 in Section 1.4). Starting from that principle we shall obtain the theorem of Kuhn-Tucker, for the case of smooth constraints and a smooth objective function. Subsequently we shall see that the numerical algorithms which came out of the method of redundant constraints can be extended to cover also nonlinear programming problems.

8.2. The principle of virtual displacements and the Kuhn-Tucker theorem

Consider a nonlinear programming problem, that is a problem of finding a vector $x = (x_1,\dots,x_n)$ which satisfies the conditions

$$f(x_1,\dots,x_n) \to \min, \tag{8.1}$$

$$\left.\begin{aligned} &g_s(x_1,\dots,x_n) \leqslant 0, \ s \in M_1 = (1,\dots,m_1), \\ &g_s(x_1,\dots,x_n) = 0, \ s \in M_2 = (m_1+1,\dots,m), \end{aligned}\right\} \tag{8.2}$$

$$x_i \geqslant 0, \ i \in N_2 = (n_1+1,\dots,n),$$

where $f(x)$, $g_1(x),\dots,g_m(x)$ are continuous functions with continuous partial derivates of the first order.

In general the solution of such a problem is not unique and there arises the difficult problem of finding the global minimum, that is the problem of finding the minimum of the function f(x) over the set of local minima for problem (8.1)-(8.2). In the following we shall mainly pay attention to the important class of problems for which the functions $f(x), g_1(x), \ldots, g_m(x)$ are convex. Such problems are called convex programming problems and for those each local minimum is also at the same time a global minimum [34, 37]. We start by recalling some definitions.

1) A set $Q \subset E_n$ of points $x = (x_1, \ldots, x_n)$ is called convex if $x^{(1)} \in Q$ and $x^{(2)} \in Q$ implies

$$\lambda x^{(1)} + (1-\lambda) x^{(2)} \in Q \text{ for } 0 \leqslant \lambda \leqslant 1.$$

2) A function f(x) is called convex on the convex domain Q if for all $x^{(1)}$ and $x^{(2)}$ in Q

$$f[\lambda x^{(1)} + (1-\lambda) x^{(2)}] \leqslant \lambda f(x^{(1)}) + (1-\lambda) f(x^{(2)})$$

for $0 \leqslant \lambda \leqslant 1$.

If equality holds only if $x^{(1)} = x^{(2)}$ then the function f(x) is called strictly convex.

3) The function f(x) is called concave if the function -f(x) is convex.

For the various criteria for convexity and assorted theorems on convex functions and convex sets the reader is referred to [32,34,37].

4) The problem (8.1)-(8.2) is called admissible if the set of vectors x satisfying the conditions (8.2) is nonempty. An admissible problem is called solvable if the function f(x) is bounded in the domain defined by (8.2) and attains its minimum on that set.

A basic property of solvable convex programming problems is that the solution set is convex and that the solution is unique in case the function f(x) is strictly convex.

In this section we shall derive necessary and sufficient conditions for minima for nonlinear programming problems starting from the principle of virtual displacements of Lagrangian analytical dynamics under the assumption that the functions $f(x_1, \ldots, x_n)$, $g_1(x_1, \ldots, x_n), \ldots, g_m(x_1, \ldots, x_m)$ are continuous and have continuous partial derivatives. In fact the function f(x) can be considered as a potential energy function for a mechanical system whose states, described by coordinates $x_1, \ldots, x_n$, are required to satisfy the equations and inequalities (8.2) which are interpreted as unilateral and bilaterial constraints imposed on the system.

Let x^* be some state of the mechanical system satisfying the constraints (i.e. a feasible state) and suppose that

$$g_s(x_1^*,\dots,x_n^*) \begin{cases} = 0 \text{ for } s \in M^{(1)} \subset M_1, \\ < 0 \text{ for } s \in M_1 \backslash M^{(1)}, \\ = 0 \text{ for } s \in M_2, \end{cases} \tag{8.3}$$

$$x_i^* \begin{cases} = 0 \text{ for } i \in N^{(2)} \subset N_2, \\ > 0 \text{ for } i \in N_2 \backslash N^{(2)}. \end{cases} \tag{8.4}$$

This means that if the system is in state x^* the unilateral constraints $g_s(x) \leqslant 0$ for $s \in M^{(1)}$ and $x_i \geqslant 0$ for $i \in N^{(2)}$ are active. Consider a state $x^* + \delta x$ near x^*. The vector $\delta x = (\delta x_1,\dots,\delta x_n)$ is called a vector of virtual displacements if the state $x^* + \delta x$ is also feasible. Therefore up to higher order small terms

$$g_s(x^*+\delta x) = g_x(x^*) + \sum_{i=1}^{n} \left.\frac{\partial g_s}{\partial x_i}\right|_{x=x^*} \delta x_i .$$

Consequently it follows from (8.3) and (8.4) that the virtual displacements $x_1,\dots,x_n$ satisfy the conditions

$$\sum_{i=1}^{n} a_{si}\delta x_i \begin{cases} \leqslant 0 \text{ for } s \in M^{(1)}, \\ = 0 \text{ for } s \in M_2, \end{cases} \tag{8.5}$$

$$\delta x_i \geqslant 0 \text{ for } i \in N^{(2)}, \tag{8.6}$$

where

$$a_{si} = \left.\frac{\partial g_s}{\partial x_i}\right|_{x=x^*}$$

The unilateral constraints which are nonactive at state x^* ($s \in M_1 \backslash M^{(1)}$, $i \in N_2 \backslash N^{(2)}$) obviously do not impose any restrictions on the virutal displacements in a neighbourhood of x^*.

Each solution of problem (8.1)-(8.2) is an equilibrium state for a mechanical system for which $f(x)$ is the potential energy and the equations and inequalities (8.2) are the constraints. For the case under consideration the principle of virtual displacements can be stated as follows:

In order that the state x^* should be an equilibrium state it is necessary and sufficient that for each feasible virtual displacement the following inequality holds

$$\sum_{i=1}^{n} \left.\frac{\partial f}{\partial x_i}\right|_{x=x^*} \delta x_i \geqslant 0,$$

where the sum above must be equal to zero for displacements δx for which no active constraints become inactive which were active at x^* and the sum above must be nonnegative in the case where the vector δx makes at least one of these constraints inactive. [19].

Denoting with

$$p_i = -\left.\frac{\partial f}{\partial x_i}\right|_{x=x^*},$$

one can write the conditions for equilibrium in the following form

$$\sum_{i=1}^{n} p_i \delta x_i \leqslant 0 \tag{8.7}$$

for all $x_1,\ldots,x_n$ satisfying conditions (8.5)-(8.6).

In this manner the principle of virtual displacements reduces the constrained minimization problem to a problem of finding feasible points x^* for which condition (8.7) holds for all solutions of the system (8.5)-(8.6). This problem can be seen as a problem of finding feasible points such that in a neighbourhood of such points the system (8.5)-(8.6) has no solutions which satisfy the conditions

$$\sum_{i=1}^{n} p_i \delta x_i > 0. \tag{8.8}$$

This formulation of the principle of virtual displacements is particularly convenient for further consideration and leads to the following theorem of Kuhn-Tucker [34,37].

THEOREM 8.1. [1)] The vector x^* is a solution of problem (8.1)-(8.2) if and only if there exists a vector w^* such that

$$w_s^* \geq 0 \text{ for } x \in M^{(1)},$$
$$L(x^*+\delta x, w^*) \geq L(x^*,w^*) \geq L(x^*,w^*+\delta w)$$

for all δx and δw satisfying the conditions $x_i^* + \delta x_i \geq 0$ for $i \in N_2$ and $w_s^* + \delta w_s \geq 0$ for $s \in M_1$.

In this theorem

$$L(x,w) = f(x) + \sum_{s \in M_1 \cup M_2} w_s g_w(x)$$

denotes the Lagrangian (function).

Proof. We shall make use of the theorems on alternatives 2.2 (see Section 2.3), which for the problem under investigation can be stated it as follows.

One only of the following alternatives is true: either the system (8.5)-(8.6) has a solution satisfying condition (8.8) or there exists a solution of the system [2)]

$$\sum_{s \in M^{(1)} \cup M_2} a_{si} w_s - p_i \begin{cases} = 0 \text{ for } i \in N^{(1)}, \\ > 0 \text{ for } i \in N^{(2)}, \end{cases} \tag{8.9}$$

$$\begin{aligned} w_s &> 0 \text{ for } s \in M^{(1)}, \\ w_s &= 0 \text{ for } s \in M_1 \backslash M^{(1)}. \end{aligned} \tag{8.10}$$

The system (8.5)-(8.8) is unsolvable in virtue of the principle of virtual displacements and as a result there does exist a solution $\tilde{w}^*$ of the system (8.9)-(8.10) which is nontrivial if the function f(x) is not a constant. We shall now verify that the m-dimensional vector w^* with the components

$$w_s^* = \begin{cases} \tilde{w}_s^* & s \in M^{(1)} \cup M_2, \\ 0 \quad \text{for } s \in M_1 \; M^{(1)} \end{cases} \tag{8.11}$$

and the n-dimensional equilibrium state vector x^* define a saddle point of the Lagrangian function

$$L(x,w) = f(x) + \sum_{s=1}^{m} w_s g_s(x).$$

This means that the following inequalities must hold

$$L(x^*+\delta x,w^*) \geqslant L(x^*,w^*) \geqslant L(x^*,w) \tag{8.12}$$

under the conditions

$$\delta x_i \geqslant 0 \text{ for } i \in N^{(2)}, \quad w_s \geqslant 0 \text{ for } s \in M_1. \tag{8.13}$$

Indeed it follows from (8.6), (8.9) and (8.11) that

$$L(x^*+\delta x,w^*) = L(x^*,w^*) + \sum_{i=1}^{n} \delta x_i =$$

$$= \sum_{s\in M^{(1)}\cup M_2} (a_{si}w_s^*-p_i) \geqslant L(x^*,w^*).$$

On the other hand, because of (8.3), (8.11), and (8.13)

$$L(x^*,w) = L(x^*,w^*) + \sum_{s=1}^{m} (w_s-w_s^*) = g_s(x^*) =$$

$$= L(x^*,w^*) + \sum_{s\in M_1\setminus M^{(1)}} w_s g_s(x^*) \leqslant L(x^*,w^*),$$

which proves the inequalities (8.12). Note that for the proof of the Kuhn-Tucker theorem just given, the convexity of the problem (8.1)-(8.2) is not required and that as a result it has been shown that the inequalities (8.12) hold in a neighbourhood of any local minimum of the function f(x) in the domain (8.2).

Let us consider as an example a nonlinear programming problem with linear constraints

$$f(x_1,\dots,x_n) \to \min, \tag{8.14}$$

$$\left.\begin{array}{l} \sum_{i=1}^{n} a_{si}x_i - \xi_i = 0, \quad s \in M, \\ \xi_s \begin{cases} \leqslant b_s \text{ for } s \in M_1, \\ = b_s \text{ for } s \in M_2, \end{cases} \\ x_i \geqslant 0 \text{ for } i \in N_2. \end{array}\right\} \tag{8.15}$$

Although the analogy of problem (8.1)-(8.2) with an equilibrium problem for a mechanical system has a rather formal character in this particular case of problem (8.14)-(8.15) one can readily picture the model. This model is different from the model of a linear programming problem as depicted in Figure 1.8 only in that the forces applied to bars of the problem

$$p_i(x_i,\ldots,x_n) = -\frac{\partial f}{\partial x_i}, \quad i = 1,\ldots,n,$$

depend on the coordinates of the state of the model. In the case under consideration the conditions for equilibrium are obviously directly analogous to (1.31)-(1.33) and they take the form

$$\left.\begin{aligned}
&\sum_{s=1}^{m} a_{si}w_s^* + \left.\frac{\partial f}{\partial x_i}\right|_{x=x^*} = 0, \quad i \in N_1\\
&\sum_{s=1}^{m} a_{si}w_s^* + \left.\frac{\partial f}{\partial x_i}\right|_{x=x^*} \begin{cases} = 0 \text{ for } x_i^* > 0, \\ \geqslant 0 \text{ for } x_i^* = 0, \end{cases} \quad i \in N_2,\\
&w_s^* \begin{cases} = 0 \text{ for } \xi_s^* < b_s, \\ \geqslant 0 \text{ for } \xi_s^* = b_s, \end{cases} \quad s \in M_1
\end{aligned}\right\} \qquad (8.16)$$

Here, precisely as in the case of linear programming problems, w_s^* is the pressure difference in the volumes $V_s^{(-)}$ and $V_s^{(+)}$ of the model of the constraints (8.15). Relations (8.16) are again, as in the case of linear programming problems, necessary and sufficient conditions for the optimality of the vector x^* and they represent the fundamental duality theorem for the class of problems under consideration. Multiplying the first n relations of (8.16) with $x_1^*,\ldots,x_n^*$ respectively and summing the results one obtains the basic equality of duality theory

$$\sum_{s=1}^{m} b_s w_s^* = \sum_{i=1}^{n} \left.\frac{\partial f}{\partial x_i}\right|_{x=x^*} x_i^*.$$

In the case of a quadratic programming problem with linear constraints, that is in the case where

$$f(x_1,\ldots,x_n) = \sum_{i=1}^{n} r_{0i}x_i + \tfrac{1}{2}\sum_{i,j=1}^{n} r_{ij}x_ix_j,$$

the equality (8.16) takes the following well known form [37].

$$\sum_{s=1}^{m} b_s w_s^* = \sum_{i=1}^{n} r_{0i}x_i^* + \sum_{i,j=1}^{n} r_{ij}x_i^*x_j^*. \qquad (8.17)$$

8.3. Numerical methods for solving nonlinear programming problems

In Chapters III, V-VII we have seen that the method of redundant constraints leads to the construction of various numerical solution algorithms and decomposition methods for linear algebra problems and linear programming problems. In these problems the redundant constraints are used to control the transition processes to an equilibrium state in the physical models of the problems under consideration. In that way a series of alternating constraints makes it possible to replace a large complicated equilibrium problem with a sequence of simpler problems of finding intermediate states each of which is in turn an equilibrium state of the model subject to additional (redundant) constraints. Below we shall see that these same redundant constraints techniques are also efficaceous in the construction of numerical algorithms for solving nonlinear programming problems.

Convex quadratic programming problems with linear constraints. Consider a nonlinear programming problem of the form

$$\left.\begin{aligned} &\sum_{i=1}^{n} r_{0i}x_i - \tfrac{1}{2}\sum_{i,j=1}^{n} r_{ij}x_ix_j \to \min,\\ &\sum_{i=1}^{n} a_{si}x_i - \xi_s = 0,\quad s = 1,\dots,m, \end{aligned}\right\} \tag{8.18}$$

$$\xi_s \begin{cases} \leqslant b_s, & s \in M_1 = (1,\dots,m_1),\\ = b_s, & s \in M_2 = (m_1+1,\dots,m), \end{cases} \tag{8.19}$$

where the second sum in the objective function is a nonnegative definite quadratic form. Clearly the physical model of problem (8.18)-(8.19) only differs from the physical model of a general linear programming problem as sketched in Figure 1.8 in that the forces $p_1,\dots,p_n$ acting on the bars are now not constant, but are functions of the state. They are determined by the formulas

$$p_i = -\frac{\partial f}{\partial x_i} = -r_{0i} - \sum_{j=1}^{n} r_{ij}x_j,\quad i = 1,\dots,n. \tag{8.20}$$

In the case under consideration one uses first of all geometric (holonomic) redundant constraints of the form

$$\left.\begin{aligned} x_i &= \text{const}, \ i = 1,\dots,\alpha-1, \ \alpha+1,\dots,n, \\ \xi_s &= \text{const}, \ s = 1,\dots,m, \\ \alpha &= 1,2,\dots, \end{aligned}\right\} \quad (8.21)$$

and, arguing as in Section 3.3 one obtains an algorithm which is analogous to the first algorithm for solving linear programming problems. In fact under redundant constraints of the type (8.21) the problem of finding the equilibrium of the physical model becomes an equilibrium problem for one single bar which is subject to the interior forces caused by the pressures of the ideal gases and an exterior force p_α which depends linearly on the coordinate x_α which defines the position of the piston (or bar).

In this manner, using exactly the same sequence of redundant constraints as was used in Section 3.3 to construct the first algorithm for solving linear programming problems numerically, one arrives at a sequence of vectors $x^{(0,1)},\dots,x^{(0,n)},\dots,x^{(1,1)},\dots,x^{(1,n)},\dots,x^{(\nu,n)},\dots$ where

$$x_i^{(\nu,\alpha)} = \begin{cases} x_i^{(\nu+1)} & \text{for } i = 1,2,\dots,\alpha-1, \\ x_i^{(\nu)} & \text{for } i = \alpha,\alpha+1,\dots,n, \end{cases}$$

$$x^{(\nu,n+1)} \equiv x^{(\nu+1,1)}, \quad \nu = 0,1,\dots$$

Consequently to arrive at state $x^{(\nu,\alpha+1)}$ from state $x^{(\nu,\alpha)}$ only the coordinate $x_\alpha^{(\nu)}$ of the state $x_\alpha^{(\nu,\alpha)}$ is changed to the value $x_\alpha^{(\nu+1)}$, which is the equilibrium value of that coordinate for the model with redundant constraints

$$x_i = \begin{cases} x_i^{(\nu+1)} & \text{for } i = 1,2,\dots,\alpha-1, \\ x_i^{(\nu)} & \text{doe } i = \alpha+1,\dots,n, \end{cases}$$

$$\xi_s^{(\nu,\alpha)} = \xi_s(x^{(\nu,\alpha)} =$$

$$\left.\begin{aligned} &= \min\{b_s, \sum_{i=1}^{\alpha-1} a_{si}x_i^{(\nu+1)} + \sum_{i=\alpha}^{n} a_{si}x_i^{(\nu)}\}, \quad s \in M_1 \\ \xi_s^{(\nu,\alpha)} &= b_s, \qquad\qquad s \in M_2. \end{aligned}\right\} \quad (8.22)$$

The equilibrium conditions by means of which $x_\alpha^{(\nu+1)}$ is determined, are obtained by substituting for the constant force p_α in the equilibrium conditions (3.13) (see Section 3.3 of Chapter III) the quantity defined by formule (8.20). Substituting the state coordinates $x^{(\nu,\alpha+1)}$ in (8.20) one obtains

$$p_\alpha(x^{(\nu,\alpha+1)}) = -r_{0\alpha} - \sum_{j=1}^{\alpha} r_{\alpha j}x_j^{(\nu+1)} - \sum_{j=\alpha+1}^{n} r_{\alpha j}x_j^{(\nu)}.$$

Comparing this with the formula for $p_\alpha(x^{(\nu,\alpha)})$

$$p_\alpha(x^{(\nu,\alpha)}) = -r_{0\alpha} - \sum_{j=1}^{\alpha-1} r_{\alpha j}x_j^{(\nu+1)} - \sum_{j=\alpha}^{n} r_{\alpha j}x_j^{(\nu)}, \tag{8.23}$$

we find

$$p_\alpha(x^{(\nu,\alpha+1)}) = p_\alpha(x^{(\nu,\alpha)}) - r_{\alpha\alpha}(x_\alpha^{(\nu+1)}-x_\alpha^{(\nu)}).$$

Substituting $p_\alpha(x^{(\nu,\alpha+1)})$ in (8.16) we obtain the equilibrium condition which determines the quantity $x_\alpha^{(\nu+1)}$:

$$p_\alpha(x^{(\nu,\alpha)})-r_{\alpha\alpha}(x_\alpha^{(\nu+1)}-x_\alpha^{(\nu)})+\tilde{q}_0 \sum_{s=1}^{m} \frac{a_{s\alpha}}{\|a_s\|} y_s(x^{(\nu,\alpha)},\xi_s^{(\nu,\alpha)})-$$

$$-(x_\alpha^{(\nu+1)}-x_\alpha^{(\nu)})\tilde{q}_0 \sum_{s=1}^{m} \frac{a_{s\alpha}^2}{\|a_s\|} \begin{cases} = 0 \text{ if } \alpha \in N_1 \\ = 0 \text{ if } \alpha \in N_2, x_\alpha^{(\nu+1)} > 0, \\ \leqslant 0 \text{ if } \alpha \in N_2, x_\alpha^{(\nu+1)} = 0, \end{cases} \tag{8.24}$$

Under the equilibrium condition (8.24)

$$y_s(x^{(\nu,\alpha)},\xi_s^{(\nu,\alpha)}) = \sum_{i=1}^{\alpha-1} a_{si}x_i^{(\nu+1)} + \sum_{i=\alpha}^{n} a_{si}x_i^{(\nu)} - \xi_s^{(\nu,\alpha)},$$

$$\|a_s\| = 1 + \sum_{i=1}^{n} |a_{si}|$$

Introducing a notation analogous to the notations used in Section 3.3:

$$\Psi_\alpha(x^{(\nu,\alpha)},\xi^{(\nu,\alpha)}) =$$

$$= x_\alpha^{(\nu)} + \frac{p_\alpha(x^{(\nu,\alpha)})-\tilde{q}_0 \sum_{s=1}^{m} \frac{a_{s\alpha}}{\|a_s\|} y_s(x^{(\nu,\alpha)},\xi_s^{(\nu,\alpha)})}{r_{\alpha\alpha}+\tilde{q}_0 \sum_{s=1}^{m} \frac{a_{s\alpha}^2}{\|a_s\|}}; \tag{8.25}$$

one obtains from the equilibrium condition a formula for calculating $x^{(\nu+1)}$

$$x_\alpha^{(\nu+1)} = \begin{cases} \Psi_\alpha(x^{(\nu,\alpha)}, \xi^{(\nu,\alpha)}), & \alpha \in N_1, \\ \Psi_\alpha(x^{(\nu,\alpha)}, \xi^{(\nu,\alpha)}) \underline{1}[\Psi_\alpha(x^{(\nu,\alpha)}, \xi^{(\nu,\alpha)}], & \alpha \in N_2. \end{cases}$$

The function $\Psi_\alpha(x^{(\nu,\alpha)}, \xi^{(\nu,\alpha)})$ turns out to be only a function of the equilibrium state vector $x^{(\nu,\alpha)}$ found during the previous step. Indeed, using the notations of Sections 3.3

$$z_s(x^{(\nu,\alpha)}) = \sum_{i=1}^{\alpha-1} a_{si} x_i^{(\nu+1)} + \sum_{i=\alpha}^{n} a_{si} x_i^{(\nu)} - b_s, \tag{8.26}$$

one verifies that in virtue of (8.22)

$$y_s(x^{(\nu,\alpha)}, \xi_s^{(\nu,\alpha)}) = y_s(x^{(\nu,\alpha)}) =$$

$$= \begin{cases} z_s(x^{(\nu,\alpha)}) \underline{1}[z_s(x^{(\nu,\alpha)})], & s \in M_1, \\ z_s(x^{(\nu,\alpha)}), & s \in M_2, \end{cases}$$

so that formula (8.25) takes the form

$$\Psi_\alpha(x^{(\nu,\alpha)}) = x_\alpha^{(\nu)} + \frac{1}{r_{\alpha\alpha} + \tilde{q}_0 \sum\limits_{s=1}^{m} \frac{a_{s\alpha}^2}{\|a_s\|}} \{p_\alpha(x^{(\nu,\alpha)}) -$$

$$- \tilde{q}_0 (\sum_{s=1}^{m_1} \frac{a_{s\alpha}}{\|a_s\|} z_s(x^{(\nu,\alpha)}) \underline{1}[z_s(x^{(\nu,\alpha)})] +$$

$$+ \sum_{s=m_1+1}^{m} \frac{a_{s\alpha}}{\|a_s\|} z_s(x^{(\nu,\alpha)}))\}. \tag{8.27}$$

The formulas

$$x_\alpha^{(\nu+1)} = \begin{cases} \Psi_\alpha(x^{(\nu,\alpha)}), & \alpha \in N_1, \\ \Psi_\alpha(x^{(\nu,\alpha)}) \underline{1}[\Psi_\alpha(x^{(\nu,\alpha)})], & \alpha \in N_2, \end{cases}$$

$\alpha = 0, 1, \ldots,$

together with formulas (8.27), (8.26), and (8.23) constitute an iterative algorithm for solving convex quadratic programming problems with linear constraints.

For a given value of the parameter $\tilde{q}_0$ the sequence of vectors $x^{(\nu,\alpha)}$, $\alpha = 1,\dots,n$, $\nu = 0,1,\dots$ converges to an equilibrium state $\bar{x}(\tilde{q}_0)$ of the physical model and for each monotonically growing sequence $\tilde{q}_0,\tilde{q}_1,\dots$ the sequence $\bar{x}(\tilde{q}_0),\bar{x}(\tilde{q}_1),\dots$ converges in virtue of Theorem 2.7 to an optimal vector of the problem (8.17)-(8.18).

General quadratic programming problems. Bilinear programming. A more difficult programming problem is the problem of minimizing a convex quadratic objective function over a convex set defined by a set of quadratic equations and inequalities. Below we shall present an approach to solving these problems which can also be extended to cover more general problems.

We consider the problem

$$\sum_{i,j=0}^{n} r_{ij}x_ix_j \to \min, \tag{8.28}$$

$$\left.\begin{aligned} &\sum_{i,j=0}^{n} a_{ij}^{(s)}x_ix_j \leqslant b_s, \quad s = 1,\dots,m, \\ &x_0 = 1, \; x_i \geqslant 0 \text{ for } i \in N_2. \end{aligned}\right\} \tag{8.29}$$

The idea of the method consists in reducing problem (8.28)-(8.29) to a series of essentially simpler problems of finding equilibria of physical models of linear programming problems. In fact, problems (8.28)-(8.29) can be rewritten in the equivalent form

$$\sum_{i,j=0}^{n} r_{ij}u_ix_j \to \min, \tag{8.30}$$

$$\left.\begin{aligned} &\sum_{i,j=0}^{n} a_{ij}^{(s)}u_ix_j \leqslant b_s, \quad s = 1,\dots,m, \\ &x_0 = u_0 = 1, \; x_i,\, u_i \geqslant 0 \text{ for } i \in N_2, \end{aligned}\right\} \tag{8.31}$$

$$u_i = x_i, \quad i = 1,\dots,n. \tag{8.32}$$

Without the constraint (8.32) this problem becomes a bilinear programming problem, and therefore everything which will be said below also applies to such problems.

The method of physical models for problems (8.30)-(8.32) can be made clear in the following way. Let $x_1^{(0)} = u_1^{(0)}$, $x_2^{(0)} = u_2^{(0)}, \ldots, x_n^{(0)} = u_n^{(0)}$ be arbitrary real numbers which are only required to satisfy the condition $x_i^{(0)} = u_i^{(0)} \geqslant 0$ for $i \in N_2$. Consider the equilibrium problem for the physical model with parameter $\tilde{q}_0$ of the following problem

$$\sum_{j=0}^{n} \left(\sum_{i=0}^{n} r_{ij} u_i^{(0)} \right) x_j \to \min, \tag{8.33}$$

$$\left. \begin{aligned} &\sum_{j=0}^{n} \left(\sum_{i=0}^{n} a_{ij}^{(s)} u_i^{(0)} \right) x_j \leqslant b_s, \quad s = 1,\ldots,m, \\ &u_0^{(0)} = x_0 = 1, \quad x_j \geqslant 0 \text{ for } j \in N_2, \end{aligned} \right. \tag{8.34}$$

$$x_j = u_j^{(0)}, \quad j = 1,\ldots,n. \tag{8.35}$$

Introducing the notations

$$\left. \begin{aligned} c_j^{(0)} &= \sum_{i=0}^{n} r_{ij} u_i^{(0)}, \quad j = 0,1,\ldots,n, \\ \alpha_{0j}^{(s)} &= \sum_{i=1}^{n} a_{ij}^{(s)} u_i^{(0)}, \quad j = 0,1,\ldots,n, \quad s = 1,\ldots,m, \end{aligned} \right\} \tag{8.36}$$

one can write problems (8.33)-(8.35) in the form

$$\sum_{j=0}^{n} c_j^{(0)} x_j \to \min, \tag{8.37}$$

$$\left. \begin{aligned} &\sum_{j=0}^{n} \alpha_{0j}^{(s)} x_j \leqslant b_s, \quad s = 1,\ldots,m, \\ &x_0 = 1, \quad x_j \geqslant 0 \text{ for } j \in N_2, \\ &x_j = u_j^{(0)}, \quad j = 1,\ldots,n. \end{aligned} \right\} \tag{8.38}$$

An equilibrium state of the physical model (of problem (8.37)-(8.38)) with parameter $\tilde{q}_0$ can be found by means of any of the algorithms of Chapter III and it is obviously not necessary to write down the corresponding formulas.

It is important to note that the (extra) conditions $x_j = u_j^{(0)}$ for $j \in N$ make problem (8.37)-(8.38) trivial. However, we are not interested in this trivial solution but in the equilibrium state of the model of problem (8.37)-(8.38) for some finite value of the parameter $\tilde{q}_0$. Therefore in an equilibrium state the conditions $x_j = u_j^{(0)}$ will no longer hold and the equilibrium vector $x^{(1)}(\tilde{q}_0)$ will be different from the vector $x^{(0)} = u^{(0)}$.

Thus, let $x_1^{(1)}(\tilde{q}_0),\ldots,x_n^{(1)}(\tilde{q}_0)$ be the coordinates of the equilibrium state vector of the physical model of problem (8.37)-(8.38). Then setting $x_i = x_i^{(1)}(\tilde{q}_0)$ in the initial problem (8.39)-(8.32) and considering the quantities $u_1,\ldots,u_n$ as the unknowns we obtain the following problem:

$$\sum_{i=0}^{n}\Big(\sum_{j=0}^{n} r_{ij}x_j^{(1)}(\tilde{q}_0)\Big)u_i \to \min, \tag{8.39}$$

$$\left.\begin{aligned} &\sum_{i=0}^{n}\Big(\sum_{j=0}^{n} a_{ij}^{(s)}x_j^{(1)}(\tilde{q}_0)\Big)u_i \leqslant b_s, \quad s = 1,\ldots,m,\\ &u_0 = x_0^{(1)} = 1, \; u_i \geqslant 0, \; i \in N_2, \end{aligned}\right\} \tag{8.40}$$

$$u_i = x_i^{(1)}(\tilde{q}_0), \; i = 1,\ldots,n. \tag{8.41}$$

Introducing the notations

$$d_i^{(1)} = \sum_{j=0}^{n} r_{ij}x_j^{(1)}(\tilde{q}_0),$$

$$\beta_{1i}^{(s)} = \sum_{j=0}^{n} a_{ij}^{(s)}x_j^{(1)}(\tilde{q}_0)$$

we write problem (8.39)-(8.41) in the form

$$\sum_{i=0}^{n} d_i^{(1)} u_i \to \min,$$

$$\sum_{i=0}^{n} \beta_{1i}^{(s)} u_i \leqslant b_s, \quad s = 1,\dots,m,$$

$$u_0 = 1, \; u_i \geqslant 0, \; i \in N_2,$$

$$u_i = x_i^{(1)}(\tilde{q}_0), \quad i = 1,\dots,n.$$

In this way we obtain a bivector $(x^{(1)}(\tilde{q}_0), u^{(1)}(\tilde{q}_0))$ which is a first approximation to the bivector $(\bar{x}(\tilde{q}_0), \bar{u}(\tilde{q}_0))$ which represents the equilibrium state of the physical model of problem (8.30)-(8.32) with parameter $\tilde{q}_0$. To proceed, the vector $u^{(1)}(\tilde{q}_0)$ must be used to construct the following problem which is analogous to problem (8.37)-(8.38).

$$\sum_{j=0}^{n} c_j^{(1)} x_j \to \min, \tag{8.42}$$

$$\left.\begin{aligned} &\sum_{j=0}^{n} \alpha_{1j}^{(s)} x_i \leqslant b_s, \quad s = 1,\dots,m, \\ &x_0 = 1, \; x_j \geqslant 0, \; j \in N_2, \\ &x_j = u_j^{(1)}(\tilde{q}_0), \quad j = 1,\dots,n, \end{aligned}\right\} \tag{8.43}$$

where, by analogy with (8.36)

$$c_j^{(1)} = \sum_{i=0}^{n} r_{ij} u_i^{(1)}(\tilde{q}_0),$$

$$\alpha_{1j}^{(s)} = \sum_{i=0}^{n} a_{ij}^{(s)} u_i^{(1)}(\tilde{q}_0),$$

$$j = 0,1,\dots,n; \quad s = 1,\dots,m.$$

Let $x^{(2)}(\tilde{q}_0)$ be an equilibrium state vector for the physical model of problem (8.42)-(8.43) and so on. The sequel is is clear and the method for finding the bivector $(\bar{x}(q_0),\bar{u}(q_0))$ consists in solving a sequence of equilibrium problems for physical models of the following (linear) problems:

1) Find the vector $x^{(k)}(\tilde{q}_0)$ which represents an equilibrium state of the physical model of the problem

$$\sum_{j=0}^{n} c_j^{(k-1)} x_j \to \min, \tag{8.44}$$

$$\left.\begin{aligned} &\sum_{j=0}^{n} \alpha_{k-1,j}^{(s)} x_j \leqslant b_s', \quad s = 1,\dots,m, \\ &x_0 = 1, \ x_j \geqslant 0, \ j \in N_2, \\ &x_j = u_j^{(k-1)}, \quad j = 1,\dots,n \end{aligned}\right\} \tag{8.45}$$

where

$$c_j^{(k-1)} = \sum_{i=0}^{n} r_{ij} u_i^{(k-1)}(\tilde{q}_0),$$

$$\alpha_{k-1,j}^{(s)} = \sum_{i=0}^{n} \alpha_{ij}^{(s)} u_i^{(k-1)}(\tilde{q}_0);$$

2) Find the vector $u^{(k)}(\tilde{q}_0)$ which represents an equilibrium state of the physical model of the problem

$$\sum_{i=0}^{n} d_i^{(k)} u_i \to \min, \tag{8.46}$$

$$\left.\begin{aligned} &\sum_{i=0}^{n} \beta_{k,i}^{(k)} u_i \leqslant b_s, \quad s = 1,\dots,m, \\ &u_0 = 1, \ u_i \geqslant 0 \text{ for } i \in N_2, \\ &u_i = x_i^{(k)}(\tilde{q}_0), \ i = 1,2,\dots,n, \end{aligned}\right\} \tag{8.47}$$

where

$$d_j^{(k)} = \sum_{j=0}^{n} r_{ij} x_j^{(k)}(\tilde{q}_0),$$

$$\beta_{k,i}^{(s)} = \sum_{j=0}^{n} a_{ij}^{(s)} x_j^{(k)}(\tilde{q}_0),$$

It will be a useful exercise for the reader to construct

the general thermodynamic potential for the problem (8.30)-(8.32). By analogy with the thermodynamic potential for a linear programming problem the result is

$$\phi(x,u) = \sum_{i,j=1}^{n} r_{ij}u_i x_j +$$

$$+ q_0 \sum_{s=1}^{m+n} [\bar{V}_s^{(+)} \ln \frac{\bar{V}_s^{(+)}}{\bar{V}_s^{(+)}+y_s} + \bar{V}_s^{(-)} \ln \frac{\bar{V}_s^{(-)}}{\bar{V}_s^{(-)}-y_s}], \qquad (8.48)$$

where $\bar{V}_s^{(+)}$ and $\bar{V}_s^{(-)}$ are constants and the quantities $y_1,\ldots,y_{m+n}$ are defined by the formulas

$$y_s = \begin{cases} z_s \underline{1}[z_s], & s = 1,\ldots,m, \\ z_s, & s = m+1,\ldots,m+n, \end{cases}$$

$$z_s = \begin{cases} \sum\limits_{i,j=0}^{n} a_{ij}^{(s)} x_j u_i - b_s, & s = 1,\ldots,m, \\ u_{s-m} - x_{s-m}, & s = m+1,\ldots,m+n- \end{cases}$$

The last sum in the expression for the thermodynamic potential (8.48) is again the Helmholtz free energy. Because each of the problems occurring in the iterative procedure described above is an equilibrium problem under stationary (i.e. time invariant) redundant constraints compatible with the initial state for the system under consideration at the time, the thermodynamic potential decreases monotonically, that is

$$\phi(x^{(0)}(\tilde{q}_0),u^{(0)}(\tilde{q}_0)) > \phi(x^{(1)}(\tilde{q}_0),u^{(1)}(\tilde{q}_0)) > \ldots$$

$$\ldots > \phi(x^{(\nu)}(q_0),u^{(\nu)}(q_0)) > \ldots .$$

Moreover in Chapter III it was shown that the thermodynamic potential also decreases monotonically with respect to a sequence of successive approximations to each intermediate equilibrium state. Therefore to solve the equilibrium problems (1) and (2) we can limit ourselves to some arbitrary number of iterations. It follows from Theorem 2.7 that the following assertions hold

$$\lim_{\nu\to\infty} x^{(\nu)}(\tilde{q}_0) = \bar{x}(\tilde{q}_0), \quad \lim_{\nu\to\infty} u^{(\nu)}(\tilde{q}_0) = \bar{u}(\tilde{q}_0),$$

$$\lim_{\tilde{q}_0\to\infty} \bar{x}(\tilde{q}_0) = \lim_{\tilde{q}_0\to\infty} \bar{u}(\tilde{q}_0) = x^*,$$

where $(\bar{x}(\tilde{q}_0), \bar{u}(\tilde{q}_0))$ is the bivector representing an equilibrium state of the physical model of the bilinear programming problem (8.30)-(8.32) and x^* is an optimal vector for problem (8.28)-(8.29).

The quadratic linear programming problem with linear constraints

$$\sum_{j=1}^{n} r_{0j}x_j + \tfrac{1}{2}\sum_{i,j=1}^{n} r_{ij}x_ix_j \to \min,$$

$$\sum_{i=1}^{n} a_{si}x_i - \xi_s = 0, \qquad s = 1,\ldots,m,$$

$$x_i \geqslant 0 \text{ for } i \in N_2,\ \xi_s \leqslant b_s,\ s = 1,\ldots,m,$$

is a special case of problem (8.28)-(8.29). It is easy to verify that in this case problem (8.46)-(8.47) splits into n simple one dimensional equilibrium problems for the physical models of the following problems:
$d_i^{(k)} u_i \to \min$, under the condition $u_i = x_i(q_0)$, $u_i \geqslant 0$ for $i \in N_2$.

Remark 1. In the case of a general quadratic programming problem with mixed (linear and quadratic) constraints it is clear that no changes in the arguments above are necessary.

Remark 2. In case $r_{ij} = 0$, $i, j = 0,1,\ldots,n$ the method described becomes a method for solving systems of equations and inequalities of the second order.

A method for reducing a general quadratic programming problem to a sequence of equilibrium problems for passive physical systems.

The idea behind the method explained below consists in replacing a general quadratic programming problem by a series of equilibrium problems for physical models of inconsistent systems of second degree equations and inequalities. In this sense the method presented is a generalization of the method for linear programming problems explained in Section 3.5. And precisely as in Section 3.5 one avoids the problem of choosing a growing sequence of model parameter values $\tilde{q}_0$, because in this case also the equilibrium state of the physical model does not depend on the size of that parameter.

Let $b_0^{(0)}$ be a number which satisfies the condition

$$b_0^{(0)} < \min \sum_{i,j=0}^{n} r_{ij} x_i x_j \tag{8.49}$$

over the domain, defined by the system of inequalities

$$\left.\begin{array}{l} \sum_{i,j=0}^{n} a_{ij}^{(s)} x_i x_j \leqslant b_s^{(0)}, \quad s = 1,\dots,m, \\ x_0 = 1, \; x_i \geqslant 0 \text{ for } i \in N_2. \end{array}\right\} \tag{8.50}$$

Writing $r_{ij} = a_{ij}^{(0)}$, consider the equilibrium problem for the physical model of the following system of second degree inequalities

$$\left.\begin{array}{l} \sum_{i,j=0}^{n} a_{ij}^{(s)} x_i x_j \leqslant b_s^{(0)}, \quad s = 0,1,\dots,m, \\ x_0 = 1, \; x_i \geqslant 0 \text{ for } i \in N_2. \end{array}\right\} \tag{8.51}$$

The system (8.51) is inconsistent by virtue of condition (8.49) and consequently the free Helmholtz energy of its physical model must be positive and assume a minimum in an equilibrium state. This follows from the fact that the Helmholtz free energy of the physical model of system (8.51) is equal to

$$F(y_0,y_1,\dots,y_m) = \sum_{s=0}^{m} \left(\bar{V}_s^{(+)} \ln \frac{\bar{V}_s^{(+)}}{\bar{V}_s^{(+)}+y_s} + \bar{V}_s^{(-)} \ln \frac{\bar{V}_s^{(-)}}{\bar{V}_s^{(-)}-y_s}\right).$$

where

$$y_s = z_s \underline{1}[z_s], \quad z_s = \sum_{i,j=0}^{n} a_{ij}^{(s)} x_i x_j - b_s, \quad s = 0,1,\dots,m,$$

and is a positive definite strictly convex function of the errors $y_0,\dots,y_m$, and in view of condition (8.49) there is no state for which all the errors are zero. Let $\bar{x}(\tilde{q}_0)$ be the vector representing the equilibrium state of the physical model of system (8.51). From the fact that the Helmholtz free energy $F(y_0,\ldots,y_m)$ is positive definite and strictly convex in $y_0,\ldots,y_m$ it follows that the equilibrium state $\bar{x}^{(0)}$ does not belong to the set (6.50) of feasible vectors of the initial problem and that moreover at the equilibrium $\bar{x}^{(0)}$ the following inequality holds

$$\sum_{i,j=0}^{n} a_{ij}^{(0)}\bar{x}_i^{(0)}\bar{x}_j^{(0)} > b_0^{(0)}. \tag{8.52}$$

This assertion is easily proved by assuming the opposite. It is also noteworthy that the gradient vector of the Helmholtz free energy of the physical model of the inequalities

$$\sum_{i,j=0}^{n} a_{ij}^{(0)}x_i x_j \leqslant b_0^{(0)},$$

$$x_0 = 1,\ x_i \geqslant 0 \text{ for } i \in N_2$$

in the point $\bar{x}^{(0)}$ is normal to the hyperplane which separates the sets (8.50) and (8.51).

The system of inequalities (8.51) is obviously equivalent to the following system

$$\left.\begin{array}{l} \sum_{i=0}^{n} a_{ij}^{(s)}u_i x_j \leqslant b_s^{(0)},\quad s = 0,\ldots,m, \\ x_0 = u_0 = 1,\ u_i = x_i,\ i = 1,\ldots,n,\ x_i \geqslant 0,\ i \in N_2 \end{array}\right\} \tag{8.53}$$

Let us consider the equilibrium problem for the physical model of system (8.53). This problem in contrast with the equilibrium problem for the physical model of (8.51) has solution bivectors $(x^{(0)},u^{(0)})$ with $x^{(0)} \neq u^{(0)}$. To find the components of the vectors $x^{(0)}$ and $u^{(0)}$ one can make use of the methods discussed above of reducing the equilibrium problem for the physical model of system (8.53) to a series of equilibrium problems for models of systems of linear inequalities. These are problems of the type (8.44)-(8.45) and (8.46)-(8.49) in the particular case that $r_{ij} = 0$. The next step of the algorithm consists of solving an equilibrium problem for the phsyical model of the system of second degree inequalities

$$\left.\begin{array}{l}\sum_{i,j=1}^{n} a_{ij}^{(s)} u_i x_j \leqslant b_s^{(1)}, \quad s = 0,1,\ldots,m,\\ u_0 = x_0 = 1, \ u_i = x_i, \ i = 1,\ldots,n, \ x_i = u_i \geqslant 0, \ i \in N_2,\end{array}\right\} \tag{8.54}$$

where by analogy with Section 3.5

$$b_0^{(1)} = \sum_{i,j=0}^{n} a_{ij}^{(0)} u_i^{(0)} x_j^{(0)},$$

$$b_s^{(1)} = b_s^{(0)}, \quad s = 1,\ldots,m.$$

Let $(x^{(1)},u^{(1)})$ be the bivector representing equilibrium for the physical model of problem (8.54). Then the following inequalities hold

$$b_0^{(0)} < b_0^{(1)} < \sum_{i,j=0}^{n} a_{ij}^{(0)} u_i^{(1)} x_j^{(1)} = b_0^{(2)}.$$

The sequel is obvious and the algorithm just presented consists of solving a series of equilibrium problems for the phsyical models of systems of inequalities of the form

$$\left.\begin{array}{l}\sum_{i,j=0}^{n} a_{ij}^{(s)} u_i x_j \leqslant b_s^{(k)}, \quad s = 0,1,\ldots,m,\\ u_0 = x_0 = 1, \ u_i = x_i, \ i = 1,\ldots,n, \ x_i = u_i \geqslant 0, \ i \in N_2,\end{array}\right\}$$

$$k = 0,1,2,\ldots, \tag{8.55}$$

where

$$b_0^{(k)} = \sum_{i,j=0}^{n} a_{ij}^{(0)} u_i^{(k-1)} x_j^{(k-1)},$$

$$b_s^{(k)} = b_s^{(0)}, \quad s = 1,\ldots,m.$$

The iterative process exposed above leads to a series of bivectors $(x^{(0)},u^{(0)})$, $(x^{(1)},u^{(1)})$,... which satisfy the conditions

$$\lim_{k\to\infty} x^{(k)} = \lim_{k\to\infty} u^{(k)} = x^*, \tag{8.56}$$

$$b_0^{(0)} < b_0^{(1)} < b_0^{(2)} < \ldots, \ \lim b_0^{(k)} = \sum_{i,j=0}^{n} r_{ij} x_i^* x_j^*, \tag{8.57}$$

where x is an optimal vector for the general quadratic programming problem (8.28)-(8.29).

Condition (8.56) holds because it is a consequence of the fact that the method proposed consists of the definition of a sequence of equilibrium problems of models corresponding to a sequence of systems of redundant constraints compatible with the previous equilibrium state. Moreover the sequence of systems of inequalities (8.55), by virtue of property (8.57), has as a limit a consistent system for which the equality of the vectors x and u at equilibrium is clear.

Polynomical programming problems. We shall call the following kind of constrained minimization problems polynomial programming problems

$$\sum_{\sigma=1}^{k} \sum_{\substack{\alpha_1,\alpha_2,\ldots,\alpha_n \\ \alpha_1+\alpha_2+\ldots+\alpha_n=\sigma}} p_\sigma^{\alpha_1,\alpha_2,\ldots,\alpha_n} U_\sigma^{(\alpha_1,\ldots,\alpha_n)}(x_1,\ldots,x_n) \to$$

$$\to \min \tag{8.58}$$

$$\sum_{\sigma=1}^{k} \sum_{\substack{\alpha_1,\ldots,\alpha_n \\ \alpha_1+\alpha_2+\ldots+\alpha_n=\sigma}} a_{s\sigma}^{(\alpha_1,\ldots,\alpha_n)} U_\sigma^{(\alpha_1,\ldots,\alpha_n)}(x_1,\ldots,x_n) \leqslant$$

$$\leqslant b_s, \quad s = 1,\ldots,m, \tag{8.59}$$

where

$$U_\sigma^{(\alpha_1,\ldots,\alpha_n)}(x_1,\ldots,x_n) = \prod_{i=1}^{n} x_i^{\alpha_i},$$

and the $\alpha_1,\ldots,\alpha_n$ are nonnegative integers.

The interior sums in the objective function and the constraint functions are overall values of the nonnegative integers $(\alpha_1,\ldots,\alpha_n)$ such that the sum of the α's is equal to σ. The general nonlinear programming problem (8.1)-(8.2) reduces to problems of type (8.58)-(8.59) if the functions $f(x)$ and $g_s(x)$, $s = 1,\ldots,m$ can be sufficiently well approximated by polynomials.

The method studied above for solving general quadratic programming problems easily extends to problems of polynomial porgramming, whose solution can also be reduced to a series of equilibrium problems for physical models of systems of linear inequalities and equalities. Indeed, let us introduce the function

$$\hat{U}^{(\alpha_1,\dots,\alpha_n)}(u_{11},\dots,u_{1,\alpha_1}, u_{21},\dots,u_{2,\alpha_2},\dots$$
$$\dots,u_{n1},\dots,u_{n,\alpha_n}) = \prod_{i=1}^{n} \prod_{j=1}^{\alpha_i} u_{ij}.$$

Then problem (8.58)-(8.59(is equivalent to the following problem

$$\sum_{\sigma=1}^{k} \sum_{\substack{\alpha_1,\dots,\alpha_n \\ \alpha_1+\dots+\alpha_n=\sigma}} p_\sigma^{(\alpha_1,\dots,\alpha_n)} \hat{U}_\sigma^{(\alpha_1,\dots,\alpha_n)} \to \min,$$

$$\sum_{\sigma=1}^{k} \sum_{\substack{\alpha_1,\dots,\alpha_n \\ \alpha_1+\dots+\alpha_n=\sigma}} a_{s\sigma}^{(\alpha_1,\dots,\alpha_n)} \hat{U}_\sigma^{(\alpha_1,\dots,\alpha_n)} \leqslant b_s, \quad s = 1,\dots,m,$$

$$u_{i1} = u_{i2} = \dots = u_{i,a_i}, \quad i = 1,\dots,n.$$

And the reduction of this problem to a series of equilibrium problems of models of linear programming problems is done by the same procedure as is followed for the general quadratic programming problem.

NOTES

1. The theorem as stated here is an extension of the well known theorem to the general case of mixed constraints.

2. The equalities $w_s = 0$, $s \in M_1 \backslash M^{(1)}$ signify that the reaction forces caused by inactive unilaterial constraints at x^* are zero.

Chapter IX

THE TANGENT METHOD

9.1. Introduction

The present chapter contains a general approach to the construction of algorithms for optimum problems, both static and dynamic. The reader will certainly notice a resemblance to the classical Newton method (tangent method), but there is an essential difference: we are looking for constrained extrema of functions or functionals by means of solving unconstrained problems. The functions involved are well-known mathematical objects and they play an important role in dynamic programming [21]. Again we shall pay a lot of attention to significant mechanical analogies.

The method of penalty functions gives rise to a quasi-static process which is realized by variations in the parameters determining the exterior conditions (see Chapter II) and the resulting solution methods for optimization methods are related to the law of the conservation of energy. Let us use another known variant of the idea of penalties where the constrained minimization problem is reduced to a sequence of free minimum problems of the form

$$\min\{\gamma f(x)+F(x)\}, \tag{9.1}$$

where $f(x)$ is the objective function and $F(x)$ a function which penalizes (violations of) the constraints. The solution $\bar{x}(\gamma)$ of (9.1) is such that $\lim_{\gamma\to\infty} \bar{x}(\gamma) = x^*$ and $\lim_{\gamma\to\infty} F(\bar{x}(\gamma)) = 0$.

It is interesting to view $\bar{x}(\gamma)$ as an equilibrium state of a mechanical system represented by a point in configuration space with a force field which is the superposition of an exterior field coming from the force function $-\gamma f(x)$ and an (elastic) field (coming from the constraints with force function $-F(x)$). Slow variations of γ from a positive value γ_0 to 0 can be naturally interpreted as giving rise to a quasi-static transformation characterized by the slowly vanishing exterior force field. The elastic (reaction) forces caused by the constraints do a certain amount of work along the path leading from the equilibrium $\bar{x}(\gamma_0)$ to the equilibrium x^*. This amount of work $F(\bar{x}(\gamma_0))$ is according to the law of the conservation of energy equal to the amount of work with opposite sign done by the exterior (vanishing) force field along the same path.

It follows that the value $F(\bar{x}(\gamma_0))$ of the penalty function at $\bar{x}(\gamma_0)$, that is the deformation energy of the constraints at the equilibrium $\bar{x}(\gamma_0)$, contains a great deal of information which can be used for solving constrained minimization problems. This is the basic idea underlying this chapter.

9.2. Constrained minimization problems

Consider a constrained minimization problem

$$\min_x \{f(x) \mid x \in \Omega\}, \tag{9.2}$$

where $f(x)$ is a convex scalar valued function of the vector variable $x = (x_1,\dots,x_n)$ and

$$\Omega = \{x \mid \phi_s(x) \leq 0,\ s = 1,\dots,m\} \tag{9.3}$$

is a convex set of admissible points for problem (9.2). Let $F(\phi_1(x),\dots,\phi_m(x))$ be a function, strictly convex outside Ω which penalizes the constraints (9.3). This could for instance be

$$F(\phi_1(x),\dots,\phi_m(x)) = \sum_{s=1}^{m} [\max(0,\phi_s(x))]^2. \tag{9.4}$$

Let us consider the one parameter family of free minimization problems

$$\min_x \Phi(x,q), \tag{9.5}$$

where

$$\Phi(x,q) = f(x) + qF(\phi_1(x),\dots,\phi_m(x)),$$

and let $\bar{x}(q)$ denote the solution of (9.5) for a fixed q. Thus the minimization of $\Phi(x,q)$ with respect to $x_1,\dots,x_n$ defines a vector function $\bar{x}(q)$ of the independent scalar variable q which by virtue of Theorem 2.5 of Chapter II has the following properties

$$\lim_{q\to\infty} \bar{x}(q) = x^*, \qquad \lim_{q\to\infty} \Phi(\bar{x}(q),q) = f(x^*), \tag{9.6}$$

with x^* the optimum vector for problem (9.2). Assuming that $\bar{x}(q)$ is known, the system of equations

$$u = f(\bar{x}(q)), \qquad v = f(\phi_1(\bar{x}(q)),\ldots,\phi_m(\bar{x}(q))) \tag{9.7}$$

is a parametrized equation for a plane curve $v = \psi(u)$, see Figure 9.1.

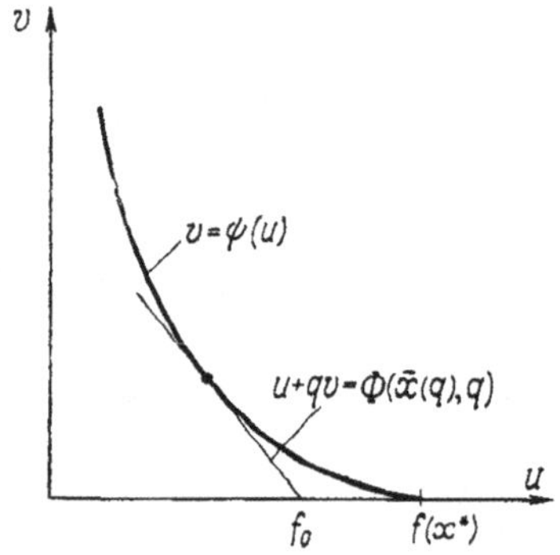

Fig. 9.1.

Consider the straight line

$$u + qv = \Phi(\bar{x}(q),q) \tag{9.8}$$

defined by the minimum of $\Phi(x,q)$. It is obvious that the straight line in question passes through the point

$$v = F(\phi(\bar{x}(q))), \; u = f(\bar{x}(q)),$$

of the curve $v = \psi(u)$, and this straight line is moreover tangent to this curve at this point. Indeed Equation (9.8) implies

$$\frac{dv}{du} = -\frac{1}{q}, \tag{9.9}$$

and it remains to check that this is also the value of the derivative $d\psi/du$. The defining equations (9.7) imply the identities

$$du = \sum_{i=1}^{n} \left.\frac{\partial f}{\partial x_i}\right|_{x=\bar{x}(q)} \frac{d\bar{x}_i}{dq}\, dq, \tag{9.10}$$

$$dv = \sum_{s=1}^{m} \left.\frac{\partial F}{\partial \phi_s}\right|_{x=\bar{x}(q)} \sum_{i=1}^{n} \left.\frac{\partial \phi_1}{\partial x_i}\right|_{x=\bar{x}(q)} \frac{d\bar{x}_i}{dq}\, dq. \tag{9.11}$$

Because $\bar{x}(q)$ minimizes (x,q) with respect to $x_1,\ldots,x_n$ in the absence of constraints it follows that at this point

$$\frac{\partial \Phi}{\partial x_i} = 0,\ i = 1,\ldots,n,$$

or

$$\left(\frac{\partial f}{\partial x_i} + q \sum_{s=1}^{m} \frac{\partial F}{\partial \phi_s} \frac{\partial \phi_s}{\partial x_i}\right)_{x=\bar{x}(q)} = 0,\ i = 1,\ldots,n. \tag{9.12}$$

Interchanging the sums in (9.11) and using (9.12) it follows that

$$dv = dq \sum_{i=1}^{n} \frac{d\bar{x}_i}{dq}\left(\sum_{s=1}^{m} \frac{\partial F}{\partial \phi_s} \frac{\partial \phi_s}{\partial x_i}\right)_{x=\bar{x}(q)} = -\frac{dq}{q} \sum_{i=1}^{n} \frac{\partial f}{\partial x_i} \frac{d\bar{x}_i}{dq}. \tag{9.13}$$

And from (9.13) and (9.10) there results

$$\frac{d\psi}{du} = -\frac{1}{q}, \tag{9.14}$$

which means the straight line (9.8) is tangent to $v = \psi(u)$, see Figure 9.1.

Now let us turn to the method itself, which consists in replacing (9.2) with a series of unconstrained minimization problems which is not dependent on the choice of a growing series of penalty parameter values $q_1, q_2, \ldots$ (as in the case of the more direct use of penalty function ideas).

Let q_0 be a fixed positive number, and $x^{(0)} = \bar{x}(q_0)$ a solution of (9.2) for $q = q_0$. This solution defines the point $u^{(0)} = f(x^{(0)})$, $v^{(0)} = F\phi(x^{(0)}))$ of the curve $v = \psi(u)$ and determines the tangent

$$u + q_0 v = \Phi(x^{(0)}, q_0) \tag{9.15}$$

at $(u^{(0)},v^{(0)})$ to this curve.

One now finds without trouble the value f_0 of the function $f(x)$ which satisfies

$$f(x^{(0)}) < f_0 < f^* = f(x^*). \tag{9.16}$$

which is reached at the point where the tangent (9.15) cuts the abscissa of the plane (u,v). Set $v = 0$ in (9.15). This gives

$$f_0 = u_0 = \Phi(x^{(0)},q_0) = f(x^{(0)}) + q_0F(\phi(x^{(0)})). \tag{9.17}$$

The calculation of f_0 (a preliminary step in the method of tangents) is followed by a series of more important steps.

The main step. Consider the unconstrained minimization problem

$$\min_x\{(f(x)-u_0)^2 + q_0F(\phi(x))\}, \tag{9.18}$$

where $q = q_0$ as above and let $x^{(1)}$ be the solution of this problem. Now consider the curve

$$(u-u_0)^2 + q_0v = R^{(1)}, \tag{9.19}$$

where

$$R^{(1)} = (f(x^{(1)})-u_0)^2 + q_0F(\phi(x^{(1)})). \tag{9.20}$$

The curve (9.19) is, as is easily checked, a parabola which is tangent to the curve $v = \psi(u)$ at the point $u^{(1)} = f(x^{(1)})$, $v^{(1)} = F(\phi(x^{(1)}))$, see Figure 9.2. Indeed the point $x^{(1)}$ satisfies the minimum conditions

$$[2(f(x^{(1)})-u_0)\frac{\partial f}{\partial x_i} + q_0\frac{\partial F}{\partial x_i}]_{x=x^{(1)}} = 0 \tag{9.21}$$

and dividing by $2(f(x^{(1)})-u_0) \neq 0$ then gives

$$\frac{\partial f}{\partial x_i} + q_1\frac{\partial F}{\partial x_i} = 0, \quad i = 1,\dots,n, \tag{9.22}$$

where

$$q_1 = \frac{q_0}{2(f(x^{(1)})-u_0)}. \tag{9.23}$$

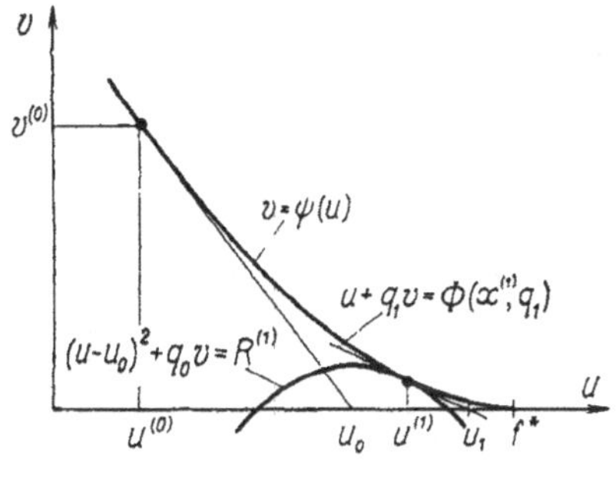

Fig. 9.2.

The equalities (9.22) are the minimum conditions for the function $f(x) + q_1F(\phi(x))$, that is the vector $x^{(1)}$ realizes at the same time the minimum of (9.18) and that of (9.2) for $q = q_1$. This result means that the point $(u^{(1)},v^{(1)})$ belongs to the curve $v = \psi(u)$ and that the straight line

$$u + q_1v = \Phi(x^{(1)},q_1) \tag{9.24}$$

is tangent to $v = \psi(u)$ at $(u^{(1)},v^{(1)})$. There is no difficulty at all in showing that it is also tangent to the parabola (9.19) at the same point.

The next approximation f_1 of the optimal value f^* is the abscissa u_1 of the intersection point of the straight line (9.24) and the u-axis

$$f_1 = u_1 = \Phi(x^{(1)},q_1),$$

where

$$f_1 = f(x^{(1)}) + \frac{q_0}{2(f(x^{(1)}-u_0)} F(\phi(x^{(1)})). \tag{9.25}$$

The next problem is

$$\min_x\{(f(x)-u_1)^2+q_0F(\phi(x))\}. \tag{9.26}$$

Its solution $x^{(2)}$ defines the parabola

$$(u-u_1)^{(2)} + q_0 v = R^{(2)}, \tag{9.27}$$

with

$$R^{(2)} = f(x^{(2)})-u_1)^2+q_0F(\phi(x^{(2)})).$$

The tangent line to (9.27) and to the curve $v = \psi(u)$ at the point $u = f(x^{(2)})$, $v = F(\phi(x^{(2)}))$ is

$$u + \frac{q_0}{2(f(x^{(2)})-u_1)} v = f(x^{(2)}) + \frac{q_0F(\phi(x^{(2)}))}{2(f(x^{(2)})-u_1)} \tag{9.28}$$

and the next approximation is

$$f_2 = f(x^{(2)}) + \frac{q_0F(\phi(x^{(2)}))}{2(f(x^{(2)})-u_1)}, \tag{9.29}$$

and it is obvious how to continue.

Thus the method of tangents consists in solving a sequence of unconstrained minimization problems of the form

$$\min_x\{[f(x)-u_v]^2+q_0F(\phi(x))\} \tag{9.30}$$

and the calculation of the terms of the sequence of numbers $f_0, f_1, \ldots$ by the formulas

$$f_\nu = f(x^{(\nu)}) + \frac{q_0F(\phi(x^{(\nu)}))}{2(f(x^{(\nu)})-u_{\nu-1})}, \tag{9.31}$$

with $x^{(\nu)}$ optimal for problem (9.30),

$$u_\nu = f_\nu, \tag{9.32}$$

$$u_0 = f(x^{(0)}) + q_0F(\phi(x^{(0)})), \tag{9.33}$$

and $x^{(0)}$ optimal for the problem

$$\min_x\{f(x)+q_0F(\phi(x))\}. \tag{9.34}$$

Under the inevitable hypothesis that $v = \psi(u)$ is convex one evidently has

$$f_\nu < f_{\nu+1}, \quad \nu = 0,1,\ldots, \tag{9.35}$$

and monotone convergence

$$\lim_{\nu\to\infty} f_\nu = f^*, \quad \lim_{\nu\to\infty} x^{(\nu)} = x^*. \tag{9.36}$$

Remark. One can also find an analogous sequence $\hat{f}_0, \hat{f}_1, \ldots$ with the property $\lim_{\nu\to\infty} \hat{f}_\nu = f^*$ which approaches the optimum from below. These points are determined by the intersection of Ou with the parabola

$$(u-\hat{f}_\nu)^2 + q_0\nu = \hat{R}^{(\nu)}, \quad \nu = 0,1,\ldots,$$

where

$$\hat{R}^{(\nu)} = (f(\hat{x}^{(\nu)})-\hat{f}_\nu)^2+q_0F(\phi(\hat{x}^{(\nu)})),$$

and $\hat{x}^{(\nu)}$ is a solution of the problem

$$\min_x\{(f(x)-\hat{f}_\nu)^2+q_0F(\phi(x))\}.$$

This being the case one has

$$\hat{f}_{\nu+1} = f_\nu + \sqrt{\hat{R}^{(\nu)}}.$$

This procedure, which was proposed in [43], converges more slowly than the one described earlier. The cause of this is the obvious inequality

$$\hat{f}_\nu < f_\nu, \quad = 1,2,\ldots (\hat{f}_0=f_0)$$

9.3. Linear programming

In the particularly important case of linear programming

$$\min_x\{\sum_{i=1}^n p_ix_i \mid \sum_{i=1}^n a_{si}x_i - b_s \leqslant 0, \; s = 1,\ldots,m\} \tag{9.37}$$

the procedure described above terminates in a finite number of steps.

Let us make use once more of the following mechanical interpretation: the linear programming problem (9.37) consists in finding the equilibrium of a system subject to unilateral constraints in a uniform force field. We have shown (see

Section 4.4 of Chapter IV) that the method of penalty functions is a way to make use of the principle of removing constraints in that it replaces the rigid constraints by elastic ones. This leads to a problem of finding the equilibrium in a force field which is a superposition of the basic force field f(x) and the elastic reaction force field defined by the deformable constraints. There is an advantage in using the following way of employing penalties

$$\min\{\gamma f(x)+F(\phi(x))\}, \tag{9.38}$$

where

$$f(x) = \sum_{i=1}^{n} p_i x_i, \quad \phi_s(x) = \sum_{i=1}^{n} a_{si} x_i - b_s, \quad s = 1,\ldots,m, \tag{9.39}$$

$$F(\phi(x)) = \sum_{s=1}^{m} [\max(0;\phi_s(x))]^2. \tag{9.40}$$

In problem (9.38) the parameter γ (together with f(x)) determines the intensity of the basic force field and the equilibrium $\bar{x}(\gamma)$ is obviously a vector valued function of the parameter γ such that $\lim_{\gamma\to\infty} \bar{x}(\gamma) = x^*$. What kind of trajectory does the point $\bar{x}(\gamma)$ describe when γ varies from a positive value γ_0 to zero? It is a broken line consisting of a number of straight line segments which join $x(\gamma_0)$ to the optimum point $x = \bar{x}(0)$ (see Chapter V). The corner points of the trajectory $x = x(\gamma)$ coincide with the points where it cuts the hyperplanes $\phi_s(x) = 0$. It is known (see Section 4.4) that at the state $\bar{x}(\gamma_0)$, $F(\phi(\bar{x}(\gamma_0))$ represents the deformation energy of the elastic constraints. Let the trajectory of $x = \bar{x}(\gamma)$, $\gamma_0 \geqslant \gamma \geqslant 0$ be the trajectory of a quasi-static process caused by the slow variation of the parameter γ from $\gamma_0 > 0$ to zero, then, given that $F(\phi(x^*)) = 0$ the amount $F(\phi(\bar{x}(\gamma_0))$ is equal, according to the law of the conservation of energy, to the amount of work A (with opposite sign) which the force γ grad $f(x) = \gamma\bar{p}$ of the fundamental field has done along the trajectory under consideration. that is one has the equality

$$A = -\int_{\bar{x}(\gamma_0)}^{x^*} \gamma(\bar{p}, d\bar{s}) = F(\phi(\bar{x}(\gamma_0))), \tag{9.41}$$

where $d\bar{s}$ is an element of the trajectory of $x = \bar{x}(\gamma)$. Noting that the scalar product

$$(\bar{p}, d\bar{s}) = -df$$

one obtains

$$A = \int_{f(\bar{x}(\gamma_0))}^{f(x^*)} \gamma df = F(\phi(\bar{x}(\gamma_0))). \tag{9.42}$$

This expression for the laws of the conservation of energy turns out to be most useful for what follows below. It does not make possible an immediate calculation of the optimal value $f^* = f(x^*)$ of the objective function, because we do not know $f = f(\bar{x}(\gamma))$, or, from this, the value of γ as a function of f.

All we know is that the curve representing the function $\gamma(f)$ is a broken convex line and that the values of γ at the corner points coincide with its values at the corner points of the trajectory $x = \bar{x}(\gamma)$ (see Chapter V). Figure 9.3 depicts the relation $\gamma = \gamma(f)$ along the trajectory $x = \bar{x}(\gamma)$ representing the quasi-static process.

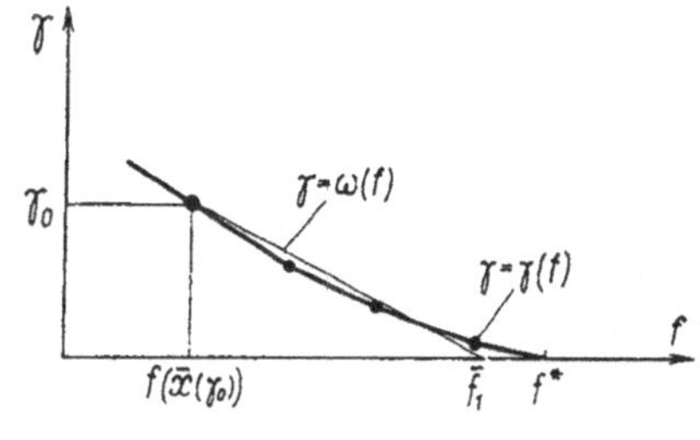

Fig. 9.3.

The property of convexity of this piece-wise linear function which corresponds to a real quasi-static process suggests the idea of replacing it by an imaginary transformation represented by a simple dependence $\gamma = w(f)$ defined by the conditions

(1) $w(f(\bar{x}(\gamma_0))) = \gamma_0$

(2) $w(f) = 0$ for $f \in (\bar{f}_1, f^*)$

(3) $w(f)$ is a linear function of the independent variable f on the segment $(f(\bar{x}(\gamma_0)), \bar{f}_1)$

(4) $\displaystyle\int_{f(\bar{x}(\gamma_0))}^{f(x^*)} \omega(f)df = F(\phi(\bar{x}(\omega_0)))$.

Condition (4) which determines $\bar{f}_1$ requires the imaginary process $\gamma = w(f)$ to respect the law of the conservation of energy (9.42). Its integral is easily calculated and one has the equality (see Figure 9.3),

$$\int_{f(\bar{x}(\gamma_0))}^{f(x^*)} \omega(f)df = \int_{f(\bar{x}(\gamma_0))}^{\bar{f}_1} \omega(f)df = \tfrac{1}{2}\gamma_0[\bar{f}_1 - f(\bar{x}(\gamma_0))].$$

Condition (4) thus determines $\bar{f}_1$:

$$\tfrac{1}{2}\gamma_0[\bar{f}_1 - f(\bar{x}(\gamma_0))] = F(\phi(\bar{x}(\gamma_0))),$$

so that

$$\bar{f}_1 = f(\bar{x}(\gamma_0)) + \frac{2F(\phi(\bar{x}(\gamma_0)))}{\gamma_0}. \qquad (9.43)$$

The inequality

$$f(\bar{x}(\gamma_0)) < \bar{f}_1 \leq f(x^*) \qquad (9.44)$$

is obvious.

It is an important fact that there is a positive number $\bar{\gamma}$ such that $\bar{f}_1 = f(x^*)$ for each value of the parameter $\gamma \in (0,\bar{\gamma})$. The interval $(0,\bar{\gamma})$ is the set of values of γ corresponding to the last segment of the broken straight line $x = \bar{x}(\gamma)$ terminating in x^*. At the equilibrium states $\bar{x}(\gamma)$, $\gamma \in (\bar{\gamma},0)$ only those unilateral constraints are deformed which are active in the position x^*. The imaginary process defined by (1)-(4) becomes for $\gamma_0 \in (\bar{\gamma},0)$ a real quasi-static process.

Thus the various numerical inconveniences attached to the use of penalty methods seem to be related to a neglect of the abundant information which is contained in the deformation energy of the constraints (the penalty function) at equilibrium.

If $\gamma_0 \notin (\bar{\gamma},0)$ that is if $\gamma_0 > \bar{\gamma}$ and hence $\bar{f}_1 < f(x^*)$, one necessarily proceeds (according to Section 9.2) to consider the free minimum problem

$$\min_x\{(f(x)-\bar{f}_1)^2+F(\phi(x))\}. \tag{9.45}$$

Let $x^{(1)}$ be the solution of this problem. One checks without difficulty that this vector is also a solution of a problem of type (9.38) for

$$\gamma = \gamma_1 = 2(f(x^{(1)})-\bar{f}_1) \tag{9.46}$$

and that $x^{(1)} = \bar{x}(\gamma_1)$ is then a point of the quasi-static transformation process $x = \bar{x}(\gamma)$. One obtains the next approximation $\bar{f}_2$ of $f(x^*)$ by repeating the procedure and thus arrives at the formula

$$\bar{f}_2 = f(x^{(1)}) + \frac{F(\phi(x^{(1)}))}{f(x^{(1)})-\bar{f}_1} \tag{9.47}$$

analogous to (9.43). Similarly one has evidently

$$\bar{f}_1 < \bar{f}_2 \leqslant f(x^*). \tag{9.48}$$

All this constitutes a finite algorithm because there is only a finite number of rectilinear segments making up the trajectory $\bar{x}(\gamma) = x$.

It is clear that the number of unconstrained minimum problems that must be solved to obtain the solution of (9.37) is at most equal to the number m of constraints. (In general it is considerably less).

Let us show that this line of arguments which is inspired by the law of the conservation of energy leads nicely to the method of tangents.

We have introduced in Section 9.2 a convex function $v = \psi(u)$ of which little is known beyond the fact that it is nonnegative and that it satisfies the conditions

$$\psi(f(x^*)) = \left.\frac{d\psi}{du}\right|_{u=f(x^*)} = 0. \tag{9.49}$$

However, in this case, because we are dealing with a linear programming problem we know a great deal more: $\psi(u)$ is not only strictly convex when $u < f(x^*)$ but it is also a regular function made up of segments of parabolas, see Figure 9.4, of which the abscissae of the corner points coincide with the corner points of the piece-wise linear function $\gamma(f)$ (see Figure 9.3).

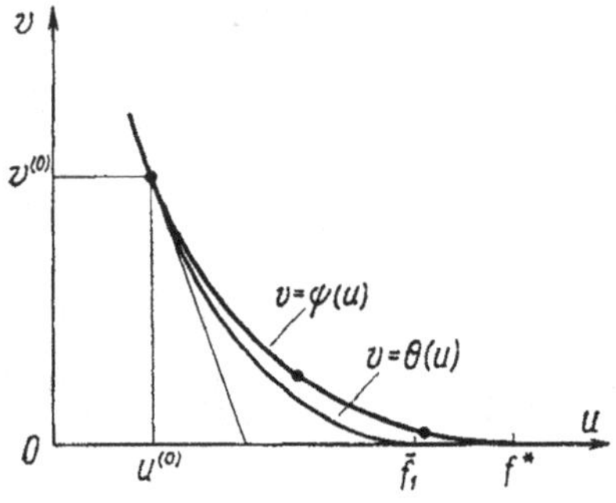

Fig. 9.4.

This assertion results from formula (9.42):

$$\psi(u) = -\int_{f(x^*)}^{u} \gamma(f)df. \tag{9.50}$$

Associated with the imaginary process $\gamma = w(f)$ defined by conditions (1)-(4) there is the function

$$f(u) = \theta(u) = -\int_{f(x^*)}^{u} \omega(f)df \tag{9.51}$$

which is, for $u \leqslant \bar{f}_1$ a parabola tangent to the curve $v = \psi(u)$ at the point (u_0, v_0) and tangent to the abscissa axis at $f = f_1$. Evidently $\theta(u) = 0$ for $u \in (\bar{f}_1, f^*)$.

Using conditions (1)-(4) one arrives easily at the equation of the curve $v = \theta(u)$, see Figure 9.4:

$$\theta(u) = \begin{cases} \frac{\gamma_0^2}{4v_0}[u-(u_0+\frac{2v_0}{\gamma_0})]^2 & \text{for } u < \bar{f}_1, \\ 0 & \text{for } u \in (\bar{f}_1, f^*) \end{cases} \tag{9.52}$$

and at the formula

$$\bar{f}_1 = u_0 + \frac{2v_0}{\gamma_0} \tag{9.53}$$

which coincides with (9.43).

If $\gamma_0 \in (0,\bar{\gamma})$ and $u \in (u^{(0)}, f^*)$ one has the same obvious identity $\theta(u) \equiv \psi(u)$.

If we want to distinguish the procedure just described from the method of tangents in its strict sense (section 9.2) we shall call this one the method of tangential parabolas.

The method in question generalizes immediately to nonlinear programming and the convexity of $\gamma(f)$ along the trajectory of the quasi-static process $x = \bar{x}(\gamma)$ is always sufficient for the convergence of the sequence $\bar{f}_1, \bar{f}_2, \ldots$. It is important to note that the speed of convergence does not decrease as one approaches the optimal value of the objective function. Indeed it then becomes possible to linearize $f(x)$ as well as the constraints and a linear problem leads to a finite number of unconstrained minimization problems, usually a number much smaller than the dimension of the problem.

9.4. Dynamic problems of optimal control

The method of tangents is nicely applicable to optimal control theory. The subject is rather large and we shall restrict ourselves to the problem

$$\min \int_{t_0}^{t_1} f_0(x_1,\ldots,x_n, u_1,\ldots,u_m,t)dt \tag{9.54}$$

over the set of piece-wise differentiable trajectories $x(t)$ and piece-wise continuous controls connected by the conditions

$$\frac{dx_i}{dt} = f_1(x_1,\ldots,x_n, u_1,\ldots,u_m,t). \tag{9.55}$$

$$x_i(t_0) = x_i^{(0)},\ x_i(t_1) = x_i^{(1)},\ i = 1,\ldots,n, \tag{9.56}$$

$$u \in \Omega \tag{9.57}$$

where Ω is a bounded set in E_m of admissible controls. For convenience with respect to what follows we shall instead of (9.54)-(9.57) be concerned with the following equivalent problem

$$\min x_0(t_1) \tag{9.58}$$

subject to (9.55)-(9.57), and the additional constraint

$$\frac{dx_0}{dt} = f_0(x_1,\ldots,x_n,\ u_1,\ldots,u_m,t), \tag{9.59}$$

$$x_0(t_0) = 0. \tag{9.60}$$

Using penalty function ideas one obtains a sequence of problems of the form

$$\min_{x\in X,\,u\in\Omega} \{\gamma x_0(t_1)+\int_{t_0}^{t_1} \sum_{i=0}^{n} [\frac{dx_i}{dt} - f_i(x,u,t)]^2 dt\}, \tag{9.61}$$

where X is the set of piece-wise differentiable trajectories $x(t) = (x_0(t),\ldots,x_n(t))$ satisfying (9.56) and (9.60). This sequence corresponds to a sequence $\gamma_0,\gamma_1,\ldots$ of values of the parameter γ such that

$$\gamma_0 > \gamma_1 > \gamma_2 > \ldots \lim_{v\to\infty} \gamma_v = 0.$$

In (9.61) one recongnizes a known variational problem with the right end point $x_0(t_1)$ free and free controls within Ω).

We now proceed to the corresponding numerical methods where we restrict ourselves to a simple iterative technique which in general gives good results. Let us look for the solution of problem (9.61) within the class of broken Euler curves for $x_i(t)$, $i = 0,\ldots,n$ and step functions for the $u_s(t)$, $s = 1,\ldots,m$.

First step of the first cycle. Choose an arbitrary continuous trajectory $x^{(0)}(t)$ which is only required to satisfy the boundary conditions (9.56) and (9.60). For instance we could take a straight line segment joining the initial and the final point. With $x_i(t) = x_i^{(0)}(t)$ fixed, problem (9.61) reduces to

$$\min_{u\in\Omega} \sum_{i=0}^{n} \left[\frac{dx_i}{dt} - f_i(x^{(0)}(t),u,t)\right]^2 \tag{9.62}$$

for each instant $t \in (t_0,t_1)$. Let $u^{(0)}(t)$ denote the solution of (9.62).

Second step of the first cycle. Set $u(t) = u^{(0)}(t)$ in the original problem (9.61). There results the variational problem

$$\min_{x}\{\gamma x_0(t_1) + \int_{t_0}^{t_1} \sum_{i=0}^{n} \left[\frac{dx_i}{dt} - f_i(x,u^{(0)}(t),t)\right]^2 dt\} \tag{9.63}$$

which can be solved in various ways, including the tangent method of Euler. Let $x^{(1)}(t)$ be the solution of (9.63). Solving (9.62) and (9.63) terminates the first cycle.

The first step of the next cycle differs from the first one only in the fact that $x^{(0)}(t)$ is replaced with the solution of (9.63), the vector valued function $x^{(1)}(t)$. It is obvious how to continue and the algorithm thus yields minimizing sequences of controls and trajectories $x^{(0)}(t), x^{(1)}(t), \ldots$.

Thus we suppose that for every fixed nonnegative γ the solution of problem (9.61) has been approximated as well as desired.

Let $\bar{x}(t,\gamma)$, $\bar{u}(t,\gamma)$ be a pair of vector valued functions which solves (9.61) for a fixed value > 0 of the parameter γ. As in Section 9.2 the system of equations

$$\begin{cases} V = \int_{t_0}^{t_1} \sum_{i=0}^{n} \left[\frac{d\bar{x}_i}{dt} - f_i(\bar{x}(t,\gamma),\bar{u}(t,\gamma),t\right]^2 dt, \\ U = \bar{x}_0(t_1,\gamma) \end{cases} \tag{9.64}$$

can be viewed as a parametrized system of equations defining a curve $V = \psi(U)$, and we have

$$\begin{cases} \psi(U) > 0 \text{ for } U < x_0^*(t_1), \\ \left.\frac{d\psi}{dU}\right|_{U=x_0^*(t_1)} = \psi(x_0^*(t_1)) = 0. \end{cases} \tag{9.65}$$

Let us apply the arguments of Section 9.2. One checks easily that the point $(U^{(0)}, V^{(0)})$, where

$$\begin{cases} U^{(0)} = \bar{x}(t_1, \gamma_0), \\ V^{(0)} = \int_{t_0}^{t_1} \sum_{i=0}^{n} [\frac{d\bar{x}_i}{dt} - f_i(\bar{x}(t,\gamma_0), \bar{u}(t,\gamma_0), t]^2 dt, \end{cases} \tag{9.66}$$

belongs to the curve $V = \psi(U)$ and that the straight line

$$\gamma_0 U + V = \gamma_0 U^{(0)} + V^{(0)} \tag{9.67}$$

is tangent to the curve $V = \psi(U)$ at $(U^{(0)}, V^{(0)})$.

Set $V = 0$ in (9.67), the result is

$$U_0 = U^{(0)} + \frac{V^{(0)}}{\gamma_0}, \tag{9.68}$$

which is a quantity satisfying

$$U^{(0)} < U_0 < x^*(t_1). \tag{9.69}$$

Finding U_0 is the preliminary step in an algorithm for solving optimal control problems. This algorithm is analogous to the one of Section 9.2.

The algorithm itself consists of solving a sequence of problems of the form

$$\min_{x \in X, u \in \Omega} \{(x_0(t_1) - u_\nu)^2 + \int_{t_0}^{t_1} \sum_{i=0}^{n} [\frac{dx_i}{dt} - f_i(x,u,t)]^2 dt ,$$

$$\nu = 0, 1, \ldots \tag{9.70}$$

where

$$U_\nu = U^{\nu)} + \frac{V^{(\nu)}}{2(U^{(\nu)} - U_{\nu-1})}, \tag{9.71}$$

$$U^{(\nu)} = \bar{x}_0^{(\nu)}(f_1), \tag{9.72}$$

$$V^{(\nu)} = \int_{t_0}^{t_1} \sum_{i=0}^{n} [\frac{dx_i}{dt} - f_i(\bar{x}^{(\nu)}(t), \bar{u}^{(\nu)}(t), t)]^2 dt, \tag{9.73}$$

where $(\bar{x}^{(\nu)}(t),\bar{u}^{(\nu)}(t))$ is the solution of problem (9.70).

This treatment of the method of tangents for optimal control problems has been very short so as to not repeat everything said in Section 9.2. To conclude we note that by means of convex tangent parabolas (see Section 9.3) one arrives at an analogous algorithm: the only significant difference is that formula (9.71) is replaced by

$$U_\nu = U^{(\nu)} + \frac{V^{(\nu)}}{U^{(\nu)}-U_{\nu-1}}.$$

Chapter X

MODELS FOR ECONOMIC EQUILIBRIUM

10.1. Introduction

In this chapter certain models for economic equilibria are considered. It is characteristic of these models that they consist of several different systems with their own objective functions, resources and budgets. The interrelations between the constituting (sub)economies take the form of exchanges. The physical analogues of such interrelations are transition processes which are acompanied by transfer or mass or energy in the transition of the physical model to an equilibrium state. Equalizations of the intensive variables then take place in the various parts of the physical system and a certain state function, called a thermodynamic potential attains its extremum. For isolated systems this potential is the entropy of Clausius or the Helmholtz free energy. An economy evolving towards equilibrium presents a similar picture: the prices which the users of a certain resource are prepared to pay for that resource become equal and a certain function which one could call the entropy of the economy reaches an extremum value. We shall consider an economy as a control system in which a central controlling agency can allocate budgets (capital) and impose resource restrictions. This is thus a partially centralized economy in which the constituting economic parts are semi-independent, but subject to exterior controls of their activities from a central controlling agency. The controlling economic centre uses an equilibrium mechanism and limits itself to controls effected by that mechanism as a result of budget allocations and to ensure stability of the transition process to an equilibrium state.

The physical analogues of this type of control of an economic system are suitable changes in the parameters defining the exterior conditions of the physical system which grant the system the possibility of spontaneous transitions to an equilibrium state corresponding to those exterior conditions. Thus we influence nature and make use of her laws by establishing exterior conditions such that the desired state is also the most probable one [1].

There also exist other models for economic equilibrium, in which, for example there are no budget restrictions and the control parameters are the elements of a matrix determining the initial allocation of resources. In such models an equilibrium state establishes itself by means of a natural exchange of resources (barter). Examples of such barter models are studied in [11] [2]. It is also possible to construct physical analogues for such models and use the method of redundant constraints to obtain numerical algorithms for solving such equilibrium problems. Unfortunately there is no space in this book to explain those methods and results.

10.2. Equilibrium problems for linear exchange models

In this section we turn to a model for an economy, analysed by Gale [29, 34, 41, 42], and we shall find that the equilibrium problem for this linear exchange model is equivalent to an equilibrium problem for a certain physical system. Studying this physical model we shall see that the basic results obtained by Gale are consequences of the principles of analytical mechanics and physics.

We consider an economy in which there are n goods $G_1,\dots,G_n$, m consumers $A_1,\dots,A_m$ and one producer. Suppose the last one produces during a given period (for example one year) X_1 units of product $G_1,\dots,X_n$ units of product G_n, Suppose, furthermore, that the quantities $P_1,\dots,P_m$ denote the incomes of the consumers $A_1,\dots,A_m$ during the same period. We assume that each consumer uses all his income to purchase products from the producer and that relative utility of one unit of good G_i for consumer A_s is given by a nonnegative number $c_i^{(s)}$. We shall call a nonnegative n-dimensional vector $x^{(s)} = (x_1^{(s)},\dots,x_n^{(s)})$ of which the components $x_i^{(s)}$ are the amount of units of good G_i purchases by consumer A_s the bundle of goods of that consumer.

By virtue of the linearity assumption of the model the utility of the bundle $x^{(s)} = (x_1^{(s)},\dots,x_n^{(s)})$ purchased by consumer A_s is given by the linear function

$$f_s(x^{(s)}) = \sum_{i=1}^{n} c_i^{(s)} x_i^{(s)} .$$

The assumption of linearity of the function $f_s(x^{(s)})$ is generally speaking not realistic and may be justified only in the case of small values $x_i^{(s)}$. Such a situation may obtain in the case where each period during which the bundle of goods $X = (X_1,\ldots,X_n)$ is produced is sufficiently short for the quantities X_i and P_s to be small.

We shall further suppose that a price vector is given $p = (p_1,\ldots,p_n)$ where p_i is the price of one unit of product G_i. This obliges the consumer A_s to choose his bundle of goods in such a way that the following conditions hold

$$\sum_{i=1}^{n} c_i^{(s)} x_i^{(s)} \to \max, \tag{10.1}$$

$$\sum_{i=1}^{n} p_i x_i^{(s)} \leqslant P_s. \tag{10.2}$$

The last inequality we shall call the budget constraint. In this manner each consumer has to solve a quite simple linear programming problem. However, it may happen that the total amount of some good $x_i^{(1)} + \ldots + x_i^{(m)}$ exceeds the total available, that is it may happen that the resource constraint

$$\sum_{s=1}^{m} x_i^{(s)} \leqslant X_i$$

is not fulfilled.

Consequently if the prices are arbitrary it may happen that the demand exceeds the supply or that the supply exceeds the demand. Such a situation necessarily affects the prices of the goods: rising prices if the demand exceeds the supply and diminishing prices if the supply exceeds the demand. In this manner there arises a price equilibrium problem in the sense of the following definition.

DEFINITION 10.1. A price vector $p = (p_1,\ldots,p_n)$ is called an equilibrium price vector if there exist bundles of goods $x^{(1)},\ldots,x^{(n)}$ satisfying conditions (10.1) and (10.2) and such that moreover the following equalities hold

$$\sum_{s=1}^{m} x_i^{(s)} = X_i, \quad i = 1,2,\ldots,n.$$

If such a choice of bundles of goods exists it is also called an equilibrium set of bundles of goods.

It is natural to suppose that the matrix $(c_i^{(s)})$ of subjective utilities has the property of having at least one strictly positive element in each row and each column. In other words each consumer desires to purchase at least one good and no good is such that no consumer at all is interested in it.

In this way an equilibrium of the economic model under consideration can be understood in the sense of the following definition.

DEFINITION 10.2. Let $C = (c_i^{(s)})$ be a nonnegative matrix which has at least one strictly positive element in each row and in each column. Let $P_1,\ldots,P_m$ be positive numbers. A nonnegative n-dimensional vector $p = (p_1,\ldots,p_n)$ is called an equilibrium price vector and the nonnegative n-dimensional vectors $x^{(1)},\ldots,x^{(m)}$ are called equilibrium bundles of goods if the following conditions are satisfied

$$\begin{aligned} &\sum_{i=1}^{n} c_i^{(s)} x_i^{(s)} \to \max, && s = 1,2,\ldots,m, \\ &\sum_{i=1}^{n} p_i x_i^{(s)} \leqslant P_s, && s = 1,2,\ldots,m, \\ &\sum_{s=1}^{m} x_i^{(s)} = X_i, && i = 1,2,\ldots,n. \end{aligned} \tag{10.3}$$

The basic result concerning the equilibrium problem under consideration for exchange economies is the equilibrium theorem of Gale [29, 41].

THEOREM 10.1. For every matrix C and system of numbers $P_1,\ldots,P_m$ satisfying the assumptions mentioned in definition 10.2 there exists an equilibrium price vector.

Now the equilibrium bundles of goods $x^{(1)*},\ldots,x^{(m)*}$ maximize the function

$$\phi(x^{(1)},\ldots,x^{(m)}) = \prod_{s=1}^{m} \Big(\sum_{i=1}^{n} c_i^{(s)} x_i^{(s)} \Big)^{P_s}$$

over the domain defined by the last n conditions of (10.3) and the equilibrium prices $p_1^*,\dots,p_n^*$ are defined by the formulas

$$p_i^* = \max_s \frac{c_i^{(s)} P_s}{\sum_{j=1}^{n} c_j^{(s)} x_j^{(s)*}} \tag{10.4}$$

In the following it will be more convenient for us to consider instead of the problem of maximizing the function $\phi(x^{(1)},\dots,x^{(m)})$, the equivalent problem

$$\sum_{s=1}^{m} P_s \ln\Big(\sum_{i=1}^{n} c_i^{(s)} x_i^{(s)}\Big) \to \max, \tag{10.5}$$

$$\left.\begin{aligned} &\sum_{s=1}^{m} x_i^{(s)} = X_i, \\ &x_i^{(s)} \geqslant 0,\ i = 1,\dots,n;\ s = 1,\dots,m. \end{aligned}\right\} \tag{10.6}$$

The problem in the formulation (10.5), (10.6) turns out to be equivalent to an equilibrium problem for some physical system and the function to be maximized turns out to be the entropy of that system and the equilibrium price vector is the vector of dual (shadow) prices (sensitivities) for the constraints (10.6). Indeed the entropy S of an ideal gas in a volume V in the case where the volume changes under constant temperature is expressed by formula (2.21) of Chapter II:

$$S = \mu R \ln V + \mu c_v \ln T + \mu a_1.$$

Choosing an integration constant a_1 satisfying the condition

$$a_1 + c_v \ln T = 0,$$

one obtains

$$S = \mu R \ln V.$$

It is now clear that the function (10.5) can be seen as the (total) entropy of an ideal gas in volumes $V^{(1)},\dots,V^{(m)}$ which are equal to

$$V^{(s)} = \sum_{i=1}^{n} c_i^{(s)} x_i^{(s)}, \quad s = 1,\dots,m,$$

and which contain respectively

$$\mu_s = \frac{P_s}{R}, \quad s = 1,\dots,m,$$

mols of an ideal gas. If, in addition, one recalls the model for a system of linear equations, described in Chapter I then one easily arrives at the physical model for problem (10.5)-(10.6) which is sketched in Figure 10.1. In Figure 10.1 on the right there are represented n systems of m cylinders of height l and base area one. The communicating volumes of the i-th system of cylinders are filled with an incompressible liquid such that the volumes on the right of the pistons contain X_i units of liquid and the volumes on the left $nl-X_i$ units. In this way the system of cylinders just described is a model for the constraints equalities (10.6), and the coordinates $x_i^{(s)}$ which indicate the positions of the pistons automatically satisfy those constraints. Further in Figure 10.1 there are shown m systems of communicating volumes $V^{(1)},\dots,V^{(m)}$ containing respectively $\mu_1 = P_1/R,\dots,\mu_m = P_m/R$ mols of an ideal gas. The communicating volumes of the s-th system are volumes of cylinders with base area equal to $c_i^{(s)}$. The right ends of these cylinders are not movable and on the left the cylinders are closed by means of movable pistons whose distance to the base is $x_i^{(1)}$.
We have thus constructed a physical system of which the admissible values of the state parameters $x_i^{(s)}$ are subject to the constraints (10.6) and whose entropy is equal to

$$S = \sum_{s=1}^{m} P_s \ln\Big(\sum_{i=1}^{n} c_i^{(s)} x_i^{(s)}\Big).$$

Consequently problem (10.5)-(10.6) is equivalent to the equilibrium problem for the physical system sketched in Figure 10.1. [3)]

In order to state the conditions for equilibrium for the model let us introduce the following notations

1) w_i^* is the pressure difference at an equilibrium state $x_i^{(s)*}$. More precisely it is the difference between the left

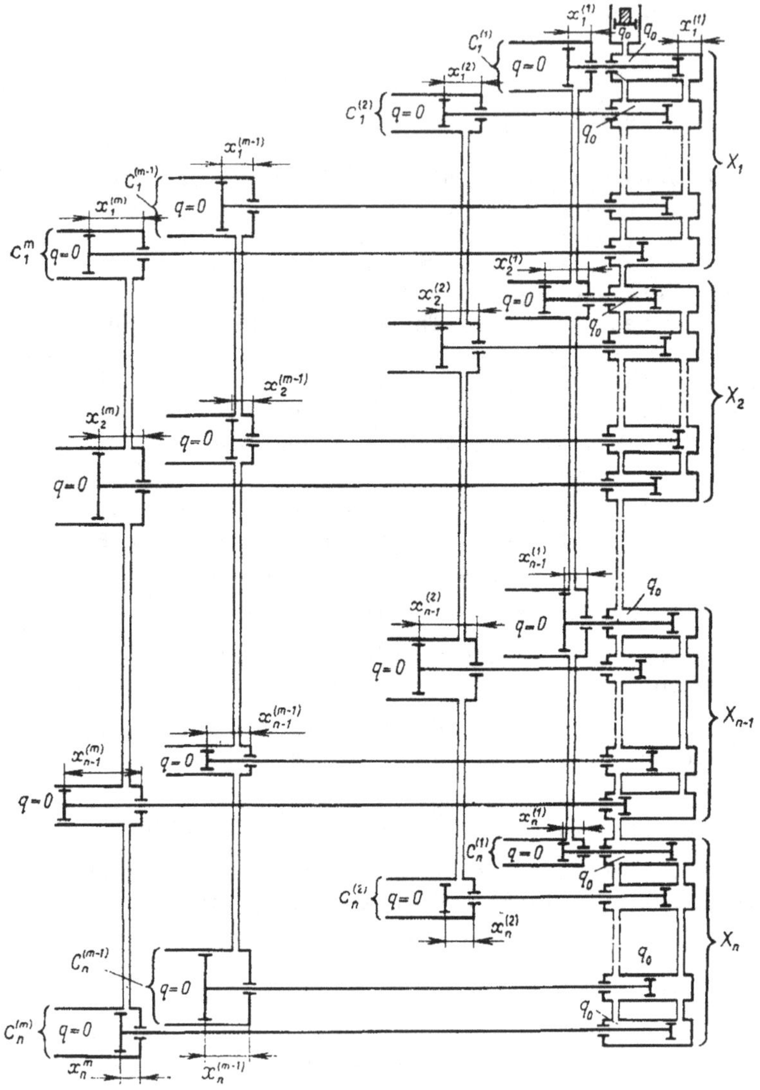

Fig. 10.1

and the right side of the pistons of the i-th system of communicating volumes filled with an incompressible liquid and which models the condition $x_i^{(s)} = X_i$.

2) $q^{(s)*}$ is the pressure in the s-th system of communicating volumes of sizes $c_1^{(s)}x_1^{(s)},\ldots,c_n^{(s)}x_n^{(s)}$ respectively which contains P_s/R mols of an ideal gas.

The conditions for equilibrium in the model obviously are the conditions on the bars connecting pistons stating that the various forces acting on them balance each other. This gives

$$w_i^* - c_i^{(s)}q^{(s)*}\begin{cases} = 0 \text{ for } x_i^{(s)*} > 0, \\ \geqslant 0 \text{ for } x_i^{(s)*} = 0, \end{cases} \tag{10.7}$$

$i = 1,\ldots,n;\ s = 1,\ldots,m.$

The quantities $q^{(1)*},\ldots,q^{(m)*}$ are connected to the equilibrium state coordinates by the Clapeyron-Mendeleev [4] equation

$$q^{(s)*}\sum_{i=1}^{n} c_i^{(s)}x_i^{(s)} = \mu_s RT.$$

Because we consider only isothermic processes we can set $T = 1$, and using the notation $\mu_s = P_s/R$ in the last equation one obtains

$$q^{(s)*} = \frac{P_s}{\sum_{i=1}^{n} c_i^{(s)}x_i^{(s)*}}. \tag{10.8}$$

Then it follows from the equilibrium conditions (10.7) that

$$w_i^*\begin{cases} = \dfrac{P_s c_i^{(s)}}{\sum_{j=1}^{n} c_j^{(s)}x_j^{(s)*}} \text{ for } x_i^{(s)*} > 0 \\ \geqslant \dfrac{P_s c_i^{(s)}}{\sum_{j=1}^{n} c_j^{(s)}x_j^{(s)*}} \text{ for } x_i^{(s)*} = 0 \end{cases} \tag{10.9}$$

or

$$w_i^* = \max_s \frac{P_s c_i^{(s)}}{\sum_{j=1}^{n} c_j^{(s)} x_j^{(s)*}} . \tag{10.10}$$

Thus the dual prices (sensitivities) $w_1^*,\dots,w_n^*$ of the constraint equations (10.6) are determined by the same formulas of Gale which determine the components of an equilibrium price vector. And therefore $p_i^* = w_i^*$, $i = 1,\dots,n$. Moreover, from the equilibrium conditions (10.7) it follows that the budget constraints hold. Indeed multiplying the relations of (10.9) with $x_1^{(s)*},\dots,x_n^{(s)*}$ respecitvely and summing we obtain the equality

$$\sum_{i=1}^{n} w_i^* x_i^{(s)} = P_s, \quad s = 1,\dots,m.$$

In order to conclude this physical demonstration of the equilibrium theorem it is necessary to show that the vectors $x^{(1)*},\dots,x^{(m)*}$ are optimal vectors for the m (linear programming) problems

$$\sum_{i=1}^{n} c_i^{(s)} x_i^{(s)} \to \max \tag{10.11}$$

under the conditions

$$\left.\begin{aligned} &\sum_{i=1}^{n} p_i x_i^{(0)} = P_s, \\ &x_i^{(0)} \geqslant 0, \quad s = 1,\dots,m. \end{aligned}\right\} \tag{10.12}$$

Indeed, by virtue of the principle of removing constraints the state $(x^{(1)*},\dots,x^{(m)*})$ remains an equilibrium state if the constraints (interrelations) (10.3) are replaced by the corresponding reaction forces $w_1^*,\dots,w_n^*$ at that state. But then clearly the system splits into m independent systems each of which remains in equilibrium for $x_i^{(s)} = x_i^{(s)*}$. Replacing the constraints by their reaction forces makes each of these physical systems active and in an equilibrium state $(x^{(1)*},\dots,x^{(m)*})$ the functions

$$\Phi_s = P_s \ln \sum_{i=1}^{n} c_i^{(s)} x_i^{(s)} - \sum_{i=1}^{n} w_i^* x_i^{(s)}, \qquad (10.13)$$

$$s = 1,\dots,m,$$

assume their maximum over their respective domains defined by $x_i^{(s)} \geqslant 0$.

On the other hand, by virtue of the principle of virtual work, in order that the state $(x^{(1)*},\dots,x^{(m)*})$ should be an equilibrium state it is necessary and sufficient that the sum of the amounts of work done by the given forces should be zero or negative for all possible displacements compatible with the constraints in a neighbourhood of the state $(x^{(1)*},\dots,x^{(m)*})$ [19], Thus it follows from the principle of virtual work that the following conditions hold

$$\sum_{i=1}^{n} (c_i^{(s)} q^{(s)} - w_i^*)\delta x_i^{(s)} \leqslant 0, \quad s = 1,\dots,m,$$

for all $\delta x_i^{(s)}$ satisfying only the conditions

$$\delta x_i^{(s)} \geqslant 0 \text{ for } x_i^{(s)*} = 0. \qquad (10.14)$$

It is known [18] that an equilibrium persists provided that the constraints in the virtual displacements $\delta x_i^{(s)}$ are compatible with that state. This is for example the case for the equalities

$$\sum_{i=1}^{n} w_i^* x_i^{(s)} = P_s, \quad s = 1,\dots,m, \qquad (10.15)$$

which are compatible with the equilibrium state as a consequence of (10.10). The quantities $\delta x_i^{(s)}$ are therefore also required to satisfy the conditions

$$\sum_{i=1}^{n} w_i^* \delta x_i^{(s)} = 0, \quad s = 1,\dots,m.$$

This last assertion means that the functions (10.13) attain their maxima for $x_i^{(s)} = x_i^{(s)*}$ also on the set of states $\{x_i^{(s)}\}$ satisfying condition (10.15) and $x_i^{(s)} \geqslant 0$. But because in that subset

$$\sum_{i=1}^{n} w_i^* x_i^{(s)} = P_s = \text{const}, \quad s = 1,\dots,m,$$

we have

$$\max P_s \ln \sum_{i=1}^{n} c_i^{(s)} x_i^{(s)} = P_s \ln \sum_{i=1}^{n} c_i^{(s)} x_i^{(s)*}$$

under conditions (10.15) and $x_i^{(s)} \geqslant 0$. This means that the vectors $x^{(1)*},\dots,x^{(m)*}$ are optimal vectors for problem (10.11)-(10.12) and that as a consequence of this

$$w_i^* = p_i^*, \quad i = 1,\dots,n,$$

that is, the dual prices of the resource constraints of problem (10.5)-(10.6) are the components of the equilibrium price vector.

10.3. An algorithm for solving numerically equilibrium problems for linear exchange economies

We shall now see that the method of redundant constraints as described in Section 3.2 makes it possible to obtain quite simply an iterative algorithm for solving equilibrium problems for the Gale exchange model. We turn our attention again to the physical model for problem (10.5)-(10.6) as depicted in Figure 10.1 and we choose any state $x_i^{(s)0}$ for the model which is compatible with the constraints (10.6). For example one could take for this state the one whose coordinates are given by [5)]

$$x_i^{(s)0} = \frac{X_i}{m}, \quad i = 1,\dots,n; \; s = 1,\dots,m.$$

The algorithm described below is the mathematical description of the spontaneous transition process to equilibrium of the physical model of Figure 10.1 under various redundant constraints imposed in the successive steps.

In the first step one isolates all the volumes of the model containing an ideal gas and one solves the n equilibrium problems in which the problem splits after these redundant constraints have been introduced. (These are of the impermeable membrane type.) The physical model of one of these problems is shown in Figure 10.2. Let $\{x_i^{(s)1}\}$ be the equilibrium state of the model subject to these extra constraints. Then for fixed values $x_i^{(s)} = x_i^{(s)1}$ one removes the impermeable membrane and determines the pressures which establish themselves in the communicating volumes $c_1^{(s)}x_1^{(s)},\ldots,c_n^{(s)}x_n^{(s)}$, $s = 1,\ldots,m$ which contain ideal gases.

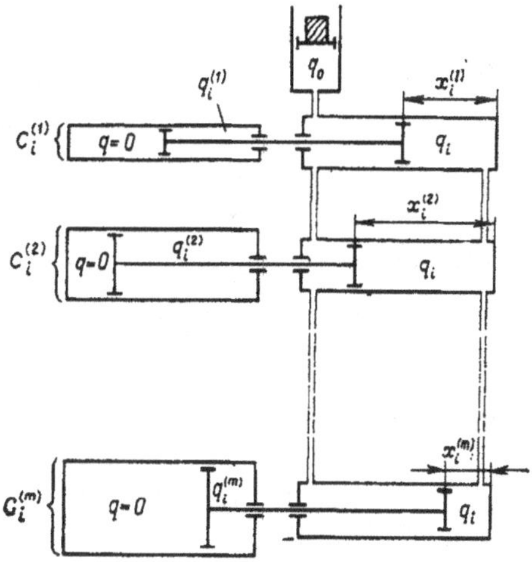

Fig. 10.2

The next step in the algorithm is essentially the same and differs only in that the initial coordinates are now the $x_i^{(s)1}$ calculated in the first step. The sequel is obvious. As results one obtains a sequence of states $\{x_i^{(s)0}\},\{x_i^{(s)1}\}$, ... which corresponds to a monotonically increasing sequence of values of the entropy of the physical model This follows from the second law of thermodynamics because in every step an equilibrium problem is solved for the model under stationary extra constraints compatible with the state determined during the previous step. Under the conditions: $x_i^{(s)0} > 0$ if $c_i^{(s)} > 0$, the limit of this sequence $\{x_i^{(s)0}\}$, $\{x_i^{(s)1}\}$ is an equilibrium state $x_i^{(s)*}$ which determines the required equilibrium bundles of goodx $x^{(1)*},\ldots,x^{(m)*}$. The price equilibrium vector is then defined by Gale's formula (10.6).

The meaning of the conditions $x_i^{(s)0} > 0$ if $c_i^{(s)} > 0$ will be made clear below.

Let us now consider the problem of the first step.

Denote with $q_i^{(s)}$ the pressures in the isolated volumes $c_i^{(s)} x_i^{(s)}$, filled with an ideal gas. Clearly until they are isolated the pressures in the volumes $c_i^{(s)} x_i^{(s)}$ in state $x_i^{(s)0}$ are, according to (10.8) equal to

$$q_i^{(s)}(x_i^{(s)0}) = q_i^{(s)0} = \frac{P_s}{\sum_{\alpha=1}^{n} c_\alpha^{(s)} x_\alpha^{(s)0}} = q^{(s)}(x_i^{(s)0}).$$

After isolation the pressures in these volumes for arbitrary $x_i^{(s)}$ are determined by the law of Boyle-Mariotte

$$q_i^{(s)}(x_i^{(s)}) = \frac{q_i^{(s)0} x_i^{(s)0}}{x_i^{(s)}} = \frac{P_s x_i^{(s)0}}{x_i^{(s)} \sum_{\alpha=1}^{n} c_\alpha^{(s)} x_\alpha^{(s)0}} \tag{10.16}$$

Denote with $x_i^{(s)1}$ the equilibrium state of the model for isolated volumes $c_i^{(s)} x_i^{(s)}$ and with $w_i^{(1)}$ the pressure difference at state $\{x_i^{(s)1}\}$ between the left and right side of the pistons in the i-th system of communicating volumes filled with an incompressible liquid.

The equilibrium conditions for the model subject to those extra constraints have by analogy with (10.7) the form

$$w_i^{(1)} - c_i^{(s)} q_i^{(s)1} \begin{cases} = 0 \text{ for } x_i^{(s)1} > 0, \\ \geqslant 0 \text{ for } x_i^{(s)1} = 0, \end{cases} \tag{10.17}$$

where according to (10.6)

$$q_i^{(s)1} = \frac{P_s x_i^{(s)0}}{x_i^{(s)1} \sum_{\alpha=1}^{n} c_\alpha^{(s)} x_\alpha^{(s)0}} \tag{10.17}$$

We remark that under the conditions

$$x_i^{(s)0} > 0 \text{ for } c_i^{(s)} > 0.$$

This obvious physical fact follows because for an ideal gas of volumes $c_i^{(s)} x_i^{(s)0} > 0$ and pressure $q_i^{(s)0}$ it is not possible to reduce the volume to zero by means of a finite force. On the other hand, it follows from the equilibrium conditions (10.17) that

$$x_i^{(s)1} = 0 \text{ for } c_i^{(s)} = 0. \qquad (10.19)$$

Indeed, suppose that $c_{i_1}^{(s_1)} = 0$. Then the corresponding equilibrium condition has the form

$$w_{i_1}^1 \begin{cases} = 0 \text{ for } x_{i_1}^{(s_1)} > 0, \\ \geqslant 0 \text{ for } x_{i_1}^{(s_1)} = 0. \end{cases}$$

The case $w_{i_1}^{(1)} = 0$ is not possible by virtue of the assumption that the matrix $(c_i^{(s)})$ has no zero columns. This means that exists at least one consumer for which the product G_{i_1} has a positive value. Obviously condition (10.19) also holds for the elements $x_i^{(s)*}$ of the matrix of optimal partition of resources. Then one has for each iteration

$$x_i^{(s)k} = 0 \text{ for } c_i^{(s)} = 0, \; k = 0,1,\ldots\,.$$

Taking this property of the optimal solution into account it is advantageous to take for the elements of the matrix $(x_i^{(s)0})$ of initial approximations numbers which are proportional to the product of the budget and the price which a given consumer assigns to the good in question

$$x_i^{(s)0} = \lambda_i P_i c_i^{(s)}. \qquad (10.20)$$

The quantities $\lambda_1,\ldots,\lambda_n$ can be found by using the condition

$$\sum_{s=1}^{m} x_i^{(s)} = X_i. \qquad (10.21)$$

From (10.20) and (10.21) it follows that

$$x_i^{(s)0} = \frac{X_i P_s c_i^{(s)}}{\sum_{\alpha=1}^{m} c_i^{\alpha} P_\alpha}$$

Now turn again to the equilibrium conditions (10.17) of the model with isolated volumes $c_i^{(s)} x_i^{(s)}$, and let us use these conditions to obtain formulas for the equilibrium state coordinates. Multiplying the inequalities (10.17) with the corresponding coordinates $x_i^{(s)1}$ of the equilibrium state one obtains the equations

$$w_i^{(1)} x_i^{(s)1} - c_i^{(s)} q_i^{(s)1} x_i^{(s)1} = 0 \tag{10.22}$$

$i = 1,\dots,n;\ s = 1,\dots,m.$

But according to (10.18)

$$q_i^{(s)1} x_i^{(s)1} = \frac{P_s x_i^{(s)0}}{\sum_{\alpha=1}^{n} c_\alpha^{(s)} x_\alpha^{(s)0}}, \tag{10.23}$$

as a result one finds from (10.22) and (10.23)

$$x_i^{(s)1} = \frac{1}{w_i^{(1)}} \frac{P_s c_i^{(s)} x_i^{(s)0}}{\sum_{\alpha=1}^{n} c_\alpha^{(s)} x_\alpha^{(s)0}}. \tag{10.24}$$

The unknown quantities $w_i^{(1)}$ are easily eliminated from (10.24), by making use of the resource constraints (10.3). From (10.3) and (10.24) it follows that

$$X_i = \sum_{s=1}^{m} x_1^{(s)1} = \frac{1}{w_i^{(1)}} \sum_{s=1}^{m} \frac{P_s c_i^{(s)} x_i^{(s)0}}{\sum_{\alpha=1}^{n} c_\alpha^{(s)} x_\alpha^{(s)0}}$$

or

$$w_i^{(1)} = \frac{1}{X_i} \sum_{i=1}^{m} \frac{P_s c_i^{(s)} x_i^{(x)0}}{\sum_{\alpha=1}^{n} c_\alpha^{(s)} x_\alpha^{(s)0}}$$

Substituting this in (10.24) one obtains formulas for calculating the required values of the coordinates of the equilibrium state

$$x_i^{(s)1} = X_i \frac{\dfrac{P_s c_i^{(i)} x_i^{(s)0}}{\sum_{\alpha=1}^{n} c_\alpha^{(s)} x_\alpha^{(s)0}}}{\sum_{\alpha=1}^{n} \dfrac{P_\sigma c_i^{(\sigma)} x_i^{(\sigma)0}}{\sum_{\alpha=1}^{n} c_\alpha^{(\sigma)} x_\alpha^{(\sigma)0}}}$$

By means of the notation

$$\xi_i^{(s)0} = \frac{P_s c_i^{(s)} x_i^{(s)0}}{\sum_{\alpha=1}^{m} c_\alpha^{(s)} x_\alpha^{(s)0}}$$

one finds a formula for calculating the quantities $x_i^{(s)1}$ in the form

$$x_i^{(s)1} = X_i \frac{\xi_i^{(s)0}}{\sum_{\sigma=1}^{m} \xi_i^{(\sigma)0}} .$$

Obviously if $\{x_i^{(s)0}\}$ is not an equilibrium state for the physical model of problem (10.5)-(10.6) then this state will also not be an equilibrium state for the same model after the redundant constraints isolating the volumes $x_i^{(s)} c_i^{(s)}$ have been introduced (the impermeable membranes). Therefore the entropy (10.5) of the physical model in states $\{x_i^{(s)0}\}$, $\{x_i^{(s)1}\}$ must satisfy the inequality

$$\sum_{s=1}^{m} P_s \ln \sum_{i=1}^{n} c_i^{(s)} x_i^{(s)0} < \sum_{s=1}^{m} P_s \ln \sum_{i=1}^{n} c_i^{(s)} x_i^{(s)1} .$$

Taking now the state $x_i^{(s)1}$ which also satisfies the constraints (10.6), as an initial state for the physical model and repeating the corresponding arguments we obtain the formulas

$$x_i^{(s)2} = X_i \frac{\xi_i^{(s)1}}{\sum_{\sigma=1}^{m} \xi_i^{(\sigma)1}},$$

$$\xi_i^{(s)1} = \frac{p_s c_i^{(s)} x_i^{(s)1}}{\sum_{\alpha=1}^{n} c_\alpha^{(s)} x_\alpha^{(s)1}}$$

defining the next approximation to the required equilibrium state or, equivalently, the solution of problem (10.5)-(10.6). It is clear how to proceed further and the iterative formulas determining the equilibrium bundles of goods $x^{(1)},\ldots,x^{(m)}$, therefore have the form

$$\left.\begin{aligned} x_i^{(s)N+1} &= X_i \frac{\xi_i^{(s)N}}{\sum_{\sigma=1}^{m} \xi_i^{(\sigma)N}}, \\ \xi_i^{(s)N} &= \frac{p_s c_i^{(s)} x_i^{(s)N}}{\sum_{\alpha=1}^{n} c_\alpha^{(s)} x_\alpha^{(s)N}}, \\ x_i^{(s)0} &= X_i \frac{p_s c_i^{(s)}}{\sum_{\sigma=1}^{n} p_\sigma c_i^{(\sigma)}} \end{aligned}\right\} \qquad (10.25)$$

$$N = 0,1,2,\ldots,\ i = 1,2,\ldots,n,\ s = 1,2,\ldots,m,$$

$$x_i^{(s)*} = \lim_{N\to\infty} x_i^{(s)N}.$$

The components of the equilibrium price vector p^* are defined by formula (10.4) or by the equalities

$$p_i^* = w_i^* = \frac{1}{X_i} \sum_{s=1}^{m} \xi_i^{(s)*},\ i = 1,\ldots,n,$$

which follow from that formula and condition 10.3. Here

$$\xi_i^{(s)*} = \lim_{N\to\infty} \xi_i^{(s)N} = \frac{P_s c_i^{(s)} x_i^{(s)*}}{\sum_{\alpha=1}^{m} c_\alpha^{(s)} x_\alpha^{(s)*}},$$

The convergence of the sequence of states defined by formula (10.25) to an equilibrium state $\{x_i^{(s)*}\}$ in which the entropy (10.5) of the physical model is maximal follows from the second principle of thermodynamics. Indeed if $\{x_i^{(s)0}\}$ is not an equilibrium state, then this same state will still not be an equilibrium state after the redundant constraints have been introduced which isolate the various volumes containing ideal gases because these extra constraints at state $\{x_i^{(s)0}\}$ do not diminish the number of degrees of freedom of the model and leave unchanged the forces acting on the bodies making up the physical model. The assertion that the limit state $\{x_i^{(s)*}\}$ is an equilibrium state is now easily demonstrated by assuming the opposite.

10.4. Discussion. The Boltzmann principle

In statistical problems of optimal planning or use of resources and also in dynamical problems of optimal control, the constraints which define the set of feasible parameters of the problem or the set of admissible controls are often not given a priori and can be changed within given limits. For the person making planning decisions or constructing a system of controls it is always important to know the possible changes in constraints, investments and gains from these changes. These bits of information which are always important can be obtained in quantitative form by considering the constraints on the choice of parameters or controls as elastic constraints, which also provides more realistic models and problems in the vast majority of cases.

The efficacy of such an approach is also indicated by the fact that the freedom of choice of parameters or controls is then complemented very usefully by freedom of choice of the parameters of the elastic constraints. Not only in technical matters but also in the great majority of economic cases the constraints are elastic and their exact models turn out to be vector fields. Let us consider as an example the linear exchange model of Gale studied in Section 10.2. The problem (10.5)-(10.6) on the equilibrium of the exchange model acquires an interesting economic meaning when the function to be maximized (10.5), is rewritten in a quite simple manner.

In fact, obviously

$$\sum_{s=1}^{m} P_s \ln(c^{(s)},x^{(s)}) = \sum_{s=1}^{m} (c^{(s)},x^{(s)}) -$$

$$-\{\sum_{s=1}^{m} [(c^{(s)},x^{(s)}) - P_s \ln(c^{(s)},x^{(s)})]\}. \quad (10.26)$$

It is easily checked that the function

$$\Phi(x^{(1)},\dots,x^{(m)}) = \sum_{s=1}^{m} [(c^{(s)},x^{(s)})-P_n \ln(c^{(s)},x^{(s)})] \quad (10.27)$$

is strictly convex and assumes it minimum under the conditions

$$(c^{(s)},x^{(s)}) = P_s, \; s = 1,\dots,m, \quad (10.28)$$

and the function

$$\Psi(x^{(1)},\dots,x^{(m)}) = \Phi(x^{(1)},\dots,x^{(m)}) - \sum_{s=1}^{m} P_s(1-\ln P_s), \quad (10.29)$$

which differs from Φ only by a constant, is moreover a positive definite function of the quantities

$$(c^{(s)},x^{(s)}) - P_s, \; s = 1,\dots,m.$$

The truth of this last statement follows from the expression

$$\Psi(x^{(1)},\dots,x^{(m)}) = \frac{1}{2}\sum_{s=1}^{m} \frac{P_s}{[P_s+\theta((c^{(s)},x^{(s)})-P_s)]^2} \times$$

$$\times [(c^{(s)}x^{(s)})-P_s]^2,$$

where $0 \leqslant \theta \leqslant 1$. This results from the Taylor formula for the function (10.29) with a Lagrangian rest term. It is now already clear that $\psi(x^{(1)},\dots,x^{(m)})$ is a finite penalty function for violations of the constraints (10.28).

And thus we arrive at the desired economic interpretation of problem (10.5)-(10.6): The equilibrium set of bundles of goods maximizes over the set defined by the resource constraints and the nonnegativity conditions the sum of the quality criteria of all consumers with finite amount of penalties for violations of the corresponding budgets.

The simple and elegant exchange model of Gale turns out to be instructive and does not represent one unique model of economic systems but can be seen as containing both rigid constraints and elastic ones.

In Section 10.2 it was shown that problem (10.5)-(10.6) is equivalent to an equilibrium problem of a simple physical system for which the function to be maximized is the entropy (of Clausius). This result, naturally, leads to a new interpretation of the economic model under consideration. At the basis of this new interpretation lies Boltzmann's fundamental theorem

$$S = k \ln W, \tag{10.30}$$

where S is the entropy, k the Boltzmann constant and W the thermodynamical probability [6] of the state defined by the matrix $(x_i^{(s)})$. Ignoring the constant k which is inessential for our constrained maximization problem and comparing (10.5) and (10.30) we arrive at the equality

$$\sum_{s=1}^{m} P_s \ln(c^{(s)}, x^{(s)}) = \ln \prod_{s-1}^{m} (c^{(s)} x^{(s)})^{P_s} = \ln W. \tag{10.31}$$

From (10.31) it follows that

1) The quantity

$$\prod_{s=1}^{m} (c^{(s)}, x^{(s)})^{P_s} = W(x^{(1)}, \ldots, x^{(m)}) \tag{10.32}$$

is the probability [7] of the state $(x_i^{(s)})$ of the economy under consideration.

2) The quantity $(c^{(s)}, x^{(s)})$ is the probability of the event that consumer s with budget P_s has available the bundle of goods $x^{(s)}$,

3) Problem (10.5)-(10.6) is equivalent to the problem

$$\max\{W(x^{(1)}, \ldots, x^{(m)}) \mid \sum_{s=1}^{m} x_i^{(s)} = X_i, \; i = 1, \ldots, n, \; x_i^{(s)} \geqslant 0\} \tag{10.33}$$

and for its solution $x^{(1)},\ldots,x^{(m)}$ (the equilibrium bundles of goods) the probability function W of the states of the economy assumes its maximum value, in accordance with the second principle of thermodynamics.

Here only one example has been treated, but, undoubtedly the result of these arguments can be employed for deeper reflections and more extensive generalizations. This conviction arises from the fact that at the basis of the arguments used there lies the very general principle of Boltzmann to the effect that

Spontaneous processes in closed physical systems are transition processes from less probable states to more probable states and a state whose probability is maximal is an equilibrium state.

It is certainly necessary to keep in mind that to apply this principle certain difficulties must be overcome. Those are concerned (as Boltzmann indicated) with finding rules defining equally probable states and the very mathematical definition of the probability of a given state of the object being studied. It is not possible to indicate the boundaries of usefulness of the principle of maximal probability. It is possible that it will turn out to be the most general one of the extremum principles of nature. It is remarkable that the possibility of applying this principle in biology, economics and sociology was pointed out in 1904 by that giant of physics Ludwig Boltzmann in his brilliant treatise 'On statistical mechanics'. Taking as starting point the principle of Boltzmann as formulated above, the second principle of thermodynamics and the additivity property of entropy, Willard Gibbs obtained formula (10.30) as a simple theorem of probability theory [8]. Concerning the value of the Boltmann principle in general M. Planck wrote: "This principle opens up a way to new methods of calculating the probability of systems in a given state; methods which will go far beyond the limits of the usual methods of thermodynamics". Using the considerations of Gibbs one can obtain an analogue of equality (10.30) in which the left side is a general thermodynamic Gibbs potential (see for example Chapter II, Section 2.2.). On the other hand this principle affirms the appropriateness of the processes taking place in nature and affirms the possibility - in principle - of discovering the unique goal towards which things evolve if we find a way of comparing states with respect to probability. And that, as Euler wrote in one of his letters to Maupertuis "... is not only most laudable but also often useful for deepening our understanding of problems".

10.5. Equilibrium of linear economic models

Below we shall study models for an economy consisting of a collection of enterprises each of which has its own goals and budget. An equilibrium for such an economy is a set of prices of the resources for which each of the participating enterprises of the economy using its budget to purchase resources maximizes its objective function. We shall see once more that the equilibrium problem for such an economy is equivalent to an equilibrium problem of some physical system. The last statement means that there exists a state function which reaches its maximum at the equilibrium state and which has the meaning of the entropy of the physical model of the economy under consideration. As in the case of the exchange model of Gale [29, 41], this function is a weighted sum of logarithms of the objective functions of the enterprises making up the economy and the weights are the budgets of these enterprises.

Now to complete the model being considered we assume the existence of a central controlling organism which uses the equilibrium mechanism and restricts its influence on the economy to budget allocations. If the aim of the central controlling agency is to maximize some global objective function then the problem of that centre is to choose that allocation of budgets, for which the entropy of the economy becomes closest to the global goal.

Thus we shall consider models for economies in which a central control is naturally combined with semi-independent economic subsystems. A fundamental property of such models is the existence of one particular function which assumes its maximum in the state for which the objective functions of all the constituting economic subsystems attain their maxima and the set of parameters on which this function depends is naturally considered the set of control parameters for the central agency.

Formulation of the problem. We consider an economy which has m types of resources $B_1,\dots,B_m$ which are present in the amounts $b_1,\dots,b_m$. There are k consumers of these resources $A_1,\dots,A_k$ and they dispose of budgets $P_1,\dots,P_k$ to obtain resources. Instead of consumers of resources one can also consider separate firms or sectors of the economy. In the following we shall say firm instead of consumer. The firm A_α, having a bundle of goods $b^{(\alpha)} = (b_1^{(\alpha)},\dots,b_m^{(\alpha)})$ at its disposal uses this for production choosing among n production processes and trying to maximize an objective (utility fun function)

$$\sum_{i=1}^{n} c_i^{(\alpha)} x_i^{(\alpha)} \to \max, \tag{10.34}$$

$$\left.\begin{aligned} &\sum_{i=1}^{n} a_{si}^{(\alpha)} x_i^{(\alpha)} \leqslant b_s^{(\alpha)}, \quad s = 1,\dots,m, \\ &x_i^{(\alpha)} \geqslant 0, \qquad\qquad i = 1,\dots,n. \end{aligned}\right\} \tag{10.35}$$

In problem (10.34)-(10.35) $x_i^{(\alpha)}$ denotes the intensity with which firm A makes use of technological process i. Let $p = (p_1,\dots,p_m)$ be a nonnegative vector of prices for the resources $B_1,\dots,B_m$. If firm A_α has initially only its budget available it must acquire resources and the problem becomes more complicated and takes the form

$$\sum_{i=1}^{n} c_i^{(\alpha)} x_i^{(\alpha)} \to \max,$$

$$\sum_{i=1}^{n} a_{si}^{(\alpha)} x_i^{(\alpha)} - b_s^{(\alpha)} = 0, \quad s = 1,\dots,m,$$

$$\sum_{s=1}^{m} p_s b_s^{(\alpha)} \leqslant P_\alpha,$$

$$x_i^{(\alpha)} \geqslant 0, \; b_s^{(\alpha)} \geqslant 0, \; s = 1,\dots,m, \; i = 1,\dots,n.$$

In this problem the intensities $x_i^{(\alpha)}$ are of course still unknown as is also the vector $b^{(\alpha)}$ of resource amounts desired by firm A_α.

Below we shall study equilibrium problems for such economies in the sense of the following definition.

DEFINITION 10.3. The set of vectors $b^{(\alpha)} = (b_1^{(\alpha)},\dots,b_m^{(\alpha)})$, $\alpha = 1,\dots,k$ with nonnegative components is called an equilibrium of bundles of goods and a nonnegative vector $p = (p_1,\dots,p_n)$ is called an equilibrium price vector if the following conditions are satisfied

(1) $f^{(\alpha)}(x^{(\alpha)}) = \sum_{i=1}^{n} c_i^{(\alpha)} x_i^{(\alpha)} \to \max$

$\alpha = 1,\dots,k,$

(2) $\sum_{i=1}^{n} a_{si}^{(\alpha)} x_i^{(\alpha)} - b_s^{(\alpha)} = 0,$

$\alpha = 1,\dots,k,\ s = 1,\dots,m,$

(3) $\sum_{s=1}^{m} p_s b_s^{(\alpha)} \leqslant P_\alpha,$

$\alpha = 1,\dots,k,$

(4) $\sum_{\alpha=1}^{n} b_s^{(\alpha)} \leqslant b_s,$

$s = 1,\dots,m.$

10.6. Physical models for economic equilibrium. The equilibrium theorem

The aim of the present section is to construct a physical model of the problem outlined in the previous section and to prove an equilibrium theorem which is a generalization of Gale's equilibrium theorem for exchange economies.

EQUILIBRIUM THEOREM 10.2. Suppose that the matrix $(c_i^{(\alpha)})$ does not contain columns or rows of zeros. Then a set of equilibrium bundles of goods $b^{(1)}, b^{(2)},\dots,b^{(k)}$ maximizes the function

$$\sum_{\alpha=1}^{k} P_\alpha \ln \xi_\alpha \qquad (10.36)$$

under the conditions

$$\xi_\alpha = \sum_{i=1}^{n} c_i^{(\alpha)} x_i^{(\alpha)},\quad \alpha = 1,\dots,k, \qquad (10.37)$$

$$\sum_{i=1}^{n} a_{si}^{(\alpha)} x_i^{(\alpha)} - b_s^{(\alpha)} = 0, \qquad (10.38)$$

$\alpha = 1,\dots,k,\ s = 1,\dots,m,$

$$\sum_{\alpha=1}^{k} b_s^{(\alpha)} \leqslant b_s, \; s = 1,\ldots,m, \; x_i^{(\alpha)} \geqslant 0, \tag{10.39}$$

and an equilibrium price vector $p = (p_1,\ldots,p_m)$ for the resources is given by the dual prices (the sensitivities) of the constraints (10.39).

To prove this theorem we shall make use of the physical properties of a model of problem (10.36)-(10.39) and it is therefore necessary to describe this model first. The model sketched in figure 10.3 consists of k blocks which are models for the systems of linear equations (10.37) and (10.38). Each of these blocks is a model of the type familiar to the reader from Chapter I, and consists of systems of communicating volumes filled with an incompressible liquid. The blocks in turn are also connected to each other by a system of communicating volumes also filled with an incompressible liquid and these systems of volumes model the m conditions (10.39). The objective function (10.36) is finally modelled as was done in the exchange model, that is by means of k volumes respectively equal to $\xi_1,\ldots,\xi_k$ and filled with an ideal gas in such amounts that the pressure in volume $\xi_\alpha = c_1^{(\alpha)}x_1^{(\alpha)} + \ldots + c_n^{(\alpha)}x_n^{(\alpha)}$ is equal to one if $\xi_\alpha = P_\alpha$. It is easy to check that for a constant absolute temperature T = 1 these volumes must contain $\mu_\alpha = P_\alpha/R$ mols of an ideal gas (R is the universal gas constant). Given all this the function (10.36) is by analogy with the one occurring in the exchange model of Gale (see Section 10.2) the entropy of the physical model described above. Because the maximal entropy is assumed in a state of equilibrium, problem (10.36)-(10.39) is equivalent to the equilibrium problem for the model sketched in Figure 10.3.

The mechanical equilibrium conditions for the model of problem (10.36)-(10.39) are the conditions that the forces acting on the various movable parts of the model cancel each other. The movable parts are the nk bars whose positions indicate the quantities $x_i^{(\alpha)}$, the mk bars whose positions determine the quantities $b_s^{(\alpha)}$ and m pistons indicating the magnitude of $\beta_1,\ldots,\beta_m$, the amount of 'left-over' resources.

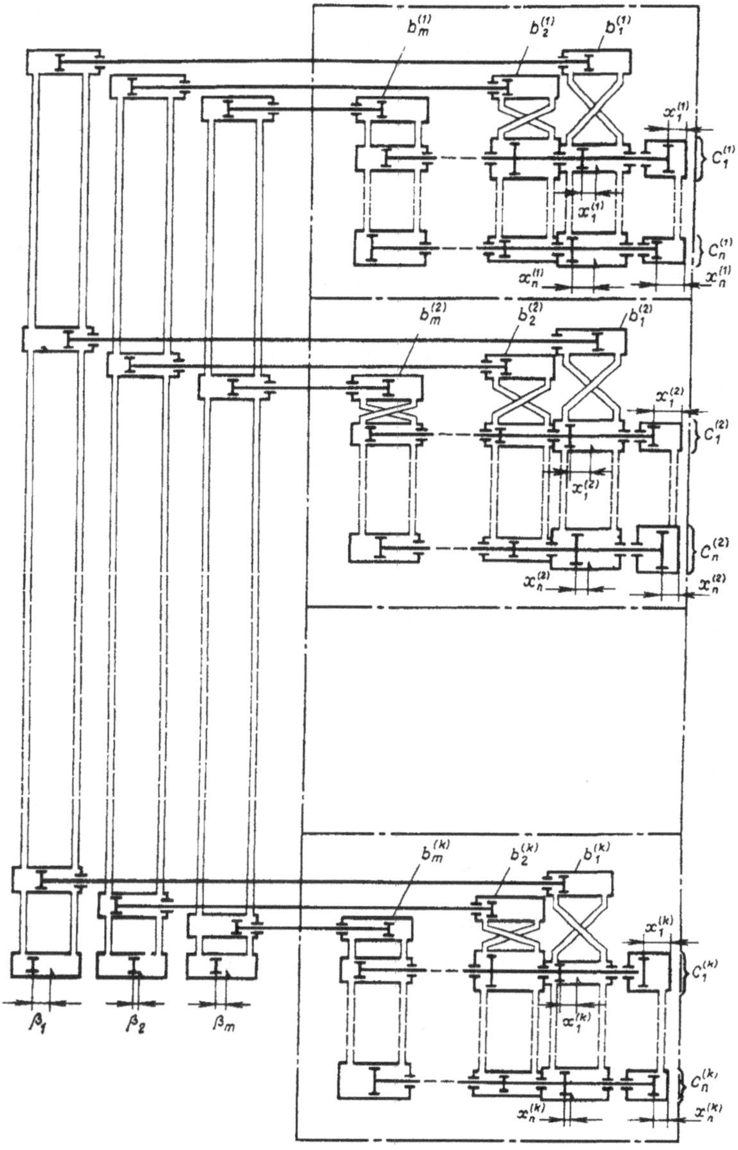
$b_m^{(1)}$
$b_2^{(1)}$
$b_1^{(1)}$
$x_1^{(1)}$
$C_1^{(1)}$
$x_1^{(1)}$
$C_n^{(1)}$
$x_n^{(1)}$
$x_n^{(1)}$
$b_m^{(2)}$
$b_2^{(2)}$
$b_1^{(2)}$
$x_1^{(2)}$
$C_1^{(2)}$
$x_1^{(2)}$
$C_n^{(2)}$
$x_n^{(2)}$
$x_n^{(2)}$
$b_m^{(k)}$
$b_2^{(k)}$
$b_1^{(k)}$
$x_1^{(k)}$
$C_1^{(k)}$
$x_1^{(k)}$
$C_n^{(k)}$
$x_n^{(k)}$
$x_n^{(k)}$
β_1
β_2
β_m

Fig. 10.3.

Let us introduce some notation:

1) $q_s^{(-)} - q_s^{(+)}$ the pressure difference between the left and right side of the pistons in the communicating volumes modelling conditions (10.39).

2) $q_s^{(\alpha)-} - q_s^{(\alpha)+}$ the pressure difference between the systems of communicating volumes $V_s^{(\alpha)-}$ and $V_s^{(\alpha)+}$ modelling conditions (10.37) and (10.38). By analogy with the models for linear programming problems (see Section 1.2) the volumes $V_s^{(\alpha)+}$ and $V_s^{(\alpha)-}$ are defined by the formulas

$$V_s^{(\alpha)-} = l(1+\sum_{i=1}^{n}|a_{si}^{(\alpha)}|) - \sum_{i=1}^{n} a_{si}^{(\alpha)}x_i^{(\alpha)} + b_s^{(\alpha)},$$

$$V_s^{(\alpha)+} = l(1+\sum_{i=1}^{n}|a_{si}^{(\alpha)}|) + \sum_{i=1}^{n} a_{si}^{(\alpha)}x_i^{(\alpha)} - b_s^{(\alpha)},$$

3) q_α the pressure in the volume $V_0^{(\alpha)}$ filled with an ideal gas. The quantity q_α is connected with the coordinates $x_1^{(\alpha)},\ldots,x_n^{(\alpha)}$ by the Clapeyron-Mendeleev state equation

$$q_\alpha\xi_\alpha = \mu_\alpha RT,$$

so that for $T = 1$ and $\mu_\alpha = R^{-1}P_\alpha$

$$q_\alpha = \frac{P_\alpha}{\xi_\alpha}. \tag{10.40}$$

It is now quite simple to write down the conditions for equilibrium of the model of problem (10.36)-(10.39). The equilibrium conditions for the nk bars whose positions indicate the coordinates $x_i^{(\alpha)}$ and whose positions are subject to the unilateral constraints $x_i^{(\alpha)} \geqslant 0$ have the form

$$\sum_{s=1}^{m}(q_s^{(\alpha)-}-q_s^{(\alpha)+})a_{si}^{(\alpha)} \begin{cases} \geqslant q_\alpha c_i^{(\alpha)} & \text{for } x_i^{(\alpha)} = 0 \\ = q_\alpha c_i^{(\alpha)} & \text{for } x_i^{(\alpha)} > 0, \end{cases} \tag{10.41}$$

$i = 1,\ldots,n,\ \alpha = 1,\ldots,k.$

The conditions for equilibrium for the bars whose

positions determine the quantities $b_s^{(\alpha)}$ and which are also subject to the unilateral constraints $b_s^{(\alpha)} \geqslant 0$ have the form

$$q_s^{(\alpha)-} - q_s^{(\alpha)+} \begin{cases} \leqslant q_s^{(-)} - q_s^{(+)} & \text{for } b_s^{(\alpha)} = 0, \\ = q_s^{(-)} - q_s^{(+)} & \text{for } b_s^{(\alpha)} > 0, \end{cases} \tag{10.42}$$

$s = 1,\ldots,m,\ \alpha = 1,\ldots,k.$

Finally the conditions for the pistons which define the magnitudes of the unconstrained variables

$$\beta_s = b_s - \sum_{\alpha=1}^{k} b_s^{(\alpha)}, \quad s = 1,\ldots,m,$$

are obviously

$$q_s^{(+)} - q_s^{(-)} \begin{cases} \geqslant 0 & \text{for } \beta_s = 0, \\ = 0 & \text{for } \beta_s > 0. \end{cases} \tag{10.43}$$

The conditions (10.41)-(10.43) are a system of conditions for equilibrium of the physical model of problem (10.36)-(10.39) and they constitute a duality theorem for the class of problems under consideration. In particular the pressure differences $q_s^{(-)} - q_s^{(+)}$ at equilibrium of the model are dual prices (sensitivities) for the constraints (10.39), that is they are the desired equilibrium prices $p_1,\ldots,p_m$ for the resources. Indeed multiplying inequalities (10.39) respectively with the $x_i^{(\alpha)}$ and summing with respect to i we obtain the k equations

$$\sum_{s=1}^{m} (q_s^{(\alpha)-} - q_s^{(\alpha)+}) \sum_{i=1}^{n} a_{si}^{(\alpha)} x_i^{(\alpha)} = q_\alpha \xi_\alpha, \quad \alpha = 1,\ldots,k,$$

and taking account of (10.38) and (10.40) we obtain from these equations

$$\sum_{s=1}^{m} b_s^{(\alpha)} (q_s^{(\alpha)-} - q_s^{(\alpha)+}) = P_\alpha, \quad \alpha = 1,\ldots,k. \tag{10.44}$$

The quantities $q_s^{(\alpha)-} - q_s^{(\alpha)+}$ are the dual prices of the constraints (10.38) for all nonnegative values of the $b_s^{(\alpha)}$

which respect the constraints (10.39). As a consequence the quantities

$$p_1^{(\alpha)} = q_1^{(\alpha)-} - q_1^{(\alpha)+}, \ldots, p_m^{(\alpha)} = q_m^{(\alpha)-} - q_m^{(\alpha)+}$$

are the respective prices which firm A_α attributes to its bundle of resources $b_1^{(\alpha)}, \ldots, b_m^{(\alpha)}$. Recall that this holds for any fixed bundle of goods $b_1^{(\alpha)}, \ldots, b_m^{(\alpha)}$ satisfying conditions (10.39). For an equilibrium set of bundles of resources equation (10.42) holds and multiplying these with the $b_s^{(\alpha)}$ and summing over s one obtains

$$\sum_{s=1}^{m} b_s^{(\alpha)}(q_s^{(\alpha)-} - q_s^{(\alpha)+}) = \sum_{s=1}^{m} b_s^{(\alpha)}(q_s^{(-)} - q_s^{(+)}).$$

From (10.44) and (10.45) it follows that

$$\sum_{s=1}^{m} b_s^{(\alpha)}(q_s^{(-)} - q_s^{(+)}) = P_\alpha, \; \alpha = 1, \ldots, k,$$

$$p_s^{(\alpha)} = p_s = q_s^{(-)} - q_s^{(+)} \text{ for } b_s^{(\alpha)} > 0, \; s = 1, \ldots, m. \tag{10.46}$$

Thus in an equilibrium state the prices attributed to a given type of resource by all firms who find that resource useful ($b_s^{(\alpha)} > 0$) are all equal, and equal to the equilibrium price of that resource. Moreover it follows from (10.42) that

$$p_s = \max_\alpha (q_s^{(\alpha)-} - q_s^{(\alpha)+}), \; s = 1, \ldots, m.$$

Condition (10.43) expresses the well known economic fact that if a resource is not completely used its price is zero.

We are now in a position to finish the proof of the equilibrium theorem. Let $\{x_i^{(\alpha)*}, b_s^{(\alpha)*}\}$ be the coordinates of an equilibrium state for the model of problem (10.36)-(10.39), and let $p_s = q_s^{(-)*} - q_s^{(+)*}$ be the dual prices of the constraints (10.39) or in other words the reaction forces produced by the unilateral constraints (10.39). The idea of the proof is now to replace these constraints by precisely these reaction forces $p_1^*, \ldots, p_m^*$ at equilibrium. According to the principle of removing constraints $\{x_i^{(\alpha)*}, b_s^{(\alpha)*}\}$ remains an equilibrium when we do this. However, the equilibrium problem thus obtained by removing constraints (10.39) splits into k equilibrium problems for the various blocks which are then separate active physical systems.

Let us consider any one of these problems. Because $x_1^{(\alpha)*},\ldots,x_n^{(\alpha)*}$, $b_1^{(\alpha)*},\ldots,b_m^{(\alpha)*}$ is an equilibrium state for the block α, then by virtue of the principle of virtual work (virtual displacements) (see Theorem 1.5 of Chapter I), in a neighbourhood of this equilibrium the following condition holds

$$\sum_{s=1}^{m} p_s^* \delta b_s^{(\alpha)} + q_\alpha^* \sum_{i=1}^{n} c_i^{(\alpha)} \delta x_i^{(\alpha)} \leqslant 0, \qquad (10.47)$$

where according to (10.40)

$$q_\alpha = (\xi_\alpha^*)^{-1} P_\alpha$$

In (10.47) there is an inequality only if the vector of virtual displacements $\delta x^{(\alpha)}$ removes at least one of the constraints $x_i^{(\alpha)} \geqslant 0$, $i = 1,\ldots,n$. If we limit our choice of variations $\delta b_s^{(\alpha)}$ to those which satisfy condition (10.46)

$$\sum_{s=1}^{m} b_s^{(\alpha)} p_s^* = P_\alpha$$

which holds in an equilibrium state for the model of (10.36)-(10.39) then we obtain

$$\sum_{s=1}^{m} p_s^* \delta b_s^{(\alpha)} = 0,$$

Therefore the necessary and sufficient condition for equilibrium (10.47) for the block α takes the form

$$P_\alpha \frac{\sum_{i=1}^{n} c_i^{(\alpha)} \delta x_i^{(\alpha)}}{\sum_{i=1}^{n} c_i^{(\alpha)} x_i^{(\alpha)}} = P_\alpha \delta(\ln \xi_\alpha) \leqslant 0. \qquad (10.48)$$

The inequality (10.48) concludes the proof of the equilibrium theorem because it follows from it that $\xi_\alpha^* \to \max$.

In this way an equilibrium state of the physical model

of the physical model of problem (10.36)-(10.39) defines the parameters

$$\{x_i^{(\alpha)*}, b_s^{(\alpha)*}, p_s^*\},\ i = 1,\ldots,n,\ s = 1,\ldots,m,\ \alpha = 1,\ldots,k,$$

satisfying all the desired requirements for equilibrium. In this manner we have once more shown that the general principles of analytical mechanics also play an important role in models of mathematical economics. And it follows from those principles that in the nondegenerate cases, because of equilibrium prices, one may ignore the constraints on resources. These will be fulfilled automatically at equilibrium.

10.7. An algorithm for solving equilibrium problems for linear economic models

The method of redundant constraints provides iterative algorithms for solving the equilibrium problem for the linear economic model described in Section 10.5. We already know (see Section 10.6) that the solution of (10.36)-(10.39) yields optimal plans $x^{(1)},\ldots,x^{(k)}$ for the firms $A_1,\ldots,A_k$ and a set of equilibrium bundles of resources $b^{(1)},\ldots,b^{(k)}$. All these quantities are parameters of the equilibrium state of the model of problem (10.36)-(10.39) depicted in Figure 10.3.

The method described below for solving equilibrium problems for linear economic models is an extension of the ideas of decomposition theory, which were studied in Chapter VII, to the equilibrium problem of (10.36)-(10.39). Indeed this problem is different from problem (7.7)-(7.9) in that the objective function (10.36) is nonlinear and as a consequence for fixed values of a resource allocation matrix $(b_s^{(\alpha)})$ one obtains k problems of the form

$$P_\alpha \ln \xi_\alpha \to \max \tag{10.49}$$

$$\left.\begin{aligned} &\sum_{i=1}^{n} c_i^{(\alpha)} x_i^{(\alpha)} - \xi_\alpha = 0,\\ &\sum_{i=1}^{n} a_{si}^{(\alpha)} x_i^{(\alpha)} - b_s^{(\alpha)} = 0,\\ &s = 1,\ldots,m, \end{aligned}\right\} \tag{10.50}$$

where the quantities $b_s^{(\alpha)0}$ are chosen so that (10.39) holds. Of course the optimal vectors for the problem (10.49)-(10.50) coincide with those of the problem of $\xi_\alpha \to \max$ under the constraints (10.50), but in the following we shall use, as we did for decomposition theory, models with containers filled with an ideal gas and for such models the equilibrium states for problem (10.49)-(10.50) and the problem $\xi_\alpha \to \max$ under (10.50) are of course different.

Here it is once more necessary to pay attention to the fact that in the method set out below we make essential use of the property of ideal gases: changes in pressure under constant temperature are possible only under corresponding changes in volume. It is that property which makes it possible to represent by means of a physical model the reallocation of resources with simultaneous equalizations of prices assgined by the firms to these resources.

The model of the economic process just indicated is a transition process of a closed physical system to an equilibrium state during which an equalization of pressures takes place and corresponding changes in volumes. The maximization of the entropy of the physical system under these processes leads to the important conclusion that there exists a state function of the economy which assumes its maximum at equilibrium for exchange economies.

Thus the method for solving (10.36)-(10.39), as in the case of decomposition theory, is a cyclic iterative process of which each cycle consists of two steps. The content of the first step is the solution of k problems (10.49)-(10.50) for a fixed allocation of resources matrix $(b_s^{(\alpha)})$ which satisfies (10.39). The second step of each cycle consists of the determination of the next approximate resource allocation matrix $(b_s^{(\alpha)})$ for fixed values of the quantities $x_i^{(\alpha)}$, which are the equilibrium coordinates of the phsyical model of the previous step. The problems of the second step are problems of reallocation of resources and they coincide with the problems we encountered in the second step of the iterative decomposition method in Chapter VII. They also split into m simple problems of reallocating each single type of resource.

Let us consider one of the first steps, that is a problem (10.49)-(10.50). Because these problems are all of the same type we leave out the label α which indicates the number of the problem and write these first step problems in the form

$$P \ln \xi \to \max, \tag{10.51}$$

$$\left.\begin{aligned} &\sum_{i=1}^{n} c_i x_i - \xi = 0, \\ &\sum_{i=1}^{n} a_{si} x_i - b_s^{(0)} = 0, \\ &s = 1,\dots,m. \end{aligned}\right\} \tag{10.52}$$

The physical model of problem (10.51)-(10.52) is one of the blocks of the model of problem (10.36)-(10.39), depicted in Figure 10.3. We shall assume that the communicating volumes of the model of the linear equations (10.52) are filled with an ideal gas in such quantities that in any state satisfying equations (10.52) the pressures in these volumes are equal to a given pressure q.

Let $x^{(0)} = (x_1^{(0)},\dots,x_n^{(0)})$ be an arbitrary positive vector. Define the number $\xi^{(0)}$ as the coordinate of the equilibrium state of the model of (10.51)-(10.52) for the fixed values $x_i = x_i^{(0)}$, $i = 1,\dots,n$. The quantity $\xi^{(0)}$ can obviously be found from the equilibrium conditions for the bars, that is from the equations

$$q_0^{(+)0} - q_0^{(-)0} = q, \tag{10.53}$$

where according to (10.40)

$$q = \frac{P}{\xi^{(0)}}$$

Moreover according to (2.32)

$$q_0^{(+)0} - q_0^{(-)0} = \tilde{q}_0 \frac{-\sum_{i=1}^{n} c_i x_i^{(0)} + \xi^{(0)}}{1 + \sum_{i=1}^{n} |c_i|} \tag{10.54}$$

where $\tilde{q}_0 = \ell^{-1} 2q_0$. From (10.53) and (10.54) one obtains a quadratic equation of which the unique positive root is equal to 9)

$$\xi^{(0)} = \frac{1}{2}\{ \sum_{i=1}^{n} c_i x_i^{(0)} + \sqrt{(\sum_{i=1}^{n} c_i x_i^{(0)})^2 + \frac{4P}{\tilde{q}_0}(1 + \sum_{i=1}^{n} |c_i|)}\}.$$

Now find, using any of the algorithms of Chapter III the equilibrium state $x^{(1)}$ of the model of system /10.52) for the fixed value $\xi = \xi^{(0)}$ and find the next approximation $\xi^{(1)}$ of ξ by the analogous formula

$$\xi^{(1)} = \frac{1}{2}\{ \sum_{i=1}^{n} c_i x_i^{(1)} + \sqrt{(\sum_{i=1}^{n} c_i x_i^{(1)})^2 + \frac{4P}{\tilde{q}_0}(1 + \sum_{i=1}^{n} |c_i|)}\}.$$

It is clear how to continue.

We note that the algorithm for solving the first step problems is very similar to the algorithm for solving linear programming problems which is based on a reduction of that problem to a series of problems of finding minimal norms of errors for systems of equations and inequalities, see Section 3.5.

Indeed after one has calculated $\xi^{(\nu)}$ by means of the formula

$$\xi^{(\nu)} = \frac{1}{2}\{ \sum_{i=1}^{n} c_i x_i^{(\nu)} - \sqrt{(\sum_{i=1}^{n} c_i x_i^{(\nu)})^2 + \frac{4P}{\tilde{q}_0}(1 + \sum_{i=1}^{n} |c_i|)}\}.$$

The problem of finding the next vector $x^{(\nu+1)}$ turns out to be the problem of minimizing the Helmhotlz free energy for the model of the system of linear equations

$$\sum_{i=1}^{a} a_{si} x_i = b_s^{(0)}, \quad s = 1,\dots,m,$$

$$\sum_{i=1}^{n} c_i x_i = \xi^{(\nu)}, \quad x_i > 0,$$

which is analogous to (3.37). The only difference is in the formula for calculating iteratively the right hand term of the (m+1)-th equation. It is therefore not necessary to write down the formulas for calculating the components of the vector $x^{(\nu+1)}$. The reader can find these formulas in Section 3.5. As was mentioned above the second step problems consist of the calculation of the next iteration $(b_s^{(\alpha)1})$ of the resource allocation matrix, and this coincides with the problems occurring in the second steps (of each cycle) of the iterative decomposition algorithm. In Section 7.2 explicit finite formulas were obtained for solving these. Because each step of the algorithm consists of a transition of some physical system to an equilibrium state it is obvious that the algorithm just described converges monotonically, independent of the number of iterations used in solving the problems of the first step (of each cycle).

10.8. A generalization of the economic equilibrium problem

Above a situation was discussed in which the firms at the beginning of a production cycle only had their budgets $P_1,\dots,P_k$ disposable and every chosen bundle of resources $(b_1^{(\alpha)},\dots,b_m^{(\alpha)})$ desired by firm A_α has to be bought against prices $(p_1,\dots,p_m)$ with funds from its own budget.

We shall now consider a more realistic situation, by assuming that at the start of a production cycle each firm A_α has available some initial bundle of resources $(\beta_1^{(\alpha)},\dots,\beta_m^{(\alpha)})$ and a budget P_α, which can be used for buying additional amounts of resources which are necessary for the firm to maintain its production at a previous level or to extend this level.

In this case the problem of firm A_α is the following:

$$\sum_{i=1}^{n} c_i^{(\alpha)} x_i^{(\alpha)} \to \max,$$

$$\sum_{i=1}^{n} a_{si}^{(\alpha)} x_i^{(\alpha)} - \beta_s^{(\alpha)} - b_s^{(\alpha)} \leqslant 0, \quad s = 1,\dots,m_1,$$

$$\sum_{i=1}^{n} a_{si}^{(\alpha)} x_i^{(\alpha)} - \beta_s^{(\alpha)} \leqslant 0, \qquad s = m_1+1,\dots,m,$$

$$\sum_{s=1}^{m_1} p_s b_s^{(\alpha)} \leqslant P_\alpha, \quad \alpha = 1,\dots,k,$$

$$x_i^{(\alpha)} \geqslant 0, \quad i = 1,\dots,n; \ \alpha = 1,\dots,k,$$

The equilibrium problem for such an economy consits in

determining the equilibrium prices $p_1,\dots,p_{m_1}$ for the resources $B_1,B_2,\dots,B_{m_1}$. That is such that the m_1-dimensional vectors $b^{(1)},\dots,b^{(k)}$ satisfy the conditions

$$\left.\begin{array}{l}
\sum_{i=1}^{n} c_i^{(\alpha)} x_i^{(\alpha)} \to \max, \\
\sum_{i=1}^{n} a_{si}^{(\alpha)} x_i^{(\alpha)} - \beta_s^{(\alpha)} - b_s^{(\alpha)} \leqslant 0,\ s=1,\dots,m_1, \\
\sum_{i=1}^{n} a_{si}^{(\alpha)} x_i^{(\alpha)} - \beta_s^{(\alpha)} \leqslant 0,\ s = m_1+1,\dots,m, \\
\sum_{s=1}^{m_1} p_s b_s^{(\alpha)} \leqslant P_\alpha,\ x_i^{(\alpha)} \geqslant 0,\ i = 1,\dots,n, \\
\sum_{\alpha=1}^{k} b_s^{(\alpha)} \leqslant R_s,\ s = 1,\dots,m_1
\end{array}\right|\ \alpha = 1,\dots,k, \tag{10.55}$$

where $R_1,\dots,R_{m_1}$ are the total amounts of resources $B_1,B_2,\dots,B_{m_1}$ which are available to the central organization for allocation.

Arguing analogously as in Section 10.6 the reader can convince himself that the equilibrium plans $x^{(1)},\dots,x^{(k)}$ of the firms and the equilibrium bundles of resources desired by the firms satisfy the conditions

$$\begin{array}{l}
\sum_{\alpha=1}^{k} P_\alpha \ln \sum_{i=1}^{n} c_i^{(\alpha)} x_i^{(\alpha)} \to \max, \\
\left.\begin{array}{l}
\sum_{i=1}^{n} a_{si}^{(\alpha)} x_i^{(\alpha)} - \beta_s^{(\alpha)} - b_s^{(\alpha)} \leqslant 0,\ s = 1,\dots,m_1, \\
\sum_{i=1}^{n} a_{si}^{(\alpha)} x_i^{(\alpha)} - \beta_s^{(\alpha)} \leqslant 0,\ x_i^{(\alpha)} \geqslant 0, \\
s = m_1 + 1,\dots,m,\ i = 1,\dots,n,
\end{array}\right|\ \alpha = 1,\dots,k \\
\sum_{\alpha=1}^{k} b_s^{(\alpha)} \leqslant R_s,\ s = 1,\dots,m_1,
\end{array} \tag{10.56}$$

and that the components of the equilibrium price vectors are the dual prices (sensitivities) of the constraints (10.56).

To prove this equilibrium theorem it is convenient to use a physical model which is analogous to the model of probelm (10.36)-(10.39). We remark however that the quantities to be determined $\beta_s^{(\alpha)}$ can both be restricted to nonnegativity constraints but can also be free of such constraints. In the last case the firms are allowed to sell their superfluous resources at equilibrium prices. The proof of the equilibrium theorem is similar to the one given in Section 10.6.

NOTES

1. Here we are referring to the famous H-theorem of Boltzmann to the effect that the entropy is proportional to the logarithm of the probability of the state: S = k ln H.
2. This model is neither a special case nor a generalization of the model of Gale as the author wrongly stated in [15].
3. It is obviously necessary for this that the exterior pressure is zero, because in the opposite case the equilibrium of the model is determined not only by the pressure forces of the gases and liquids in the closed volumes, but also by the forces coming from the exterior pressures.
4. Or law of Boyle-Mariotte.
5. Below we shall indicate other formulas for determining the coordinates of an initial state which is essentially closer to the equilibrium state to be found.
6. In contradistinction to mathematical probabilities, thermodynamic probabilities need not be smaller than one; it can be simply the number of different configurations or complexes (L. Boltzmann).
7. In this case the quantity W is analogous to a physical probability
8. The demonstration of Gibbs is set out in [62].
9. The negative root of the quadratic equation obtained from (10.53) and (10.54) does not satisfy conditions (10.52).

Chapter XI

DYNAMIC ECONOMIC MODELS

11.1. Introduction

This chapter is essentially concerned with an extension to some dynamical mathematical models of economics of the idea of physical simulations. We shall restrict ourselves to a study of the von Neumann-Gale model [29, 41] and this will enable us to illustrate the possibilities of this method of physical models. We shall consider an economy in which the production of goods can change with time and is described by a series of state vectors corresponding to (the results of) a series of production cycles. The state X(s) of the economy will be the bundle of goods which can be used during the (s+1)-th cycle.

The intensities with which various technological processes available to the economy are used are naturally considered to be the control parameters for the evolution process (of the economy). We shall study the problem of shortest time movement (evolution) of the economy from a given initial state X(0) to some state $X \geqslant \bar{X}$ where $\bar{X}$ is a given state vector; see Figure 8.1.

The method of solving this shortest time problem consists in reducing it to a parametric minimization problem of some definite functional. And, as we shall see, this functional will turn out to be the Helmholtz free energy of an ideal gas in the various volumes of a certain physical model of the problem under consideration. And, using the methods of redundant constraints, we shall obtain an iterative algorithm for solving numerically this problem which converges monotonically on physical grounds. A good initial approximation reduces the calculation times substantially, and it is natural to make use for this purpose of von Neumann trajectories (rays) of maximal balanced growth, [39, 42, 44]. Therefore we shall first consider the problem of maximal (balanced) growth rates and describe an algorithm for solving this which consists of a sequence of linear programming problems. In the final section of this chapter we shall consider the problem of finding a minimal time balanced growth trajectory.

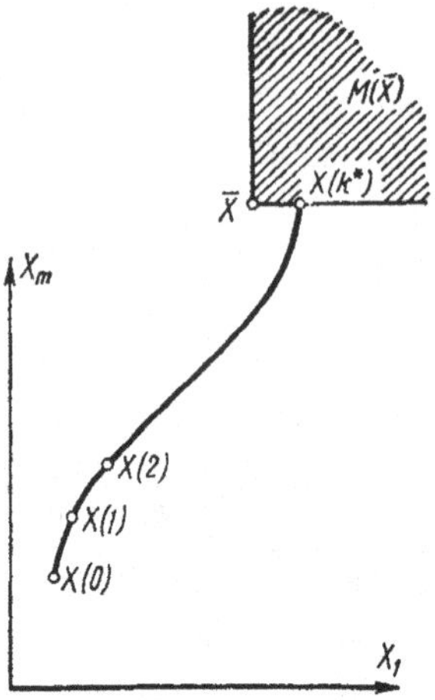

Fig. 11.1.

11.2. The Von Neumann-Gale model. Growth rates and interest rates

We shall consider a linear economic model in which there are m (kinds of) goods (products), $G_1,\ldots,G_m$ and n technological production processes $P_1,\ldots,P_n$. The technological production process P_i is characterized by a pair of m dimensional vectors a_i and b_i of which $a_i = (a_{1i},\ldots,a_{mi})$ describes the input amounts and $b_i = (b_{1i},\ldots,b_{mi})$ the output amounts of goods $G_1,\ldots,G_m$ if the process is used at intensity one. The choice of intensity units for each of the technological processes is arbitrary. We shall denote with x_i the intensity with which production process P_i is used. As the model is linear this means that process P_i will use up the bundle of goods $(a_{1i}x_i,\ldots,a_{mi}x_i)$ and produce the bundle $(b_{1i}m_i,\ldots,b_{mi}x_i)$. The matrices with nonnegative elements A and B

$$A = \begin{Vmatrix} a_{11} & a_{12} & \cdots & a_{1n} \\ a_{21} & a_{22} & \cdots & a_{2n} \\ \cdots & \cdots & \cdots & \cdots \\ a_{m1} & a_{m2} & \cdots & a_{mn} \end{Vmatrix}, \quad B = \begin{Vmatrix} b_{11} & b_{12} & \cdots & b_{1n} \\ b_{21} & b_{22} & \cdots & b_{2n} \\ \cdots & \cdots & \cdots & \cdots \\ b_{m1} & b_{m2} & \cdots & b_{mn} \end{Vmatrix}$$

will be called the input and outpur matrix respectively. We shall assume in the following that the matrices A and B satisfy the following properties [29, 41].

1) for all i at least one of the numbers $a_{1i},\dots,a_{mi}$ is strictly positive; that is each technological production process consumes at least one kind of good

2) for all j at least one of the numbers $b_{j1},\dots,b_{jn}$ is strictly positive; that is for every good there is at least one technological production process which produces that good.

In the following (A,B) will denote the model defined by the matrices A and B.

We shall proceed to discuss the important notion of a rate of growth for an expanding economy. For a given vector of intensities $x = (x_1,\dots,x_n)$ of use of the technological production processes $P_1,\dots,P_n$, the sums

$$\sum_{i=1}^{n} a_{ji}x_i \quad\text{and}\quad \sum_{i=1}^{n} b_{ji}x_i$$

are respectively the total input and total outpur of good G_j. The quantity

$$\alpha_j(x) = \frac{\sum_{i=1}^{n} b_{ji}x_i}{\sum_{i=1}^{n} a_{ji}x_i}$$

is called the technological growth rate of the j-th good. Here the amount of input of good G_j is supposed to be nonzero. For a given intensity vector x we can now define a nonnegative number $\alpha(x)$ as follows:

$$\alpha(x) = \min_j \alpha_j(x).$$

One has of course

$$\sum_{i=1}^{n} b_{ji}x_i \geqslant \alpha(x) \sum_{i=1}^{n} a_{ji}x_i, \quad j = 1,\dots,m,$$

and the number $\alpha(x)$ is called the growth rate of model (A,B) at intensity x.

For the model (A,B) the problem of finding the maximal growth rate thus becomes

$$\left.\begin{array}{l}\alpha \to \max, \\ \sum_{i=1}^{n} b_{ji}x_i \geqslant \alpha \sum_{i=1}^{n} a_{ji}x_i,\ j = 1,\ldots,m, \\ x_i \geqslant 0,\ i = 1,\ldots,n.\end{array}\right| \qquad (11.1)$$

The maximum rate $\alpha = \alpha^*$, when it exists, is called the technological growth rate of model (A,B), and the corresponding vector x is called the optimal intensity vector. For models satisfying conditions 1) and 2) above there does exist a positive growth rate. A proof of this assertion can be found, for example, in [29]. One can also easily obtain a proof by means of the alternatives Theorem 2.2 of Section 2.3. Because the growth rate $\alpha(x)$ does not change if x is replaced by kx, where $k > 0$ is a scalar one completes the constraints in (11.1) with the equality

$$\sum_{i=1}^{n} x_i = 1. \qquad (11.2)$$

Now introduce a price vector $p = (p_1,\ldots,p_m)$, $p_i \geqslant 0$ of prices for the goods $G_1,\ldots,G_m$. For this given price system the quantities

$$\sum_{j=1}^{m} a_{ji}p_j \text{ and } \sum_{j=1}^{m} b_{ji}p_j$$

are respectively the total costs and total income of activity P_i used at unit intensity. The quotient

$$\beta_i(p) = \frac{\sum_{j=1}^{m} b_{ji}p_j}{\sum_{j=1}^{m} a_{ji}p_i}$$

under the assumption $\sum_j a_{ji}p_j > 0$ is the income from P_i for unit cost and it is therefore a measure of rentability of this particular activity or production process. The quantity

$$\beta(p) = \max_i \beta_i(p)$$

is called the interest rate of the model (A,B) for these (fixed) prices. The problem of finding the (minimal) interest rate of the model (A,B) is now the problem of finding that price system p for which the quantity $\beta(p)$ becomes minimal.

This gives the problem

$$\left.\begin{aligned} &\beta \to \min \\ &\sum_{j=1}^{m} b_{ji}p_j \leqslant \beta \sum_{j=1}^{m} a_{ji}p_j, \quad i = 1,\ldots,n, \\ &\sum_{j=1}^{m} p_j = 1, \; p_j \geqslant 0, \qquad j = 1,\ldots,m. \end{aligned}\right\} \tag{11.3}$$

The scalar β^* which solves (11.3) is called the interest rate of model (A,B) and the corresponding p^* is called the optimal price vector. The existence of a strictly positive interest rate (if conditions 1) and 2) are satisfied) is established in exactly the same way as the existence of a strictly positive growth rate [29]. We shall later see that always $\beta^* \geqslant \alpha^*$ with a strict inequality possible only if the model (A,B) is reducible. This means that the matrix A can be put in the form (by column and row permutations)

$$A = \left\| \begin{array}{c|c} A_1 & 0 \\ \hline \multicolumn{2}{c}{A_2} \end{array} \right\| \tag{11.4}$$

where 0 is a submatrix consisting of zeros. The reducibility of a model means that there exists a sub collection of goods which can be produced using only those goods and no others. One has for example $\beta^* > \alpha^*$ is the model (A,B) is the union of two sub-models (A_1,B_1), (A_2,B_2) with different growth rates. Then one has clearly

$$(A,B) = \left\{ \left\| \begin{matrix} A_1 & 0 \\ 0 & A_2 \end{matrix} \right\|, \left\| \begin{matrix} B_1 & 0 \\ 0 & B_2 \end{matrix} \right\| \right\},$$

and as a result A is of the form (11.4).

11.3. A method for solving the problem of maximum growth rates

The conditions (11.1)-(11.2) which define the growth rate can be stated as a nonlinear programming problem

$$\left.\begin{aligned}&\alpha \to \max\\ &\sum_{i=1}^{n} b_{ji}x_i \geqslant \alpha \sum_{i=1}^{n} a_{ji}x_i, \quad j = 1,\ldots,m\\ &\sum_{i=1}^{n} x_i = 1,\ x_i \geqslant 0, \qquad i = 1,\ldots,n\ .\end{aligned}\right\} \tag{11.5}$$

The method of solution to be described below reduces this problem to a sequence of linear programming problems.

Consider the problem

$$\left.\begin{aligned}&u \to \max,\\ &u - \sum_{i=1}^{n} b_{ji}x_i \leqslant 0,\ j = 1,\ldots,m,\\ &\sum_{i=1}^{n} x_i = 1,\ x_i \geqslant 0,\ i = 1,\ldots,n.\end{aligned}\right\} \tag{11.6}$$

Because of conditions (1) and (2) there exists a solution $(x^{(0)},u^{(0)})$ of problem (11.6) which satisfies the conditions: $x^{(0)}$ is a semi-positive vector and $u^{(0)}$ is a scalar > 0. Define the number $\alpha^{(0)}$ as follows

$$\alpha^{(0)} = \min_j \left\{ \frac{\sum_{i=1}^{n} b_{ji}x_i^{(0)}}{\sum_{i=1}^{n} a_{ji}x_i^{(0)}} \right\},$$

Then $\alpha^{(0)}$ is positive and finite. To check this assertion it suffices to verify that

$$\sum_{i=1}^{n} a_{ji}x_i^{(0)} > 0,\ j = 1,\ldots,m,$$

for some j, and for that it suffices to show that

$$\left.\begin{aligned}&\sum_{i=1}^{n} a_{ji}x_i = 0,\ i = 1,\ldots,m,\\ &\sum_{i=1}^{n} x_i = 1,\ x_i \geqslant 0,\ i = 1,\ldots,n,\end{aligned}\right\} \tag{11.7}$$

has no solutions. Indeed the system of Equations (11.7) has

as its complementary alternative system the system of inequalities

$$\sum_{j=1}^{m} a_{ji}w_j \geqslant -w_{m+1} > 0,\ i = 1,\dots,n.$$

(see Theorem 2.2 of Section 3). This can also be written in the form

$$\sum_{j=1}^{m} a_{ji}w_j + w_{m+1} \geqslant 0,\ i = 1,\dots,n,\ w_{m+1} < 0,$$

and this last system obviously has a solution in view of condition 1. It now follows from Theorem 2.2 that (11.7) has no solutions.

Now proceed by considering the linear programming problem

$$\left.\begin{aligned} & u \to \max \\ & u - \sum_{i=1}^{n} (b_{ji} - \alpha^{(0)} a_{ji})x_i \leqslant 0,\ j = 1,\dots,m \\ & \sum_{i=1}^{n} x_i = 1,\ x_i \geqslant 0,\ i = 1,\dots,n. \end{aligned}\right\} \qquad (11.8)$$

This problem also has a solution $(x^{(1)}, u^{(1)})$ which satisfies the conditions: $u^{(1)} \geqslant 0$ and $x^{(1)}$ is a semipositive vector. This follows from the fact that the scalar $u = 0$ and the semipositive vector $x^{(0)}$ are feasible for (11.8). Continue by defining $\alpha^{(1)}$ by the formula

$$\alpha^{(1)} = \min_j \left\{ \frac{\sum_{i=1}^{n} b_{ji}x_i^{(1)}}{\sum_{i=1}^{n} a_{ji}x_i^{(1)}} \right\}.$$

It is easy to show that

$$\alpha^{(1)} \geqslant \alpha^{(0)}. \qquad (11.9)$$

Indeed the property $u^{(1)} \geqslant 0$ and the constraints of (11.8) imply

$$\sum_{i=1}^{n} b_{ji}x_i^{(1)} - \alpha^{(0)} \sum_{i=1}^{n} a_{ji}x_i^{(1)} \geqslant 0, \text{ or } \alpha^{(0)} \leqslant \frac{\sum_{i=1}^{n} b_{ji}x_i^{(1)}}{\sum_{i=1}^{n} a_{ji}x_i^{(1)}},$$

$$j = 1,\dots,m,$$

and from this (11.9) follows in view of the definition of $\alpha^{(1)}$. It is obvious how to continue further, so that the calculation of the growth rate reduces to solving a sequence of linear programming problems of the form

$$\left.\begin{aligned} & u \to \max, \\ & u - \sum_{i=1}^{n} (b_{ji} - \alpha^{(k)} a_{ji})x_i \leqslant 0, \; j = 1,\dots,m, \\ & \sum_{i=1}^{n} x_i = 1, \; x_i \geqslant 0, \; i = 1,\dots,n, \end{aligned}\right\} \tag{11.10}$$

$$\begin{aligned} \alpha^{(k)} &= \min_j \left\{ \frac{\sum_{i=1}^{n} b_{ji}x_i^{(k)}}{\sum_{i=1}^{n} a_{ji}x_i^{(k)}} \right\}, \; k = 1,2,\dots, \\ \alpha^{(0)} &= 0. \end{aligned} \tag{11.11}$$

The solution of problem (11.10) consists of a semipositive vector $x^{(k+1)}$ and a scalar $u^{(k+1)} \geqslant 0$.

In this manner we find a nondecreasing sequence of numbers $\alpha^{(0)} \leqslant \alpha^{(1)} \leqslant \alpha^{(2)} \leqslant \dots$ of which the limit will be the growth rate $\alpha^* = \lim_{k\to\infty} \alpha^{(k)}$ which we are looking for. Moreover if for some integer k the equality $\alpha^{(k)} = \alpha^{(k+1)}$ holds then $\alpha^{(k)} = \alpha^*$, and $x^{(k)} = x^*$. Just suppose the opposite, i.e. that $\alpha^{(k)} = \alpha^{(k+1)}$ but $\alpha^* = \alpha^{(k)} + \Delta\alpha$ with $\Delta\alpha > 0$. Clearly then $(x^{(k)},0)$ and $(x^*,0)$ will be respective optimal solutions of problem (11.10) and the problem

$$\left.\begin{array}{l} u \to \max, \\ u - \sum_{i=1}^{n} (b_{ji} - \alpha^* a_{ji}) x_i \leqslant 0, \ j = 1,\ldots,m, \\ \sum_{i=1}^{n} x_i = 1, \ x_i \geqslant 0, \ i = 1,\ldots,n. \end{array}\right\} \tag{11.12}$$

Substituting $\alpha^* = \alpha^{(k)} + \Delta\alpha$ in (11.12) gives us the system of inequalities

$$\sum_{i=1}^{n} (b_{ji} - \alpha^{(k)} a_{ji}) x_i^* \geqslant \Delta\alpha \sum_{i=1}^{n} a_{ji} x_i^*, \ j = 1,\ldots,m.$$

Let $M_1 \subset M = \{1,\ldots,m\}$ be the subset of those indices j for which the equality

$$\sum_{i=1}^{n} a_{ji} x_i^* = 0, \ j \in M_1, \tag{11.13}$$

holds. We note that the subset M_1 may be empty. Above it was shown that the system (11.7) admits no solutions and the set $M \backslash M_1$ for which the condition

$$\sum_{i=1}^{n} a_{ji} x_i^* > 0, \ j \in M \backslash M_1, \tag{11.14}$$

holds, is consequently non empty. Therefore (x^*, α^*) is also an optimal solution of the problem

$$\left.\begin{array}{l} \alpha \to \max \\ \sum_{i=1}^{n} b_{ji} x_i \geqslant \alpha \sum_{i=1}^{n} a_{ji} x_i, \ j \in M \backslash M_1, \\ \sum_{i=1}^{n} x_i = 1, \ x_i \geqslant 0, \ i = 1,\ldots,n, \end{array}\right\} \tag{11.15}$$

because it follows from (11.13) that for $x = x^*$

$$\sum_{i=1}^{n} b_{ji} x_i^* \geqslant \alpha \sum_{i=1}^{n} a_{ji} x_i^* = 0, \ j \in M_1,$$

so that the corresponding constraints of (11.15) are inessential. Setting $x = x^*$, $\alpha = \alpha^* = \alpha^{(k)} + \Delta\alpha$ in (11.15) we obtain

$$\sum_{i=1}^{n} b_{ji}x_i^* - \alpha^{(k)} \sum_{i=1}^{n} a_{ji}x_i^* \geqslant \Delta\alpha \sum_{i=1}^{n} a_{ji}x_i^*, \; j \in M\backslash M_1,$$

where, by virtue of (11.14)

$$\min_{j \in M\backslash M_1} \Delta\alpha \sum_{i=1}^{n} a_{ji}x_i^* > 0. \tag{11.16}$$

The vector x^* is therefore an admissible vector for problem (11.10) and the optimal solution $(x^{(k+1)},u^{(k+1)})$ of this problem satisfies, as a consequence of (11.16), the conditions

$$u^{(k+1)} \geqslant \min_{j \in M\backslash M_1} \Delta\alpha \sum_{i=1}^{n} a_{ji}x_i^* > 0.$$

This last inequality contradicts the hypothesis because the equality $\alpha^{(k)} = \alpha^{(k+1)}$ can only hold if $u^{(k+1)} = 0$.

The existence of a finite limit of the monotone nondecreasing sequence $\alpha^{(0)},\alpha^{(1)},\ldots$ follows from the fact proved above that (11.7) admits no solutions. Indeed for all k at least one of the scalar products

$$(a_1,x^{(k)}),\ (a_2,x^{(k)}),\ldots,(a_m,x^{(k)})$$

must be strictly positive. Thus the problems (11.10)-(11.11) define an algorithm for determining the growth rate of the model (A,B).

11.4. Duality and problems of growth rates and interest rates

J. von Neumann showed that the two problems of determining the growth rate and the interest rate of (A,B) are in duality and the established fundamental duality relations which are in the words of Gale [29] remarkably reminiscent of the duality theorems of linear programming. In the preceding section we saw that the nonlinear problem of finding the growth rate and the optimal intensity vector reduces to the solving of a sequence of linear programming problems. Obviously the same can be said about the problem of finding the interest rate. These results suggest that it may be possible to obtain the von Neumann duality relations within the framework of linear programming.

The problem (11.3) of finding the interest rate leads just like problem (11.1)-(11.2) to a series of linear programming problems. These are of the form

$$\left.\begin{aligned} &\lambda \to \min, \\ &\sum_{j=1}^{m} (b_{ji} - \beta^{(k)} a_{ji}) p_j \leqslant \lambda, \ i = 1,\dots,n, \\ &\sum_{j=1}^{m} p_j = 1, \ p_j \geqslant 0, \ j = 1,\dots,m, \end{aligned}\right\} \quad (11.17)$$

where the sequence $\beta^{(0)}, \beta^{(1)}, \dots$ is defined by the recursive formula

$$\beta^{(k)} = \max_i \left\{ \frac{\sum_{j=1}^{m} b_{ji} p_j^{(k)}}{\sum_{j=1}^{m} a_{ji} p_j^{(k)}} \right\}, \ k = 1,2,\dots,$$

$$\beta^{(0)} = 0.$$

In the same way as in the case of growth rates one shows that the formula

$$\beta^* = \lim_{k \to +\infty} \beta^{(k)},$$

determines the magnitude of the interest rate of the model (A,B).

Let us write down formally the sequence of linear programming problems dual to (11.10):

$$v \to \min,$$

$$v - \sum_{j=1}^{m} (b_{ji} - \alpha^{(k)} a_{ji}) r_j \geqslant 0, \ i = 1,\dots,n,$$

$$\sum_{j=1}^{m} r_j = 1, \ r_j \geqslant 0, \ j = 1,\dots,m,$$

and also the sequence of linear programming problems dual to the sequence of problems (11.17):

$$\mu \to \max,$$

$$\mu - \sum_{i=1}^{n} (b_{ji}-\beta^{(k)}a_{ji})\xi_i \leqslant 0,\ j = 1,\dots,m,$$

$$\sum_{i=1}^{n} \xi_i = 1,\ \xi_i \geqslant 0,\ i = 1,\dots,n,$$

By virtue of the duality theorem of linear programming, see Chapter I, the following relations hold for all k

$$u^{(k)} = v^{(k)},$$

$$r_j^{(k)} \begin{cases} = 0 \text{ for } u^{(k)} < \sum_{i=1}^{n} (b_{ji}-\alpha^{(k-1)}a_{ji})x_i^{(k)}, \\ \geqslant 0 \text{ for } u^{(k)} = \sum_{i=1}^{n} (b_{ji}-\alpha^{(k-1)}a_{ji})x_i^{(k)}, \end{cases}$$

$$x_i^{(k)} \begin{cases} = 0 \text{ for } v^{(k)} > \sum_{i=1}^{m} (b_{ji}-\alpha^{(k-1)}a_{ji})r_s^{(k)}, \\ \geqslant 0 \text{ for } v^{(k)} = \sum_{j=1}^{m} (b_{ji}-\alpha^{(k-1)}a_{ji})r_j^{(k)}, \end{cases}$$

$$\lambda^{(k)} = \mu^{(k)}$$

$$\xi_i^{(k)} \begin{cases} = 0 \text{ for } \sum_{j=1}^{m} (b_{ji}-\beta^{(k-1)}a_{ji})p_j^{(k)} \leqslant \lambda^{(k)} \\ \geqslant 0 \text{ for } \sum_{j=1}^{m} (b_{ji}-\beta^{(k-1)}a_{ji})p_j^{(k)} = \lambda^{(k)} \end{cases}$$

$$p_j^{(k)} \begin{cases} = 0 \text{ for } \mu^{(k)} < \sum_{i=1}^{n} (b_{ji}-\beta^{(k-1)}a_{ji})\xi_i^{(k)} \\ \geqslant 0 \text{ for } \mu^{(k)} = \sum_{i=1}^{n} (b_{ji}-\beta^{(k-1)}a_{ji})\xi_i^{(k)} \end{cases}$$

and as a consequence the following limit relations hold

$$\lim_{k\to\infty} u^{(k)} = \lim_{k\to\infty} v^{(k)} = \lim_{k\to\infty} \mu^{(k)} = \lim_{k\to\infty} \lambda^{(k)} = 0, \quad (11.18)$$

$$r^*_j \begin{cases} = 0 \text{ for } \sum_{i=1}^{n} (b_{ji} - \alpha^* a_{ji}) x^*_i > 0, \\ \geqslant 0 \text{ for } \sum_{i=1}^{n} (b_{ji} - \alpha^* a_{ji}) x^*_i = 0, \end{cases} \tag{11.19}$$

$$x^*_i \begin{cases} = 0 \text{ for } \sum_{j=1}^{m} (b_{ji} - \alpha^* a_{ji}) r^*_j \quad 0, \\ \geqslant 0 \text{ for } \sum_{j=1}^{m} (b_{ji} - \alpha^* a_{ji}) r^*_j = 0, \end{cases} \tag{11.20}$$

$$\xi^*_i \begin{cases} = 0 \text{ for } \sum_{j=1}^{m} (b_{ji} - \beta^* a_{ji}) p^*_j < 0 \\ \geqslant 0 \text{ for } \sum_{i=1}^{m} (b_{ji} - \beta^* a_{ji}) p^*_j = 0 \end{cases} \tag{11.21}$$

$$p^*_j \begin{cases} = 0 \text{ for } \sum_{i=1}^{n} (b_{ji} - \beta^* a_{ji}) \xi^*_i > 0, \\ \geqslant 0 \text{ for } \sum_{i=1}^{n} (b_{ji} - \beta^* a_{ji}) \xi^*_i = 0, \end{cases} \tag{11.22}$$

where

$$x^* = \lim_{k\to\infty} x^{(k)}, \quad r^* = \lim_{k\to\infty} r^{(k)},$$

$$\xi^* = \lim_{k\to\infty} \xi^{(k)}, \quad p^* = \lim_{k\to\infty} p^{(k)}.$$

From (11.18)-(11.22) there follows the following theorem of von Neumann:

THEOREM 11.1. If the model (A,B) satisfies the two conditions 1) and 2) then there exists a semi-positive n dimensional vector x^* and a semi-positive m dimensional vector p^* and a number γ such that

$$\sum_{i=1}^{n} b_{ji} x^*_i \geqslant \gamma \sum_{i=1}^{n} a_{ji} x^*_i,$$

and if $\sum_{i=1}^{n} b_{ji} x^*_i > \gamma \sum_{i=1}^{n} a_{ji} x^*_i$ then $p^*_j = 0$;

$$\sum_{j=1}^{m} b_{ji}p_j^* \leqslant \gamma \sum_{j=1}^{m} a_{ji}p_j^*,$$

and if $\sum_{j=1}^{n} b_{ji}p_j^* < \gamma \sum_{j=1}^{m} a_{ji}p_j^*$ then $x_i^* = 0$.

In fact it follows from (11.19) and (11.20) and from (11.21) and (11.22) that the properties mentioned in Theorem 11.1 are satisfied by (x^*,r^*,α^*) or by (ξ^*,p^*,β^*).

From the relations (11.18)-(11.22) the following assertion also results

THEOREM 11.2. For a model (A,B) which satisfies conditions 1) and 2) the following inequality holds

$$\beta^* \leqslant \alpha^*$$

The strict inequality $\beta^* < \alpha^*$ can only hold in the case of a reducible model.

Proof. The duality conditions (11.20) mean that a semi-positive vector $r^* = (r_1^*,\dots,r_m^*)$ exists, such that

$$\sum_{j=1}^{m} (b_{ji}-\alpha^* a_{ji})r_j^* \leqslant 0, \quad i = 1,\dots,n.$$

Consequently α^* and r^* from an admissible point for the problem of interest rates and the definition of β^* then implies $\beta^* \leqslant \alpha^*$. Let us now see when $\beta^* < \alpha^*$ is possible. Let $\alpha^* = \beta^* + \varepsilon$ with $\varepsilon > 0$. Then from the relation

$$\sum_{i=1}^{n} (b_{ji}-\alpha^* a_{ji})x_i^* \geqslant 0, \quad j = 1,\dots,m,$$

which is satisfied by the growth rate and the optimal intensity vector, there follows

$$\sum_{i=1}^{n} (b_{ji}-\beta^* a_{ji})x_i^* \geqslant \varepsilon \sum_{i=1}^{n} a_{ji}x_i^*, \quad j = 1,\dots,m. \tag{11.23}$$

On the other hand, condition (11.19) means that there exists a non empty subset M_1 of the index set M such that

$$\sum_{i=1}^{n} (b_{ji}-\alpha^* a_{ji})x_i^* \begin{cases} = 0, & j \in M_1, \\ > 0, & j \in M\setminus M_1. \end{cases}$$

Let us show that the semi-positive vector x^* cannot satisfy the conditions

$$\sum_{i=1}^{n} a_{ji}x_i^* > 0 \text{ for } j \in M_1. \tag{11.24}$$

In the opposite case it would follow from (11.23) that

$$\begin{aligned} &\sum_{i=1}^{n} (b_{ji}-\beta_{ji}^*)x_i^* > 0, \ j = 1,\dots,m, \\ &\sum_{i=1}^{n} x_i^* = 1, \ x_i^* \geqslant 0, \ i = 1,\dots,n. \end{aligned} \tag{11.25}$$

However, these inequalities, if satisfied, would mean that β^* is not the interest rate. Therefore the inequality (11.24) cannot hold and there exists a non empty subset $M_2 \subset M$ satisfying $M_2 \cap M_1 \neq 0$ and such that

$$\sum_{i=1}^{n} a_{ji}x_i^* = 0 \text{ for } j \in M_2.$$

The last equality means that

$$a_{ji}x_i^* = 0, \ i = 1,\dots,n, \ j \in M_2$$

which is only possible in the case where A has the form (8.4), i.e. in the case where (A,B) is reducible. This proves the theorem.

11.5. The minimal time problem

In this section we study the possibility of using physical models for dynamic economic processes by considering a minimal time problem in the framework of the classical von Neumann model. The method of solving this is based on the usual idea for minimal time problems of converting them to a parametric problem of minimizing some definite function which is zero on the real trajectories (that is on the trajectories satisfying the conditions of the problem).

This is a rather general method of treating problems in the theory of optimal control. It is interesting to compare this method with a similar method of N. N. Krasovskii of reducing these problems to a minimization problem. As both methods admit important generalizations we restrict ourselves to minimal time problems for steering a control system from a given initial state $x^{(\alpha)}$ to another given state $x^{(\beta)}$. Let us first of all state the problem.

Given the differential equations describing how the control system evolves

$$\frac{dx}{dt} = f(x,u), \tag{11.26}$$

an initial state $x^{(\alpha)}$ and a final state $x^{(\beta)}$ for the phase vector $x(t)$, and given the constraints on the magnitude of the controls $u(t)$

$$\kappa[u] \leqslant \mu, \tag{11.27}$$

It is desired to find the time $t_\beta^{(0)}$ and a corresponding possible control $u^{(0)}(t)$, $t_\alpha \leqslant t \leqslant t_\beta^{(0)}$ such that the following conditions are satisfied

1) The solution $x^{(0)}(t)$ of the equation

$$\frac{dx}{dt} = f(x^{(0)}(t),u^{(0)}(t)), \tag{11.28}$$

which satisfies $x^{(0)}(t_\alpha) = x^{(\alpha)}$ also satisfies the requirement $x^{(0)}(t_\beta^{(0)}) = x^{(\beta)}$.

2) The following condition holds

$$\kappa[u^{(0)}(t)] \leqslant \mu, \ t_\alpha \leqslant t \leqslant t_\beta^{(0)}.$$

3) For all other controls $u(t)$ satisfying (11.27) and any vector function $x(t)$ satisfying (11.26) and the conditions

$$x(t_\alpha) = x^{(\alpha)}, \ x(t_\beta) = x^{(\beta)}, \tag{11.29}$$

one has $t_\beta \geqslant t_\beta^{(0)}$.

The solution $x^{(0)}(t)$ of Equation (11.28) and the control $u^{(0)}(t)$ are called optimal or time-optimal and the number $t_\beta^{(0)} - t_\alpha$ is the optimal transition time for the process.

The idea behind the method of N. N. Krasovskii is the following. Let t_β be some fixed time instant. Ignore the constraint (11.27) and assume the existence of some non-empty set $U(t\ )$ of controls which take the system from state $x^{(\alpha)}$ at time t_α to state $x^{(\beta)}$ at time t_β. Let us look for that control $\tilde{u}(t)$ which minimizes the magnitude $\kappa[u]$. It must be remembered that $\tilde{u}(t)$ belongs to the set of controls which take state $x^{(\alpha)}$ to state $x^{(\beta)}$, and that (11.26) and (11.29) are satisfied by the trajectory.

In this manner there corresponds to each value t_β a control which takes the system from $x^{(\alpha)}$ to $x^{(\beta)}$ at time t_β and which minimizes the magnitude (as measured by κ) of all the controls which do this. Solving this problem for all $t_\beta \geqslant t_\alpha$ we find a function

$$\omega(t_\beta) = \min_{u(t) \in U(t_\beta)} \kappa[u].$$

Obviously the number $t_\beta^{(0)}$ defining the minimal transition time $t_\beta^{(0)} - t_\alpha$ is equal to the smallest number for which $\omega(t_\beta) \leqslant \mu$.

Thus the method of N. N. Krasovskii reduces the time optimal control problem to a parametric problem of minimizing the magnitude of the controls over the set of solutions of (11.26) satisfying condition (11.29). It is important that this parametric problem does not impose constraints on the controls.

Now let us describe another way of reformulating the time optimal control problem. Let t_β again be some fixed number satisfying $t_\beta > t_\alpha$. Let $C(t_\beta)$ be the set of continuously differentiable functions satisfying (11.29). Further let $I(x(t),u(t))$ be a convex, definite functional for all continuously differentiable $x(t) \in C(t_\beta)$ and piecewise continuous controls $u(t)$, which is zero if and only if the pair $x(t)$, $u(t)$, satisfies equation (11.26) and condition (11.29).

Now consider the following problem

$$I(x(t),u(t)) \to \min$$

under the conditions

$$x(t) \in C(t_\beta), \quad \kappa[u(t)] \leqslant \mu.$$

Essentially this problem consists of finding a function $\tilde{x}(t) \in C(t_\beta)$ and an admissible control $\tilde{u}(t)$ which minimize the penalty on the violations of conditions (11.26) and (11.29). Let

$$\bar{I}(t_\beta) = \min_{x(t)\in C(t_\beta),\kappa[u(t)]\leq\mu} I(x(t),u(t)) = I(\tilde{x}(t),\tilde{u}(t)),$$

then we can obviously write

$$I(t_\beta) \begin{cases} = 0 \text{ for } t_\beta \geq t_\beta^{(0)}, \\ > 0 \text{ for } t_\beta < t_\beta^{(0)}. \end{cases}$$

It is easy to verify that the function $\bar{I}(t_\beta)$ is monotonically decreasing in the interval $(t_\alpha, t_\beta^{(0)})$ and that it is identically zero for all $t_\beta \geq t_\beta^{(0)}$. Consequently the minimal transition time of the process is given by the smallest value of the argument t_β for which $\bar{I}(t_\beta)$ becomes zero.

Obviously the method described above is applicable both to linear and to nonlinear minimal time problems. It is also clear that if the constraints on the possible control choices are dependent on time and the phase coordinates of the control systems this will not introduce fundamental new difficulties for this method.

To use the method it is convenient to find first a given increasing sequence of numbers $\tau_0, \tau_1, \ldots$ in which τ_0 satisfies $t_\alpha < \tau_0 < t_\beta^{(0)}$. To find such a τ_0 is not generally difficult. The sequence $\tau_0, \tau_1, \ldots$ now corresponds to a sequence of values $\bar{I}(\tau_i)$ which is strictly decreasing to some value for $t_\beta \geq t_\beta^{(0)}$. To proceed further one can use dichotomy algorithms or chord methods.

These methods will be applied in the following section where we consider a time-optimal control problem for the growth of an economy. Here the control constraints are dependent (in an explicit way) on the state vector of economy.

11.6. A time optimal control problem for economic growth

Let us return to the model for an economy described by a pair of matrices A and B as was described in Section 11.2. The production processes of the economy under consideration are supposed to be used in cycles. At the start of cycle s the state of the economy is given by the m dimensional vector X(s-1) which becomes transformed to the vector X(s) (of the same dimension) at the end of cycle s. This vector X(s) of goods $X(s) = (X_1(s),\dots,X_m(s))$ is the bundle of goods available at the start of cycle (s+1). Thus the economy being studied is characterized by the state vector X of goods evolving in discrete time.

We shall assume that there is given an initial bundle of goods X(0). Then the input vector for the first cycle is Ax(1) where x(1) is the n-dimensional vector of intensities with which the production processes are used during the first cycle. Clearly the choice of x(1) is restricted by $Ax(1) \leqslant X(0)$, and $x(1) \geqslant 0$, which means that the total amounts of inputs cannot exceed the available amounts of goods. The production result at the end of the first cycle is the vector of goods Bx(1) and a bundle of goods not used in the first production cycle may have been conserved.

Assuming that the next cycle may also use the goods not used during the first cycle the vector X(1) will be given by the formula

$$X(1) = Bx(1) + X(0) - Ax(1),$$

where X(0) - Ax(1) is the vector of unused resources.

In the case where the possibility is allowed [1)] of using only part of the resources not used during the prèvious cycle the equation for X(1) must be written in the form

$$X(1) = Bx(1) + K(x(0)-Ax(1)),$$

where K is a diagonal matrix with diagonal elements k_{ii} satisfying $0 \leqslant k_{ii} \leqslant 1$.

We shall assume that some of the resources not used in a given production cycle can be conserved and may be used later. Conserving resources obviously can be seen as a (rather trivial) production process and these can be incorporated by enlarging the matrices A and B) [2)]. Thus the equation giving the state of the economy at the end of the first cycle takes the form

$$X(1) = Bx(1),$$

where the choice of the vector x(1) is restricted to the nonnegative solutions of the equation $Ax(1) = X(0)$.

In this manner the simple von Neumann model for an economy becomes quite significant and one can restrict attention to economic evolution processes which can be described by the system of equations and inequalities

$$\left.\begin{aligned} X(s) &= Bx(s), \\ Ax(s) &= X(s-1), \\ x(s) &\geqslant 0, \; s = 1,2,\ldots, \end{aligned}\right\} \tag{11.30}$$

where $X(0)$ is a given initial state vector.

Let $\bar{X}$ be a state of the economy which it is desired to reach. Further let $M(\bar{X})$ be the set of vectors in E_m satisfying

$$M(\bar{X}) = \{X;\; X-\bar{X} \geqslant 0\}, \tag{11.31}$$

where, as usual, the inequality $X - \bar{X} \geqslant 0$ means $X_i - \bar{X}_i \geqslant 0$, $i = 1,\ldots,m$. Condition (11.31) describes the set of states for which there is no less than desired of each good. Therefore we shall not consider below the problem of time-optimal control for eaching $\bar{X}$ but the time optimal-control problem of reaching some point in the domain $M(\bar{X})$, see Figure 11.1. This gives the following formulation of the time optimal control problem.

Time optimal control problem. To find the smallest natural number k^* for which there exist vector functions $X(s)$ and $x(s)$ defined on the natural numbers $s = 1,2,\ldots$ satisfying condition (11.30) and the condition $X(s) \in M(\bar{X})$ for $s \geqslant k^*$.

This means the time optimal control problem consists in determining the smallest integer k for which there exist solutions to the following system of equations and inequalities

$$\left.\begin{aligned} X(s) &= Bs(x), \\ Ax(s) &= X(s-1), \\ x(s) &\geqslant 0, \; s = 1,2,\ldots,k, \\ X(k) &\geqslant \bar{X}. \end{aligned}\right\} \tag{11.32}$$

11.7. A physical model for solving time optimal control problems

The reader will have no difficulty at all in finding a physical model for the system of linear equations and inequalities (11.32). In Figure 11.2 a block-scheme of such a physical model for (11.32) is shown. It consists of a chain of identical blocks of which number s corresponds to the transformation $X(s-1)$ into $X(s)$. The detailed structure of one of the blocks is shown in Figure 11.3.

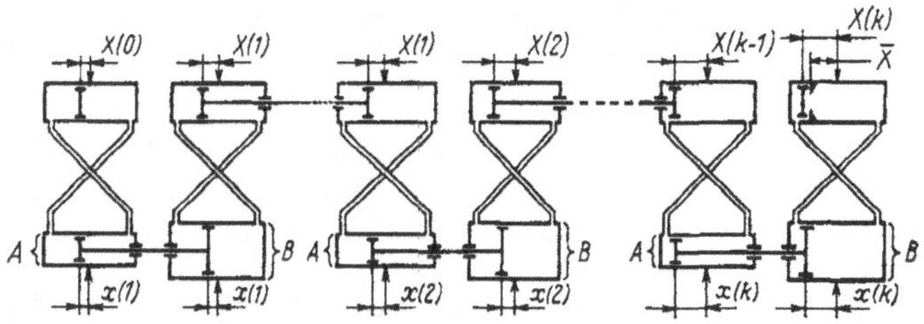

Fig. 11.2.

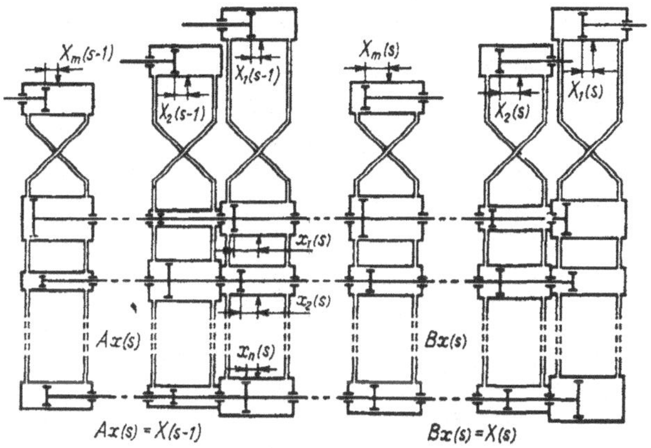

Fig. 11.3.

Such models with containers filled with an ideas gas were considered in Chapter III, and there the reader will also find algorithms to determine its equilibrium states (cf. Sections 3.3 and 3.4 of Chapter III). Here we know that an equilibrium state is a solution of the system being modelled if and only if the Helmholtz free energy of the ideal gas in the containers of the model is equal to zero. This property of the physical model is the essential fact underlying the method of solution to be described below. Indeed if the scalar k^* and the collection of vectors $\{X^*(s), x^*(s), s = 1,\ldots,k^*\}$ are a solution of the time optimal control problem then that means that there are no solutions of (11.32) for $k < k^*$ and that means that the equilibrium states of the model have strictly positive Helmholtz free energy. Thus one can say that the Helmholtz free energy of the physical model of (11.32) at equilibrium measures for a fixed k the impossibility of solving this system. The Helmholtz free energy of the model of system (11.32) is a positive definite function of the state parameters $X_1(s),\ldots,X_m(s)$, $s = 1,\ldots,k$, the control parameters $x_1(s),\ldots,x_n(s)$, $s = 1,\ldots,k$ and the parameter k, as is visible from expression (2.26), (see Section 2.3 of Chapter II) for the Helmholtz free energy and its minimum is, as a function of the parameter k, monotonically decreasing in the interval $1 \leqslant k \leqslant k^*$ [3]) and it is identically equal to zero for $k \geqslant k^*$. A graph of the function $F_{min}(k)$ is depicted in Figure 11.4.

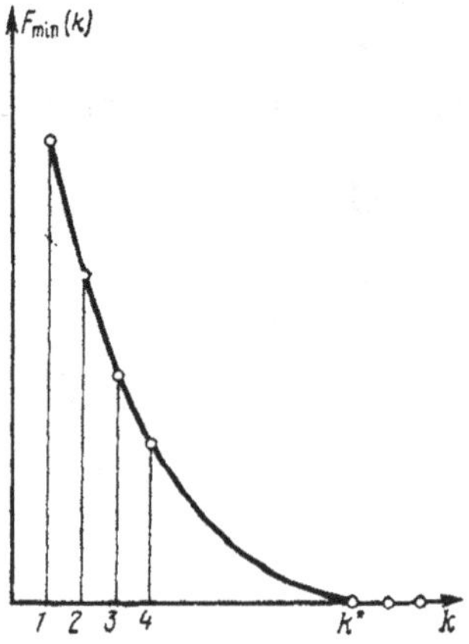

Fig. 11.4.

Thus solving the time optimal control problem becomes the determination of that integer k^* which satisfies the conditions

$$F_{min}(k) \begin{cases} > 0 \text{ for } k < k^*, \\ = 0 \text{ for } k \geqslant k^*. \end{cases}$$

One can propose several calculation procedures for solving this problem, which reduce it to a finite number of equilibrium problems. The simplest one is the following

(1) Assume that the set $M(\bar{X})$ can be reached in one cycle. This optimistic hypothesis, obtained from (11.32) by setting $k = 1$, is equivalent to assuming that the following system is solvable.

$$\begin{aligned} Bx(1) &= X(1), \\ Ax(1) &= X(0), \\ x(1) &\geqslant 0,\ X(1) \geqslant \bar{X}, \end{aligned} \tag{11.33}$$

where $\bar{X}$ and $X(0)$ are given vectors.

Let $X^{(1)}(1)$ and $x^{(1)}(1)$ be the equilibrium state of the physical model of (11.33). The coordinates of this equilibrium can be calculated by means of formulas (3.15), (3.19) of Section 3.3 or formulas (3.31), (3.36) of Section 3.4 of Chapter III.

Let $F_{min}(1)$ be the amount of Helmholtz free energy at equilibrium of the physical model of (11.33) at equilibrium. If $F_{min}(1) = 0$, then

$$k^* = 1,\ X^*(1) = X^{(1)}(1),\ x^*(1) = x^{(1)}(1)$$

and the time optimal control problem is solved. In the opposite case $F_{min}(1) > 0$ and consequently $k^* > 1$.

In this last case we continue with the next problem which assumes that the set $M(\bar{X})$ can be reached in two cycles ($k^* = 2$).

(2) Assuming that $k^* = 2$ leads to an equilibrium problem for the physical model of the system

$$\left.\begin{aligned} Bx(2) &= X(2), \\ Ax(2) &= X(1), \\ Bx(1) &= X(1), \\ Ax(1) &= X(0), \\ x(1) &\geqslant 0,\ x(2) \geqslant 0,\ X(2) \geqslant \bar{X}. \end{aligned}\right\} \tag{11.34}$$

Let $X^{(2)}(1)$, $X^{(2)}(2)$, $x^{(2)}(1)$, $x^{(2)}(2)$ denote the vectors describing the equilibrium state of the physical model of system (11.34) and let $F_{min}(2)$ be the amount of Helmholtz free energy at equilibrium. Then again one has the alternatives $k^* = 2$ if $F_{min}(2) = 0$ and $k^* > 2$ if $F_{min}(2) > 0$.

It is obvious how to continue and solving the time optimal control problem thus means finding the solutions of a sequence of equilibrium problems for physical models of the systems of linear equations and inequalities which are obtained from (11.32) for $k = 1,2,\ldots,k^*$ with k^* determined by the condition $F_{min}(k^*) = 0$.

Because the physical model of a system (11.32) for some k differs from the model for k - 1 by the inclusion of a component which does not depend on k it follows that $F_{min}(k)$ is not only monotonically decreasing but also concave; that is for all $k \geqslant 1$ the inequality

$$F_{min}(k+1) \leqslant \frac{1}{2}[F_{min}(k)+F_{min}(k+2)] \qquad (11.35)$$

holds. Here the inequality (11.35) is strict for $1 \leqslant k < k^*$. The property (11.35) enables us to speed up the solution process of the time optimal control problem substantially be making use of extrapolation methods of secant type. Indeed having solved the equilibrium problem for an integer k_1 and also for the next integer $k_1 + 1$ we can find the next (relevant) value of the parameter k by finding the point Δ_1 of intersection of the line through $(k_1, F_{min}(k_1))$ and $(k_1+1, F_{min}(k_1+1))$ with the abscissae axis. It is easy to check that Δ_1 is given by the formula

$$\Delta_1 = k_1 + \frac{F_{min}(k_1)}{F_{min}(k_1)-F_{min}(k_1+1)} .$$

Obviously it is not then necessary to solve the equilibrium problem for the physical models of system (11.32) for the intermediate parameter values k with $k_1+1 < k < \Delta_1$, and one proceeds immediately with the equilibrium problem for $k = k_2$ where k_2 is the smallest integer greater than or equal to Δ_1. A sketch of this procedure for solving the time optimal control problem is given in Figure 11.5.

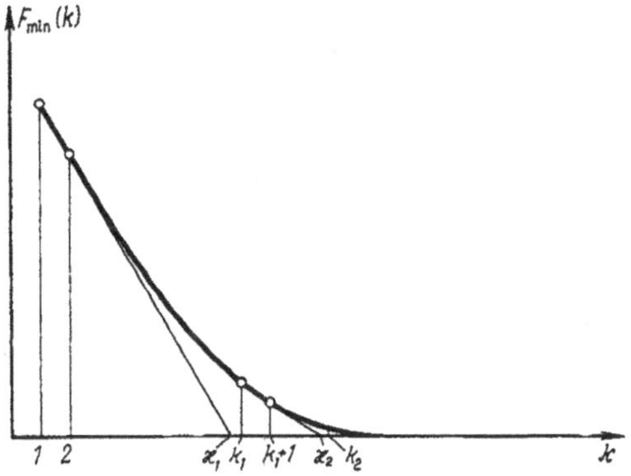

Fig. 11.5.

11.8. Decomposition for time optimal control problems

The structure of the physical model for the sytem (11.32) (for a given k) lends itself naturally to a decomposition algorithm for solving the equilibrium problem. We again resort to the method of redundant constraints. In fact, finding the equilibrium of the physical model in question can be reduced to solving several substantially less complicated problems. Let $X^{(0)}(1), X^{(0)}(2), \ldots, X^{(0)}(k)$ be a set of nonnegative vectors which can be chosen arbitrarily. Although this choice is arbitrary it greatly influences the necessary computation time and it follows that it pays to adopt some reasonable choice, depending on the circumstances. One can for example take the set of vectors defined by the formulas

$$X^{(0)}(s) = X(0) + \frac{s}{k}(\bar{X}-X(0)), \quad s = 1,\ldots,k.$$

Now consider the equilibrium problem for the physical model of (11.28) under the additional redundant constraints

$$X(s) = X^{(0)}(s), \quad s = 1,\ldots,k, \qquad (11.36)$$

that is the bars determining the positions of the pistons corresponding to the vectors X(1), X(2),...,X(k) are fixed.

From (11.32) and Figure (11.2) we see that under these

additional constraints the problem decomposes into k isolated problems of finding the equilibria of models of systems of the type (11.30) (see Figure 11.3) with identical matrices of coefficients. The equilibrium states $x^{(0)}(s)$, $s = 1,\dots,k$ of the models of these isolated systems can be found by means of the algorithms of Chapter III. In this manner the result of the first step of the decomposition algorithm yields k vectors $x^{(0)}(s)$, $s = 1,\dots,k$ which are the equilibrium states of the physical models of systems of type (11.30) for fixed vectors $X(s) = X^{(0)}(s)$. The next step of the algorithm consists in imposing the redundant constraints

$$x(s) = x^{(0)}(s),\quad s = 1,\dots,k, \tag{11.37}$$

and removing the previous redundant constraints (11.36).

Under the redundant constraints (11.37) the model decomposes into mk simple isolated models. And m(k-1) of these are models of two equations with one unknown $X_j(s)$. These are of the form

$$X_j(s) = (b_j, x^{(0)}(s)),$$
$$X_j(s) = (a_j, x^{(0)}(s-1)),$$

where the right hand side is a scalar product. The remaining m models determine the quantities $X_1(k),\dots,X_m(k)$ at equilibrium. They are models of the relations

$$X_j(k) = (b_j, x^{(0)}(k)),$$
$$X_j(k) \geqslant \bar{X}_j,$$

and from this it clearly follows that

$$X_j(k) = \begin{cases} (b_j, x^{(0)}(k)) & \text{for } (b_j, x^{(0)}(k)) \geqslant \bar{X}_j, \\ \bar{X}_j & \text{for } (b_j, x^{(0)}(k)) < \bar{X}_j. \end{cases}$$

We shall not pause to write down the explicit formulas for calculating the quantities $X_j^{(1)}(s)$, $j = 1,\dots,m$, $s = 1,\dots,k-1$, the remaining coordinates of the equilibrium state of the model under the redundant constraints (11.37). These formulas are easily obtained from the iterative formulas for solving systems of equations of Chapter III. The next step of the algorithm is similar to the first one and is different only in that the redundant constraints (11.36) are replaced by the redundant constraints.

$$X^{(s)} = X^{(1)}(s), \quad s = 1,\dots,k.$$

It is obvious how to proceed further, and the algorithm just described yields a sequence of vectors

$$\{X^{(\alpha)}(s), x^{(\alpha)}(s), \quad s = 1,\dots,k\}, \quad \alpha = 0,1,\dots$$

of which the limit is the required equilibrium state of the model of the system (11.32). To prove this last assertion one proceeds by assuming the opposite.

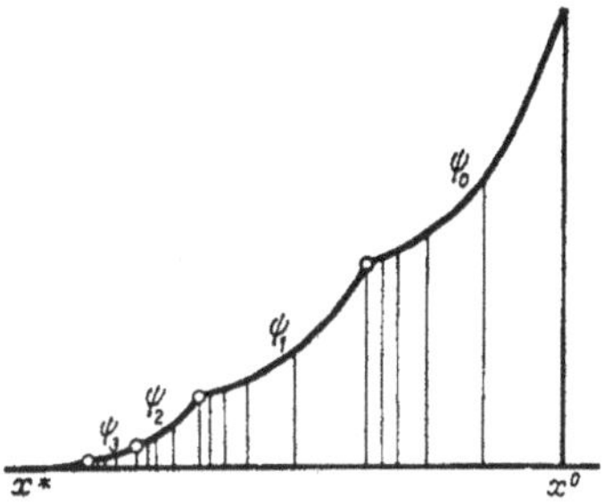

Fig. 11.6.

11.9. Rays of optimal balanced growth problems

Another interesting example of optimal time control of an economy is the problem of minimal time evolution from a given initial state $X^{(0)}$ to a direction of maximal balanced growth. We shall assume that the initial assortment of goods $X^{(0)} = (X_1^{(0)},\dots,X_m^{(0)})$ is such that the result of a production cycle can be a strictly positive vector (all coordinates > 0). This means that there exists a nonnegative vector $x = (x_1,\dots,x_n)$ of intensities for using the technological production processes which satisfies

$$\left.\begin{aligned} &\sum_{i=1}^{n} b_{si}x_i > 0, \quad s = 1,\dots,m, \\ &\sum_{i=1}^{n} a_{si}x_i \leqslant X_s^{(0)}, \quad s = 1,\dots,m. \end{aligned}\right\} \tag{11.38}$$

As before we shall consider the dynamics of the evolving economy to be taking place in discrete time, that is as a sequence of production cycles of which the inputs constitute an assortment of goods

$$\{\sum_{i=1}^{n} a_{si}x_i\}, \quad s = 1,\dots,m,$$

and the outputs are an assortment of goods

$$\{\sum_{i=1}^{n} b_{si}x_i\}, \quad s = 1,\dots,m.$$

If, as was indicated in Section 11.6 the list of technological production processes is augmented by a set of stocking processes the dimensions of the matrices A and B increase by m columns and the output vector of the k-th production cycle will coincide [4] with the input vector of production cycle number k + 1.

$$\left.\begin{aligned} &\sum_{i=1}^{n+m} a_{si}x_i^{(k)} = X_s^{(k)}, \quad s = 1,\dots,m, \\ &\sum_{i=1}^{n+m} b_{si}x_i^{(k+1)} = X_s^{(k+1)}, \quad s = 1,\dots,m. \end{aligned}\right| \tag{11.39}$$

Then the system of equations

$$\left.\begin{aligned} &\sum_{i=1}^{n+m} a_{si}x_i^{(k)} = \sum_{i=1}^{n+m} b_{si}x_i^{(k-1)}, \quad s = 1,\dots,m, \ k = 1,2,\dots, \\ &\sum_{i=1}^{n+m} a_{si}x_i^{(0)} = X_s^{(0)}, \quad s = 1,\dots,m, \ x_i^{(k)} \geqslant 0, \end{aligned}\right\} \tag{11.40}$$

determines the set of possible evolution trajectories of the economy.

The problem which we consider in this section consists in determining the smallest integer k* and the set of control vectors $x^{(0)}, x^{(1)}, \ldots, x^{(k)}$ which satisfy the following conditions

$$\sum_{i=1}^{n} b_{si} x_i^{(k^*)} = \alpha^* \sum_{i=1}^{n} a_{si} x_i^{(k^*)}, \qquad (11.41)$$

$$\sum_{i=1}^{n} a_{si} x_i^{(k^*)} = \sum_{i=1}^{n+m} b_{si} x_i^{(k^*-1)}, \qquad (11.42)$$

$$\sum_{i=1}^{n+m} a_{si} x_i^{(k)} = \sum_{i=1}^{n+m} b_{si} x_i^{(k-1)}, \qquad (11.43)$$

$$s = 1,\ldots,m; \; k = 1,2,\ldots,k^*-1$$

$$\left. \begin{array}{l} \sum_{i=1}^{n+m} a_{si} x_i^{(0)} = X_s^{(0)}, \; s = 1,\ldots,m, \\ x_i^{(\alpha)} \geqslant 0, \; i = 1,\ldots,n+m, \; \alpha = 0,1,\ldots,k^*-1 \end{array} \right\} \qquad (11.44)$$

$$x_{n+1}^{(k^*)} = x_{n+2}^{(k^*)} = \ldots = x_{n+m}^{(k^*)} = 0, \qquad (11.45)$$

where α^* is the maximal balanced growth rate of the economy. At method for finding α^* was described in Section 11.3. It is easy to interpret conditions (11.41)-(11.45) economically. Condition (11.41) means that starting with cycle k* the economy evolves with maximal growth. Conditions (11.42) and (11.45) mean that for cycle k* there are no redundant goods, i.e. no goods in stock. Conditions (11.43) and (11.44) ensure that the evolution paths are admissible (admissible controls).

A solution method. The problem of finding the minimal k under conditions (11.41)-(11.43) is clearly a time-optimal control problem and to solve it one can make effective use of the parametrization methods described in Section 11.7. Indeed let us assume that k* = 0. That is correct if there exists a solution of the system

$$\left.\begin{aligned} &\sum_{i=1}^{n} b_{si}x_i^{(0)} = \alpha^* \sum_{i=1}^{n} a_{si}x_i^{(0)}, \\ &\sum_{i=1}^{n} a_{si}x_i^{(0)} = X_s^{(0)}, \; x_i^{(0)} \geqslant 0, \end{aligned}\right\} \tag{11.46}$$

$$s = 1,\ldots,m.$$

In the opposite case the hypothesis $k^* = 0$ is incorrect and consequently $k^* > 0$.

If system (11.46) is not solvable then suppose that $k^* = 1$. That hypothesis is correct if there exists a solution of the system

$$\left.\begin{aligned} &\sum_{i=1}^{n} b_{si}x_i^{(1)} = \alpha^* \sum_{i=1}^{n} a_{si}x_i^{(1)}, \\ &\sum_{i=1}^{n} a_{si}x_i^{(1)} = \sum_{i=1}^{n+m} b_{si}x_i^{(0)}, \\ &\sum_{i=1}^{n+m} a_{si}x_i^{(0)} = X_i^{(0)}, \; s = 1,\ldots,m, \\ &x_{n+1}^{(1)} = x_{n+2}^{(1)} = \ldots = x_{n+m}^{(1)} = 0, \end{aligned}\right\} \tag{11.47}$$

and it is incorrect if the system (11.47) admits no solution.

It is obvious how to proceed further and the process continues until a value $k = k^*$ is found for which the system (11.41)-(11.45) has a solution. In this manner the method leads to a finite sequence of problems on the solvability of systems of linear equations of the form (11.41)-(11.45). A criterion for solvability of systems of the form (11.41)-(11.45) can be stated as follows. There exists a nonnegative solution of (11.41)-(11.45) if the Helmholtz free energy at equilibrium of its physical model is zero, and if the Helmholtz free energy at equilibrium is positive then the system (11.41)-(11.45) is unsolvable.

In Chapter III there were explained several algorithms for solving equilibrium problems for models of systems of equations and inequalities. Clearly any of these algorithms may be used to calculate the coordinates of the vectors $x^{(0)*},\ldots,x^{(k)*}$ of the equilibrium state of the physical model of (11.41)-(11.45) for each integer value of the parameter k. The physical model of the problem of minimal time evolution to a ray of maximal balanced growth is similar to the model (see Figure 11.2) of the problem of time-optimal control and it also consists of a chain of cinematically connected models whose states define the respective vectors $x^{(0)}, x^{(1)}, \ldots, x^{(k)}$. Therefore it seems unnecessary to write down the iterations for determining the coordinates of the equilibrium state of the model of (11.41)-(11.45). For a reader who is well acquainted with the results of Chapter III this will not be difficult. Again as in the case of time optimal control the Helmholtz free energy $F_{min}(k)$ at equilibrium is a monotonically decreasing function of k for $k < k^*$ and identically equal to zero for $k \geqslant k^*$.

In the problem under consideration as in the time optimal control problem one can significantly speed up the solution process by making use of secant methods. Applying this secant method permits one to pass from the problems for k = 1 and k = 2 directly to $k = k_1$ where k_1 is the nearest point to the right of the intersection of the line through $(1, F_{min}(1))$ and $(2, F_{min}(2))$ with the abscissae axis.

NOTES

1. Consumption, spoilage, devaluation.
2. It is not difficult to see how the matrices A and B of inputs and outputs must be enlarged. To A there is added an m×m unit matrix and to B a diagonal matrix with diagonal entries $k_{11}, \ldots, k_{mm}$ and zeros elsewhere.
3. This assumes that there is an activity intensity vector x such that Bx = Ax, which is often a reasonable assumption (translator's Note).
4. One assumes that the various forms of labour are among the collection of goods and occur in the input vectors and that consumption is an input vector for a process yielding the various kinds of labour as outputs.

Chapter XII

OPTIMAL CONTROL PROBLEMS

This book is mainly concerned with static optimization problems and mathematical programming problems. Only a few sections of Chapter IX (Tangent methods), and Chapter XI (Dynamic economic models) were concerned with dynamic control problems.

The search for analogies between constrained minimization problems for functions of m variables and equilibrium problems of mechanical systems has been fruitful, so it is not unreasonable to expect similar analogies between optimal control problems and problems of dynamics. The size of this volume prevents us from giving these aspects as much attention as they deserve and as a result this last chapter only contains a few remarks on this topic.

As one is dealing with problems of minimizing functionals it is clear that the method of variations is a generalization of the principle of virtual displacements and it is therefore hardly surprising that the inventor of the method of multipliers extended this method to a class of variational problems subject to additional conditions. It is known that during the evolution of the calculus of variations from isoperimetric problems to optimal control problems multiplier techniques have proved useful.

Other realizations of the idea of removing constraints are however no less general than multiplier methods and they can be used with success in constrained minimizing of functional problems and in optimal control. In this way one can avoid numerous technical difficulties of a numerical nature which adhere to the Lagrange method.

Consider the wellknown optimal control problem.

$$\min \int_{t_0}^{t_1} f_0(x_1,\dots,x_n,\ u_1,\dots,u_m,t), \tag{12.1}$$

under the conditions

$$\frac{dx_i}{dt} = f_1(x_1,\dots,x_n,\ u_1,\dots,u_m,t), \tag{12.2}$$

$$x_i(t_0) = x_i^{(0)}, \quad x_i(t_1) = x_i^{(1)}, \quad i = 1,\dots,n, \qquad (12.3)$$

$$u \in \Omega, \qquad (12.4)$$

where Ω is a convex set of admissible controls, and let us use some concretizations of the principle of removing constraints just mentioned; mainly the idea of displacing the constraints.

A solution by means of penalties consists in replacing (12.1.-12.4.) with a sequence of problems

$$\min_{u,x} \int_{t_0}^{t_1} \{f_0(x,u,t)+\frac{1}{2} q_v[\rho^2(u,\Omega)+ \sum_{i=1}^{n} (\frac{dx_i}{dt} - f_i(x,u,t))^2]\}dt,$$

$$v = 0,1,\dots, \qquad (12.5)$$

where $q_0, q_1, \dots$ is an arbitrary monotically increasing sequence of positive numbers which tends to infinity and $\rho(u,\Omega)$ is the distance in E^m between a point of coordinates $(u_1,\dots,u_m)$ and and the set Ω.

The unknown functions $x_1,\dots,x_n$ satisfy (12.3) and $u_1,\dots,u_n$ are not subject to any conditions. One can also replace the sequence (12.5) with

$$\min_{x} \int_{t_0}^{t_1} \min_{u\in\Omega}\{f_0(x,u,t) +\frac{1}{2} q_v \sum_{i=1}^{n} [\frac{dx_i}{dt} - f_i(x,u,t)]^2\}dt,$$

$$v = 0,1,\dots \qquad (12.6)$$

It is interesting to apply to (12.6) the cyclic algorithm of Section 9.4. Chapter IX sketches a variant for solving (12.1)-(12.4) which is more efficient and we shall therefore close here our discussion of this very simple variant of penalty function ideas and proceed with a method of displacing deformable constraints. The guiding idea behind this method was discussed in Section 4.6.

The differential equations (12.2) which desribe the evolution of the system can be written as non-holonomic [1)] constraints for a mechanical system for which the functional to be minimized is the analogue of the Hamiltonian action. [2)]

The fact that this analogue is a degenerate case (in that $f_0(x,u,t)$ does not depend on the velocities) does not affect the results and the reader need not be worried about this. There is also something else to note. The formulation (12.1)-(12.4) could make one think that the quality criterion has to be of the form (12.1). However, it can be taken to be of a more general form.

$$\int_{t_0}^{t_1} F(x_1,\ldots,x_n,\ \dot{x}_1,\ldots,\dot{x}_n,\ u_1,\ldots,u_m,t)dt \qquad (12.7)$$

And the corresponding minimization problem (obtained by adding (12.2)-(12.4)) can be reduced to the form (12.3)-(12.4) by eliminating the derivatives $\dot{x}_1,\ldots,\dot{x}_n$ in (12.7) by means of the evolution equations (12.2). This particular elimination is obviously not necessary and it does not necessarily simplify the problem. It is also possible that the form of the evolution equation is different from (12.2); that is they may not be in a form of an expression for $\dot{x}_1$. For reasons of available space we shall limit our discussion to the case of (12.1)-(12.4).

We shall now make everywhere the following hypothesis: the non holonomic constraints are deformable and the equations (12.2) only express the geometric form of the non holonomic constraints when they are not under pressure.

It is from now on impossible to work solely in the configuration space of dim n (see Chapter V); this must be replaced by a 2n dimensional space containing also velocity coordinates or impulse coordinates called phase space by Gibbs [30, 40].

Let $(x_1(t),\ldots,x_n(t),\ \dot{x}_1(t),\ldots,\dot{x}_n(t))$ be a trajectory which the system describes in phase space $\{(x,\dot{x})\}$. We shall call the time function.

$$y_i(t) = x_i - f_i(x(t),u(t),t), \qquad (12.8)$$

a deformation of the (corresponding) non holonomic constraints along this trajectory. The integral

$$\frac{1}{2}\, q \int_{t_0}^{t_1} \sum_{i=1}^{n} [\dot{x}_1 - f_i(x,u,t)]^2 dt \qquad (12.9)$$

is now - by analogy with (4.34) - interpreted as the deformation energy of the non holonomic constraints during (t_1-t_0). The factor q characterizes the degree of elasticity of the constraints.

We are dealing with an optimal control problem and the procedure of displacing the deformable non holonomic constraints now consists of replacing the problem (12.1)-(12.4) with rigid constraints with a problem (12.6) where the geometric form of the deformable unpressurized non holonomic constraints is defined by the equalities.

$$\dot{x}_i - f_i(x,u,t) + \phi_i(t) = 0, \; i = 1,\dots,n \tag{12.10}$$

where the functions $\phi_1(t),\dots,\phi_n(t)$ are called displacement functions for the non holonomic constraints and the functions

$$\psi_i(t) = \dot{x}_i - f_i(x,u,t) + \phi_i(t), \; i = 1,\dots,n, \tag{12.11}$$

where $\psi_i(t) = y_i(t) + \phi_i(t)$, $i = 1,\dots,n$, determine, by analogy with (12.8), the deformations of the non holonomic displaced constraints along the trajectory $(x(t),\dot{x}(t))$.

The problem (12.6) with deformable constraints (12.10) looks as follows

$$\min_x \int_{t_0}^{t_1} \min_{u\in\Omega}\{f_0(x,u,t) + \tfrac{1}{2}q \sum_{i=1}^{n} [\dot{x}_i-f_i(x,u,t)+\phi_i(t)]^2\}dt \tag{12.12}$$

with the conditions $x(t_0) = x^{(0)}$ and $x(t_1) = x^{(1)}$.

The essential difference between the two problems is that the parameter q in (12.12) is given so that there is no need to consider an increasing sequence of values of q. Therefore we do not have to solve a sequence of problems but one single problem (12.12) with n unknown functions $\phi_1(t),\dots,\phi_n(t)$. This problem is equivalent to the original problem (12.1)-(12.4) if there exists a vector function $\phi^*(t) = (\phi_1^*(t),\dots,\phi_n^*(t))$ such that the solution $(x^*(t), u^*(t))$ of (12.12) for $\phi(t) = \phi^*(t)$ respects conditions (12.2), that is if the deformation of the non holonomic constraints along the trajectory $x = x^*(t)$ is for $u = u^*(t)$ equal to the amount by which they have been displaced. Let $\bar{u}(t)$ be a piece-wise continuous vectorial control function $(\bar{u}(t) \in \Omega)$ which is admissible. Consider problem (12.12) for $u_s = \bar{u}_s(t)$, $s = 1,\dots,m$ given. That is consider the problem

$$\min_x \int_{t_0}^{t_1} \{f_0(x,\bar{u}(t),t) + \tfrac{1}{2}q \sum_{i=1}^{n} [\dot{x}_i - f_i(x,\bar{u}(t),t) + \phi_i(t)]^2\}dt \tag{12.13}$$

over the set of all piece wise regular curves x(t) which pass through $x^{(0)}$ and $x^{(1)}$. The Euler-Lagrange differential equations for this problem are

$$\frac{d(y_i+\phi_i)}{dt} = \frac{1}{q}\frac{\partial f_0}{\partial x_i} - \sum_{\alpha=1}^{n} (y_\alpha+\phi_\alpha)\frac{\partial f_\alpha}{\partial x_i}, \quad i = 1,\ldots,n \tag{12.14}$$

where the $y_1,\ldots,y_n$, the deformations of (12.2) following the solution of (12.13), are given by (12.8).

Set $y_i = 0$, $i = 1,\ldots,n$, in (12.14) to obtain the conditions for the optimal displacement $\phi_i = \phi_i^*$

$$\frac{d\phi_i^*}{dt} = \frac{1}{q}\frac{\partial f_0}{\partial x_i} - \sum_{\alpha=1}^{n} \phi_\alpha^* \frac{\partial f_\alpha}{\partial x_i}, \quad i = 1,\ldots,n. \tag{12.15}$$

The conditions $y_i = 0$, $i = 1,\ldots,n$, mean that the extremal will satisfy (12.2) for $\phi_i = \phi_i^*$.

The reader will have noted the analogy between the system of Equations (12.15) and the Pontriagin maximum principle for the conjugate variables. However, one should not forget that, physically, the quantities $\phi_1,\ldots,\phi_n$ are different from Lagrange multipliers in the same sense that constraints are different from deformations in elasticity theory. The analogy between the differential equations comes from a law of nature which says that there is a linear relation between these quantities. We shall omit a number of results which derive from this approach to describe a numerical procedure for solving optimal control problems.

Take as an initial approximation a zero displacement of the constraints (12.2) and consider problem (12.12) for $\phi(t) \equiv 0$. The solution can be obtained by the method of Section 9.4 which reduces the problem

$$\min_x \int_{t_0}^{t_1} \min_{u\in\Omega}\{f_0(x,u,t) + \tfrac{1}{2}q \sum_{i=1}^{n} [\dot{x}_i - f_i(x,u,t)]^2\}dt \tag{12.16}$$

to a sequence of problems ($v = 0,1,\ldots$)

$$\min_{u\in\Omega}\{f_0(x^{(\nu)}(t),u,t) + \tfrac{1}{2}q \sum_{i=1}^{n} [\dot{x}_i^{(\nu)} - f_i(x^{(\nu)}(t),u,t)]^2 \tag{12.17}$$

with given vector function $x^{(\nu)}(t)$ and

$$\min_x \int_{t_0}^{t_1} f_0(x,u^{(\nu)}(t),t) + \tfrac{1}{2}q \sum_{i=1}^{n} [\dot{x}_i - f_i(x,u^{(\nu)}(t),t)]^2 dt$$

$$x(t_0) = x^{(0)},\ x(t_1) = x^{(1)} \tag{12.18}$$

where the given vector function $u^{(\nu)}(t)$ is obtained as solution of (12.17). Here $x^{(0)}(t)$ is an arbitrary vector functions which is only required to satisfy the boundary conditions (12.3). It could for example be simply the function defined by the equation

$$x^0(t) = x^{(0)} + \frac{x^{(1)} - x^{(0)}}{t_1 - t_0}(t - t_0).$$

Let $(\bar{x}^{(0)}(t), \bar{u}^{(0)}(t))$ be the solution of problem (12.16) obtained by the technique just indicated or obtained in some other way. Denote with

$$\bar{y}_i^{(0)}(t) = \dot{\bar{x}}^{(0)} - f_i(\bar{x}^{(0)}(t), \bar{u}^{(0)}(t), t),\ i = 1,\ldots,n \tag{12.19}$$

the deformations of the non holonomic constraints (12.2) along the trajectory $x = \bar{x}^{(0)}(t)$ for $u = \bar{u}^{(0)}(t)$. One checks easily that $\bar{y}_1^{(0)}(t),\ldots,\bar{y}_n^{(0)}(t)$ satisfy for $\phi \equiv 0$ the system of differential equations (12.14). The form of these equations suggests taking as the next approximation the displacement functions of (12.2)

$$\phi_i^{(1)}(t) = \bar{y}_i^{(0)}(t),\ i = 1,2,\ldots,n. \tag{12.20}$$

Now consider again (12.17) and (12.18) to find a solution $(\bar{x}^{(1)}(t), \bar{u}^{(1)}(t))$ of problem (12.12) for $\phi(t) = \phi^{(1)}(t)$.

Thus the solution algorithm for the optimal control problem looks as follows

$$\min_{u \in \Omega} \Phi(x^{(\nu)}(t), u, t, \phi^{(\alpha)}(t)) = \Phi(x^{(\nu)}(t), u^{(\nu)}(t), t, \phi^{(\alpha)}(t)) \tag{12.21}$$

$$\min_x \int_{t_0}^{t_1} \Phi(x, u^{(\nu)}(t), t, \phi^{(\alpha)}(t))dt =$$
$$= \int_{t_0}^{t_1} \Phi(x^{(\nu+1)}(t), u^{(\nu)}(t), t, \phi^{(\alpha)}(t))dt \tag{12.22}$$

with $x(t_0) = x^{(0)}$, $x(t_1) = x^{(1)}$, and $u^{(\nu)}(t)$ the solution of (12.21). In these last two equalities

$$\Phi(x,u,t,\phi) = f_0(x,u,t) + \tfrac{1}{2}q \sum_{i=1}^{n} [\dot{x}_1 - f_1(x,u,t) + \phi_1]^2.$$

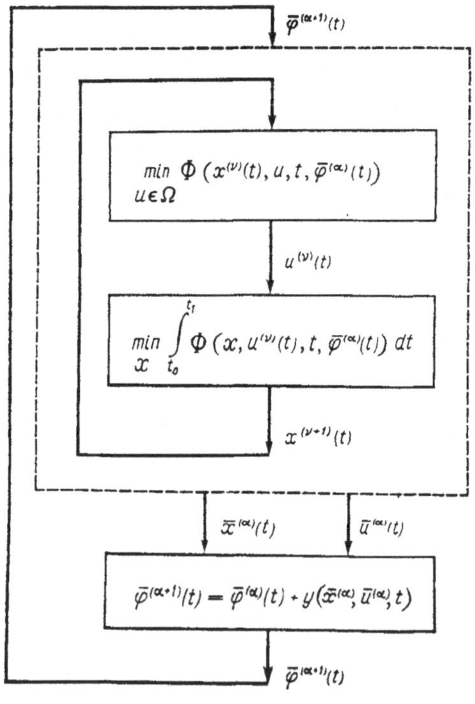

Fig. 12.1.

The algorithm is composed of inner loops $\nu = 0,1,\ldots,$ and outer loops $\alpha = 0,1,\ldots$. Each inner loop contains two problems (12.21) and (12.22) and each outer loop consists of a sequence of ν's resulting in two vector functions $\bar{x}^{(\alpha)}(t)$ and $\bar{u}^{(\alpha)}(t)$ solving (12.12) for given $\phi_i = \phi_i^{(\alpha)}(t)$, $i = 1,\ldots,n$ and giving the next approximation to the optimal displacement functions of (12.2)

$$\phi_i^{(+1)}(t) = \phi_i^{(\alpha)}(t) + y_i(\dot{\bar{x}}^{(\alpha)},\bar{x}^{(\alpha)},\bar{u}^{(\alpha)},t),\ i = 1,\ldots,n.$$

Figure (12.1) is a flow diagram for the algorithm.

NOTES

1. A constraint is non holonomic if it cannot be expressed by means of a finite equation $\phi(x_1,\ldots,x_n,t) = 0$ and if the corresponding condition involves derivations of the variables.

2. The Hamiltonian action is the integral $G = \int_{t_0}^{t_1} (T-V)dt$, where T is the kinetic energy of the mechanical system and V its potential energy. The principle of Hamilton says that the evolution of the system in the time interval (t_0,t_1) will be such that this integral assumes an extremal value. It is interpreted as the most general variational principle underlying dynamics. It can be obtained as a consequence of the principle of virtual displacements in the d'Alembert-Lagrange form. The reader who would like to improve his understanding of the variational principles of mechanics and the extension of the principle of Hamilton to nonconservative and non holonomic systems is advised to consult [30].

BIBLIOGRAPHY [1]

[1] Vilcevskii, N. O. and Razumikhin, B. S.: 1966, 'Mechanical Models and Solution Methods for General Linear Programming Problems', Avtomatika i Telemekhanika No. 4.

[2] Volkonskii, V. A.: 1965, 'Optimal Planning in High Dimensional Situations (Iterative Methods and Decomposition Principles)', Economics and Mathematical Methods 1, No. 2.

[3] Germeier, Yu. B.: 1971 , Introduction to Operations Research, Nauka.

[4] Danilov-Danilyan, V. I.: 1967, 'High Dimensional Problems and Iterative Methods for Optimal Planning', in: Collection of Algorithms and Programs for Solutions in an E.C.M., Statistika.

[5] Kantorovic, L. V.: 1939, Mathematical Methods for Organization and Planning, Izd-vo LGU.

[6] Kantorovic, L. V.: 1957, 'On Methods for Analysing some External Planning Problems', Dokl. Akad. Nauk S.S.S.R. 115, No. 3.

[7] Krasovskii, L. V.: 1968, Theory of Controlled Motion, Nauka.

[8] Levitin, E. S. and Polyak, B. T.: 1966, Minimization Methods in the Presence of Constraints, Ž. Vyc. Mat. i. Mat. Fiz. 6, No. 5.

[9] Moiseev, N. N.: 1971, Numerical Methods in the Theory of Optimal Systems, Nauka.

[10] Pervozvanskaya, T. N. and Pervozvanskii, A. A.: 1966, 'Search Algorithms for the Optimal Allocation of Central Resources', Izv. Akad. Nauk,S.S.S.R., Tekhn. Kibernetika No 3.

[11] Pittel, V. G.: 1969, 'On Exchange Models', Economics and Mathematical Methods, No. 6.

[12] Polterovic, V. M.: 1969, Block Methods for Concave Programming and their Economic Interpretation, Economics and Math. Methods, No. 6.

[13] Putilov, K. A.: 1971, Thermodynamics, Nauka.

-----

1) The bibliographical items 1-18 are all in Russian.

[14] Razumikhin, B. S.: 1967, 'Iteration Solution Methods and Decomposition Problems in Linear Programming', Avt. i Telem. No. 3.

[15] Razumikhin, B. S.: 'The Method of Physical Models in Mathematical Programming and Economics I-VI', Avt. i Telem. Nos 3, 4, 6, 11, 1972; Nos 2, 4, 1973.

[16] Razumikhin, B. S.: 1976, 'Principles of Analytical Mechanics and Optimal Control Problems I, II', Avt. i Telem. Nos 2, 3.

[17] Tikhonov, A. N.: 1965, 'On Methods for Regularization of Optimum Problems', Dokl. Akad. Nauk S.S.S.R. 162, No. 4.

[18] Cetaev, N. G.: 1955, 'Stability of Motion', Gos. Izd-vo Tekhn.-Teor. Lit., Moscow.

[19] Appell, P.: 1952-1955, Traité de mécanique rationelle, Paris.

[20] Arrow, K. J., Hurwicz, L. and Uzawa, H.: 1958, Studies in linear and nonlinear programming, Stanford Univ. Press, Stanford (California).

[21] Bellman, R.: 1957, Dynamic Programming, Princeton Univ. Press, N.Y.

[22] Bernoulli, J.: 1742, Opera omnia, t. III, Lausanne-Geneva.

[23] Courant, R., 1943, 'Variational Methods for the Solution of Problems of Equilibrium and Vibrations', Bull. Am. Math. Soc. 49, 1-23.

[24] Dantzig, G. B.: 1963, Linear Programming and Extensions, Princeton (New Jersey), Princeton Univ. Press.

[25] Dantzig, G. and Wolfe, F.: 1960, 'Decomposition Principle for Linear Programs', Operat. Res. 8 No. 1.

[26] Dennis, J. B.: 1959, Mathematical Programming and Electrical Networks, The Massachesetts Institute of Technology and John Wiley and Sons, New York.

[27] Fermi, E.: 1956, Thermodynamics, New York, Dover.

[28] Fiacco, A. and McCormick, G.: 1968, Nonlinear Programming: Sequential Unconstrained Minimization Techniques, John Wiley and Sons, New York, London, Sydney, Toronto.

[29] Gale, D.: 1960, The Theory of Linear Economic Models, New York (a.o.), McGraw-Hill.

[30] Goldstein, H.: 1950, Classical Mechanics, Cambridge (Mass.), Addison-Wesley.

[31] Goldstein, E. and Youdine, D.: 1973, Problèmes particuliers de la programmation linéaire, Ed. de Moscou.

[32] Hardy, G. H., Littlewood, J. E. and Polya, P.: 1934, Inequalities, Cambridge Univ. Press.

[33] Kantorovitch, L. V.: 1963, 'Calcul économique et utilisation des resources', Coll. Finances et Economie appliquée, Dunod XV.

[34] Karlin, S.: 1959, Mathematical Methods and Theory in Games, Programming and Economics, London.

[35] Kornai, J. and Liptak, T.: 1962, Kétszintü tervezés, Publications of the Mathematical Institute of the Hungarian Academy of Sciences, 7, 557-621.

[36] Kubo, R.: 1968, Thermodynamics, North-Holland Publ. Co., Amsterdam.

[37] Künzi, H. P. and Krelle, W.: 1963, Nichtlineare Programmierung, Springer Verlag, Berlin-Göttingen-Heidelberg.

[38] Lagrange, J. L.: 1888, Mécanique analytique, t.I., Paris.

[39] Lancaster, K.: 1968, Mathematical Economics. New York-London, Collier-MacMillan.

[40] Lanczos, C.: 1953, The Variational Principles of Mechanics, Toronto, Univ. of Toronto Press.

[41] Linear Inequalities and Related Systems, Kuhn, H. W. and Tucker, A. W. (eds.), Princeton, New Jersey, Princeton Univ. Press, 1956.

[42] Morishima, M.: 1964, Equilibrium, Stability and Growth, Oxford.

[43] Morrison, D. D.: 1968, 'Optimization by Least Squares', SIAM Journal on Numerical Analysis, 5, No. 1, 83-88.

[44] Nikaido, N.: 1968, Convex Structures and Economic Theory, Academic Press, New York and London.

[45] Optimization Techniques with Applications to Aerospace Systems, George Leitzmann (ed.), New York Academic Press, London, 1962.

[46] Pontriagin, L., Boltianski, V., Gamkrelidze, R. and Miščenko. E.: 1974, Théorie mathématique des processus optimaux, Ed. de Moscou.

[47] Powell, M. I. D.: 1969, 'A Method for Nonlinear Constraints Constraints in Minimization Problems', in R. Fletcher (ed.), Optimization, Academic Press, New York.

[48] Szymanowski, J.: 1970, 'A Comparison of Several Methods of Constrained Minimization', Report of Institute of Automatics, Warsaw University of Technology.

[49] Von Neumann, J. and Morgenstern, O.: 19545-1946, A model of General Economic Equilibrium, Rev. Econ. Studies, 13 No. 1, 1-9.

[50] Von Neumann, J. and Morgenstern, O.: 1953, Theory of Games and Economic Behaviour, Princeton Univ. Press.

[51] Wierzbicki, A. P.: 1971, 'A Penalty Function Shifting Method in Constrained Static Optimization and its Convergence Properties', Arch. Automat. and Telemech. 16, No. 4, 395-416.

[52] Young, L. Ch.: 1969, Lectures on the Calculus of Variations and Optimal Control Theory, Philadelphia (a.o.), Saunders.

[53] Ford, L. R. and Fulkerson, D. R.: 1962, Flows in Networks. Princeton University Press, Princeton, New Jersey.

[54] Razumikhin, B. S. and Razumikhin. Yu. B., 1980, 'Problems Concerning Maximal Flows Through Networks. Physical models and solution methods (Russian). Models and methods for optimization, Collection of papers VNII on operations reserach 3, Moscow.

[55] Ter Haar, D.: 1961, 'Elements of Hamiltonian Mechanics', International Series of Monographs in Natural Philosophy 34.

[56] Arnold, V. I.: 1974, Mathematical Methods of Classical Mechanics (Russian), Nauka, Moscow.

[57] Hestenes, M. R. and Stifel, E.: 1952, 'Methods of Conjugate Gradients for Solving Systems, J. Res. Nat. Bur. Standards 49.

[58] Hestenes, M. R.: 1969, 'Multiplier and Gradient Methods', J. Optimiz. Th. Appls 4, No. 3.

[59] Bertsekas, D. P.: 1976, 'Multiplier Methods', A survey, Automatica 12.

[60] Fletcher, R.: 1975, 'An Ideal Penalty Function', J. Inst. Math. Appl. 15.

[61] Numerical Methods for Constrained Optimization, P. E. P. E. Cill and W. Murrey (eds.), Academic Press, London-New York-San Francisco, 1974.

[62] Chambadal Par, P.: 1963, Evolution et Applications d'entropie. Dunod, Paris.

[63] Razumikhin, B. S.: 1980,'Elastic Constraints and the Method of Penalty Functions', Models and methods for optimization (Russian). Collection of papers VNII on Operations Research 3, Moscow.

INDEX